COMPETITIVE ADVANCES IN METALS AND PROCESSES

SOCIETY FOR THE ADVANCEMENT OF
MATERIAL AND PROCESS ENGINEERING

1st INTERNATIONAL SAMPE METALS AND METALS PROCESSING CONFERENCE

VOLUME 1

COMPETITIVE ADVANCES IN METALS AND PROCESSES

Hyatt Hotel
Cherry Hill, New Jersey
August 18-20, 1987

Edited By
Roy J. Cunningham
Mel Schwartz

Additional copies of this publication may be obtained from
Society for the Advancement of Material and Process Engineering

SAMPE Business Office
P.O. Box 2459
Covina, California 91722

ISBN 0-938994-39-5

1ST INTERNATIONAL SAMPE METALS AND METALS PROCESSING CONFERENCE

SPONSORED BY: WILMINGTON/PHILADELPHIA CHAPTER

General Chairman	Roy J. Cunningham, Boeing Vertol Company
Program Chairman	Mel Schwartz, Sikorsky Aircraft
Financial Chairman	Dick Landt, SPS Technologies, Inc.
Arrangements Chairman	Glenn Kuebeler, Hercules, Inc.
Publicity Chairman	Bob Golick, Hysol/Dexter Corporation
Registration Chairman	Vicki Sholtes, Boeing Vertol Company
Publication Chairman	Vic Mazzio, Boeing Vertol Company
Audio Visual Chairman	Jim Kachelries, Boeing Vertol Company
Paper Evaluation Chairman	Andy Blackwood, Structure Probe, Inc.

INTERNATIONAL OFFICERS

President	Gary W. Valentine
Past President	Charles J. Hurley
1st Vice President	Gayle B. Wadsworth
2nd Vice President	Jerry Bauer
3rd Vice President	Brian Wilson
Secretary	Robert Martinez
Treasurer	W. H. Face, Jr.

CHAPTER OFFICERS

Chairman	P. R. Langston
1st Vice Chairman	Vicki Lynn Sholtes
2nd Vice Chairman	Edward Dejong
Secretary	Marygail K. Hutchins
Treasurer	James C. Shuart
Sr. International Director	Frank Ko

SESSIONS AND SESSION CHAIRMEN

ALUMINUM LITHIUM PROCESSING

Dave Divecha, Naval Surface Weapons Center

CLEANLINESS AND ADVANCES IN STEEL AND NICKEL ALLOY PROCESSING

Lou Lherbier, Cytemp

METAL ALLOY DEVELOPMENT

Stan Silverstein, Sikorsky Aircraft Company

COATINGS AND CORROSION

Frank Hermanek, Alloy Metals, Inc.

LIGHT ALLOY FABRICATION

Larry Mueller, ALCOA Laboratories

METALS FABRICATION I

Thomas T. Bales, NASA Langley Research Center

METALS FABRICATION II

Paul Baumgardner, AVCO-Bell Helicopter/Textron

POWDER METALLURGY

Bob Witt, Grumman Aerospace Systems

COST REDUCTION THRU HEAT TREATING AND PROCESSING

Bill Jones, Vacuum Furnace Systems Corporation

CONTENTS

PAPERS AND AUTHORS

PAPERS AND AUTHORS

1st International SAMPE Metals Conference
August 18-20, 1987

MICROSTRUCTURES AND PROPERTIES OF Al-Li-Mg ALLOYS

T. H. Sanders, Jr. and H. R. Last
School of Materials Engineering
Georgia Institute of Technology
Atlanta, GA 30332-0245
and
S. C. Jha
Marko Materials, Inc.
144 Rangeway Rd.
North Billerica, MA 01862

ABSTRACT

The effects of Mg and Li content on the mechanical behavior in the Al-Mg-Li-0.3Mn alloy system were determined. Nine alloy combinations ranging from 1.4 to 3.1 Li and 2.0 to 4.9 Mg were evaluated. The primary strengthening agent, δ' (Al_3Li) was confirmed. It was observed that increasing both Mg and Li increased strength, but the effect of Li was greater. Ductility and toughness were affected by grain boundary precipitation and the development of precipitate free zones (PFZs). Alloys with the best combinations of strength and ductility were associated with compositions which did not lead to large volume fractions of equilibrium phases at interface or interphase boundaries.

1. INTRODUCTION

The Al-Mg-Li alloy system has a potential of providing high strength in heavy sections, good resistance to stress-corrosion cracking, and high fracture toughness.(1,2) More importantly, this alloy system also has the potential of providing a unique combination of high modulus of elasticity and low density.

Work in the Soviet Union with additions of Zr, Mn, Ti, and Cr resulted in the development of the Al-Mg-Li alloy 01420. (3-9) The alloy, which contains 4.0-7.0% Mg and 1.5-2.6% Li, has reportedly been cast in ingots up to 78.cm in diameter and has been fabricated into extrusions, forgings, sheet, and plate. (10) Strengths of the extruded bars and panels exceed strengths predicted from work with ternary alloys by approximately 70MPa, but strengths of the other products are more in line with expected strengths of ternary alloys. Alloy 01420 also has good fatigue strength in air and corrosive media, and it is weldable. (4)

The objective of this investigation was to determine the Mg and Li contents that provided the optimum combinations of strength, toughness, modulus, density, castability, and fabricability.

2. EXPERIMENTAL

2.1. Ingot Casting

To determine the effects of Mg and Li content on the mechanical behavior in the Al-Mg-Li system various target compositions presented in Table 1 were cast as 15.25cm diameter D.C. ingots. Because Li readily oxidizes in air, special precautions were followed during melting and casting. All components except Li were melted as 40 Kg charges in a plumbago crucible using a high frequency induction furnace. The melt was fluxed with Cl_2 to remove hydrogen gas and was then covered with a molten KCl-LiCl salt mixture. Elemental Li was added to the melt as 100 g charges. The melt was poured into the pouring trough under a molten KCl-LiCl salt mixture and was then transferred to the mold under an inert gas atmosphere. A single charge produced one 15.25 cm diameter x 75 cm long ingot.

2.2. Fabricating

Rods 5 cm in diameter were fabricated by hot extrusion from the ingots. The ingots were preheated 12 hours at 727K (850°F) plus 12 hours at 788K (960°F) in an argon atmosphere and were air cooled. The billets were induction reheated to 644K (700°F) and direct extruded from a 16.2cm diameter cylinder to the 5cm rod-diameter resulting in an extrusion ratio of 10.

2.3 Solution Heat Treating Practices

Solution heat treatment (SHT) temperatures investigated were in the range of 700K (800°F) to 783K (950°F) in steps of 13.9K (25°F). Specimens of the as-extruded products were SHT at the various temperatures for one hour, cold water quenched, and aged 48 hours at 464K (375°F). Single, longitudinal tension specimens were removed from the center of each sample and were tested. In addition, specimens were removed from as-extruded samples which had been artificially aged. The SHT temperature chosen was the lowest temperature which produced the greatest strength upon aging at 464K. These SHT temperatures are summarized in Table 2.

2.4 Determination of Precipitation Heat Treatment Practices

Samples were removed from the extrusions, SHT one hour at the temperatures given in Table 2, quenched in cold water, and artificially aged for one to 168 hours at 422K (300°F) to 477K (400°F) in increments of 13.9 (25°F). Rockwell "B" hardness values as a function of time and temperature were determined.

Based on the results of the hardness measurements, three aging practices were chosen to evaluate the mechanical properties of the alloys. The first practice was designed to produce maximum or near maximum strength for each alloy. The second practice was chosen to produce 90 percent of maximum strength by overaging. A third practice was used on alloys which exhibited what appeared to be two distinct peaks in hardness. These aging practices are summarized in Table 3. Longitudinal and transverse tensile and notched tensile tests were run on each alloy for each aging condition according to Table 3.

2.5 Density

Densities were determined by water displacement technique on machined 2.5cm x 2.5cm x 5cm specimens.

2.6 Moduli of Elasticity

Elastic moduli were determined for samples aged to peak strength using duplicate longitudinal specimens with strain gages 180° apart.

2.7 δ' Precipitate Distribution

To determine the effect of aging time and temperature on the distribution of the metastable δ' phase in the Al-Mg-Li system, three alloys were prepared by melting 99.95% pure Al, 99% pure Li and 99.9% pure Mg in a glove box under argon atmosphere. Approximately 100 gms. of the molten alloy was chill cast in a copper book mold. The cast ingots were homogenized at 823K for 48 hours in a salt bath. The ingots were then quenched in cold water, and hot and cold rolled to about 5mm thickness. Small, 15mm x 15mm coupons were cut from the alloy sheets, and all the samples were SHT at 823K for 0.5 hours. The chemical composition of the alloys were determined by atomic absorption spectroscopy. Table 4 shows the chemical composition of the three alloys prepared by the method described above. All the alloys contained 0.3% Mn, which was added for grain size control in the wrought microstructure.

Aging studies on the Al-Li-Mg alloys were performed by aging alloy coupons at 473K for times ranging from 1 hour to 2 weeks. After aging at 473K, the samples were cold water quenched. The heat treated samples were mechanically ground to thin sheets, and 3mm diameter discs were punched from the thin sheets for TEM sample preparation. TEM samples were prepared using a jet polishing technique, in a 25% HNO_3 - methanol solution cooled to 253K at 10 volts. TEM studies were carried out in a JEOL 200CX electron microscope, and particle size and grain boundary PFZ widths were measured as a function of the aging time.

3. RESULTS and DISCUSSION

3.1 Ingot Casting

Ingots were cast successfully for all of the alloys except alloy 3 (3.4 percent Li). Dye-checked ingot slices showed no porosity. Three separate attempts were made to cast alloy 3; however, all three ingots cracked. One of these ingots was extruded to develop information on the effects of composition on properties.

3.2 Fabricating

The breakout pressure was insensitive to Li content but increased considerably with increasing Mg content. This observation is consistent with other high strength alloys, since magnesium appears to have a greater effect on high temperature flow stress in aluminum compared to other alloying additions.

The extrusions were inspected for rear-end defects by removing 30 cm lengths, etching the cross-section, and visually examining the cross-section for defects. The defects were confined to the rear 30 cm portion of all the extrusions. Slices were also removed from locations 60 cm from the front and 30 cm from the rear of the extrusions and etched to reveal the macro grain structure. All alloys except alloy 4 showed fine grain structure after macroetching. In the case of alloy 4 the as-cast microstructure contained large columnar grains. This grain structure carried through the extrusion, resulting in a large grain structure in the product. This grain structure arose from differences in the casting practice and not from variations in composition.

3.3. SHT Practices

SHT practices for the extrusions were selected on the basis of results of tensile tests on samples which had been removed from extrusions of alloys 1 through 9, which had been SHT between 700K (800°F) to 783K (950°F) and artificially aged 48 hours at 464K (375°F).

In general, the SHT temperature producing the highest tensile properties increased with increasing solute content. For alloys with Li contents less than approximately 2.4 wt% Li, as-extruded samples that were artificially aged developed strengths as high as strengths of samples that were SHT prior to artificial aging. This result indicates that these alloys were not only solutionized during extrusion but that they are also quench-insensitive (extrusions were still-air cooled).

3.4 Determination of Aging Temperature

Representative Rockwell B hardness curves are illustrated in Figure 1.

The response to artificial aging depended significantly on Li and moderately on Mg content and can be summarized as follows:

1. Maximum hardness produced by aging increased considerably with increasing Li content, but increased only slightly with increasing Mg content.
2. Aging temperature producing maximum hardness decreased with increasing Mg content.
3. Alloys 1, 2, 3, and 5 showed what appeared to be a two-stage aging phenomenon.

3.5 Tensile Properties

The heat treatment and aging practices listed in Tables 2 and 3 were used to evaluate the different alloys. Longitudinal yield strength versus Li content (in atomic percent) for each Mg level is plotted in Figure 2. The yield strength increased considerably with increasing Li content but increased only slightly with increasing Mg content. No significant differences in tensile properties were apparent in samples aged by Practice 1 as compared to those aged by Practice 3. (See Tables 5 and 7).

Specific tensile strength (strength/density) for the highest strength Al-Mg-Li alloys was the same as that of 7050-T76 extrusions.

3.6 Densities and Elastic Moduli

The effect of composition on density is illustrated in Figure 3. The density was primarily affected by the Li content although both Mg and Li additions decreased density. As indicated in Figure 4, elastic modulus for these alloys ranged from approximately 75-81 GPa, was independent of Mg content, but increased with increasing Li content.

The ratio of elastic modulus to density was up to 30% greater than that of alloy 7050.

3.7 Toughness

Notch tensile strength/yield strength ratios are plotted against yield strength in Figure 5. All of the alloys developed combinations of toughness and strength below the band for commercial aluminum alloys. The solid line in Figure 5 represents the trend observed in alloy 7075. It was also noted that transverse toughness was decidedly lower than longitudinal toughness.

Longitudinal toughness was affected by yield strength, composition, and aging practice. Toughness of the peak-aged, high solute alloys and of the low solute alloys in the peak aged or overaged condition conformed to a relationship of the form:

$$NTS/YS = a - b\ (Y.S.)$$

+ P (Composition), as indicated in Figure 5. Toughness generally decreased with increasing Li content (Figure 5). Overaging was detrimental to the toughness of alloys containing 3.5% or more Mg. The magnitude of the decrease in toughness with overaging increased with increasing Mg and Li.

Contrary to the effect on longitudinal toughness, composition had no apparent direct effect on transverse toughness. However, yield strength and aging practice did have an effect. Overaging decreased the combination of strength and toughness in the transverse direction.

Microprobe analyses of fractured longitudinal and transverse notched tensile specimens indicate that Fe, Si, Ti, or the minor alloying element, Mn, did not contribute significantly to the low toughness. Light and scanning electron microscopy indicated that fracture was intergranular/intersubgranular.

3.8 Precipitate Free Zone Growth Kinetics

In the ternary Al-Mg-Li alloys that were cast to evaluate the influence of Mg content on the coarsening and PFZ growth behavior, the Mg content was sufficiently low, so that the precipitation of ternary Al-Li-Mg phase did not occur. In the initial stages of aging at 200°C, δ' precipitates formed mostly on the grain boundaries. The grain boundaries were also associated with δ' PFZs, as previously observed in binary Al-Li alloys.(11)

Figure 6 shows the PFZ half width plotted against the square root of the aging time at 200°C for alloys A and B. A least square line superimposed on the experimental data points is also included on the plots. The PFZ in the ternary alloys grows parabolically with the aging time as observed for the binary Al-Li alloys. The PFZ growth rate constants K_p, as defined by the following equation, were complied on the basis of the experimental data.

$$h = K_p\sqrt{t}$$

Table 11 summarizes the measured values of K_p for alloys A and B at 200°C, along with the extrapolated values for corresponding binary Al-Li alloys. From the data it appears that addition of Mg increases the PFZ growth kinetics over what would be expected in a binary in Al-Li alloy.

Figure 7 shows the plots of cube of average δ' particle size as a function of aging time at 200°C for alloys A and B. The data plot linearly showing that the coarsening of the δ' precipitates in the Al-Li-Mg alloys follows the Ostwald ripening kinetics observed in binary Al-Li alloys.(12) The data were analyzed on the basis of the following equation:

$$\bar{R}^3 - \bar{R}_o^3 = K_c t$$

where $\bar{R}$ is the average particle size at aging time, t and R_o is the particle size at the outset of coarsening. The coarsening constants K_c and R_o were evaluated by linear regression analysis of the data. Table 12 shows the values of K_c and R_o for the Al-Li-Mg alloys aged at 200°C. The estimated values of the coarsening rate constants for binary Al-Li alloys with the same Li content as in the ternary alloys, are also included in the table. It is observed that the coarsening rate in the Mg containing Al-Li alloys is faster than the rate in a corresponding binary Al-Li alloy.

4. SUMMARY OF OBSERVATIONS

1. Al-Mg-Li alloys containing up to 3.0 Li and 4.8 Mg can be cast and extruded.

2. Extrusion pressure was independent of Li content but progressively increased with increasing Mg.

3. Strength and elastic modulus increased with increasing Li, while density decreased.

4. The addition of Mg accelerates the kinetics of δ' precipitation and the growth of the δ' PFZ.

5. Mg contributes a component to the strength by a combination of solid solution strengthening and increasing the volume fraction of δ' by reducing the Li solubility.

6. The low toughness of these alloys is associated with the intergranular/intersubgranular failure process stimulated by the grain boundary precipitates and the PFZ.

5. REFERENCES

1. I.N. Fridlyander, V.F. Shamray, and N.V. Shiryaeva, Translated from Izvestiia Akademiia Nauk SSSR, Metally,, (2), pp. 153-158, 1965.

2. Undated Fulmer Research Institute advertising flyer.

3. I.N. Fridlyander, S.M. Ambartsumyan, N.V. Shiryaeva, and R.M. Gabidullin, Translated from Metallovedenie i Termicheskaya Obrabotka Metallov, No. 3, pp. 50-52, March 1968.

4. V.G. Kovrizhnykh, V.A. Tikhomirov, N.I. Saitseva, and M.V. Ermanok, Translated from Metallovedenie i Termicheskaya Obrabotka Metallov, No. 2, pp. 20-21, February 1969.

5. M.I. Paretskii, E.V. Giarsintov, and M.G. Veitsman, Translated from Metallovedenie i Termicheskaya Obrabotka Metallov, No. 3, pp. 52-54, March 1968.

6. I.N. Fridlyander, V.S. Sandler, and T.I. Nikol'skaya, Translated from Metallovedenie i Termicheskaya Obrabotka Metallov, No. 5, pp. 2-5, May 1971.

7. M.E. Drits, E.S. Kadaner, and N.I. Turkina, Izvestiia Akademiia Nauk SSSR, Metally, pp. 231-235, July-August 1971.

8. British Patent No. 1,172,736.

9. F.S. Novik and N.I. Kolobnev, Translated from Metallovedenie i Termicheskaya Obrabotka Metallov, No. 4, pp. 33-35, April 1969.

10. Dimitri Vvidenski, Foreign Science Bulletin, Vol. 4, No. 12, pp. 35-43, December 1968.

11. S.C. Jha, T.H. Sanders, Jr., and M.A. Dayananda, Acta Metallurgica, 35, pp. 473-482, 1987.

12. K. Mahalingan, B.P. Gu, G.L. Liedl, and T.H. Sanders, Jr., Acta Metallurgica, 35, 483-498, 1987.

6. ACKNOWLEDGEMENT

H.R. Last is currently supported through a National Science Foundation Graduate Fellowship.

Table 1. Remelt Analyses * of the Al-Mg-Li Extrusions

Alloy	Mg	Li	Mn
1	2.03	2.30	0.28
2	1.99	2.76	0.28
3	1.94	3.14	0.28
4	3.51	1.80	0.28
5	3.50	2.28	0.28
6	3.55	2.71	0.27
7	4.89	1.37	0.28
8	4.89	1.81	0.27
9	4.84	2.33	0.27

* Compositions were determined by atomic absorption. The Fe + Si < 0.1 for all compositions.

Table 2. Optimum Solution Heat Treatment Temperatures

Alloy	Solution Heat Treatment Temperature, K
1	727
2	755
3	783
4	727
5	783
6	783
7	727
8	727
9	783

Table 3. Aging Practices for Material Evaluation

Alloy	Practice 1	Practice 2	Practice 3
1	18 hr @ 464K	72 hr @ 477K	--
2	72 hr @ 464K	72 hr @ 477K	18 hr @ 450K
3	48 hr @ 464K	72 hr @ 477K	--
4	48 hr @ 450K	--	--
5	36 hr @ 463K	48 hr @ 477K	72 hr @ 436K
6	48 hr @ 463K	48 hr @ 477K	--
7	72 hr @ 422K	--	--
8	48 hr @ 436K	36 hr @ 477K	--
9	72 hr @ 450K	48 hr @ 477K	--

Practice 1 selected to produce maximum or near maximum hardness.
Practice 2 selected to produce 90 percent of maximum hardness by overaging.
Practice 3 selected to produce high hardness by aging to initial hardness peak.

Table 4. Chemical composition of Al-Li-Mg Alloys

Designation	wt%Li	at%Li	wt%Mg	at%Mg
Alloy A	3.37	11.9	1.66	1.7
Alloy B	3.53	12.4	2.35	2.4
Alloy C	4.80	16.4	2.50	2.4

Table 5. Longitudinal Tensile and Notch Tensile Properties of Aged Al-Mg-Li Alloys, Practice 1.

Alloy	T.S.		Y.S.		NTS		
	MPa	(ksi)	MPa	ksi	MPa	(ksi)	NTS/YS
1	449	(65.2)	411	(59.6)	510	(74.0)	1.25
2	487	(70.7)	399	(57.9)	477	(69.3)	1.20
3	482	(70.0)	423	(61.4)	383	(55.6)	0.91
4	412	(59.8)	337	(48.9)	491	(71.2)	1.46
5	493	(71.5)	432	(62.6)	521	(75.6)	1.21
6	524	(76.0)	455	(66.0)	440	(63.8)	0.97
7	517	(60.5)	286	(41.6)	475	(69.0)	1.66
8	425	(69.0)	367	(53.3)	441	(64.0)	1.20
9	517	(75.0)	443	(64.3)	434	(63.0)	0.98

Table 6. Longitudinal Tensile and Notch Tensile Properties of Aged Al-Mg-Li Alloys, Practice 2

Alloy	T.S.,		Y.S.		NTS,		
	MPa	(ksi)	MPa	(ksi)	MPa	(ksi)	NTS/YS
1	399	(57.9)	255	(37.0)	424	(61.6)	1.67
2	402	(58.3)	246	(35.7)	374	(54.2)	1.52
3	406	(59.0)	249	(36.2)	336	(48.8)	1.35
4		--		--	--	--	--
5	446	(64.7)	294	(42.6)	413	(60.0)	1.41
6	465	(67.6)	306	(44.4)	327	(47.5)	1.07
7		--	--	--		--	
8	435	(63.2)	266	(37.8)	301	(43.6)	1.16
9	437	(63.4)	284	(41.2)	271	(39.4)	0.95

Table 7. Longitudinal Tensile and Notch Tensile Properties of Aged Al-Mg-Li Alloys, Practice 3

Alloy	T.S. MPa (ksi)	Y.S., MPa(ksi)	NTS, MPa (ksi)	NTS/YS
2	474 (68.8)	--	474 (68.8)	--
5	464 (67.3)	435 (63.2)	517 (75.1)	1.19

Table 8. Transvere Tensile and Notched Tensile Properties of Aged Al-Mg-Li Alloys, Practice 1.

Alloy	T.S. MPa (ksi)	Y.S. MPa (ksi)	NTS MPa (ksi)	NTS/YS
1	368 (53.4)	251 (36.4)	359 (52.2)	1.43
2	417 (60.4)	297 (43.2)	271 (39.3)	0.91
3	409 (59.4)	331 (48.0)	211 (30.6)	0.64
4	364 (52.8)	232 (33.8)	396 (57.4)	1.70
5	412 (59.8)	286 (41.6)	326 (47.2)	1.14
6	421 (61.1)	314 (45.6)	243 (35.3)	0.77
7	366 (53.0)	227 (32.9)	395 (57.3)	1.74
8	367 (53.2)	271 (39.3)	276 (40.1)	1.02
9	386 (56.0)	313 (45.4)	205 (29.7)	0.65

Table 9. Transvere Tensile and Notched Tensile Properties of Aged Al-Mg-Li Alloys, Practice 2.

Alloy	T.S. MPa	(ksi)	Y.S. MPa	(ksi)	NTS MPa (ksi)	NTS/YS
1	346	(50.2)	218	(31.7)	323 (46.9)	1.48
2	368	(53.4)	225	(32.6)	251 (36.5)	1.12
3	278	(40.3)	241	(35.0)	136 (19.8)	0.56
4		--		--	--	--
5	391	(56.8)	245	(35.6)	302 (43.9)	1.23
6	396	(57.4)	269	(39.0)	213 (30.9)	0.79
7		--		--	--	--
8	317	(46.0)	225	(32.7)	248 (36.0)	1.10
9	327	(47.4)	265	(38.4)	162 (23.5)	0.61

Table 10. Transvere Tensile and Notched Tensile Properties of Aged Al-Mg-Li Alloys.

Alloy	Practice 3 Aged to Max Strength						
	T.S. MPa	T.S. (ksi)	Y.S. MPa	Y.S. (ksi)	NTS MPa	NTS (ksi)	NTS/YS
2	379	(55.0)	278	(40.4)	400	(58.1)	1.44
5	407	(59.0)	272	(39.6)	357	(51.8)	1.31

Table 11. PFZ growth rates for Al-Li-Mg and Al-Li alloy.

Alloy Composition	PFZ Growth Rate K_p (cm^2/sec)
Al-3.37 wt%Li-1.66 wt%Mg	5.76×10^{-15}
Al-3.37 wt%Li	2.89×10^{-15}
Al-3.53 wt%Li-2.35 wt%Mg	5.99×10^{-15}
Al-3.53 wt%Li	3.05×10^{-15}

Table 12. The parameters K_c and R_o for Al-Li-Mg alloys aged at 200°C.

Alloy Composition	K_c(m^3/sec)	R_o(A)
Al-3.37 w%Li-1.66w%Mg	48.93×10^{-30}	170
Al-3.37w%Li	33.62×10^{-30}	-
Al-3.53w%Li-2.35w%Mg	55.55×10^{-30}	229
Al-3.53w%Li	36.53×10^{-30}	-

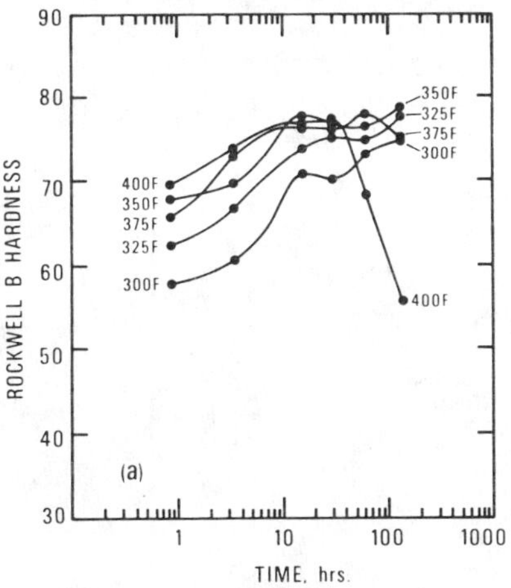

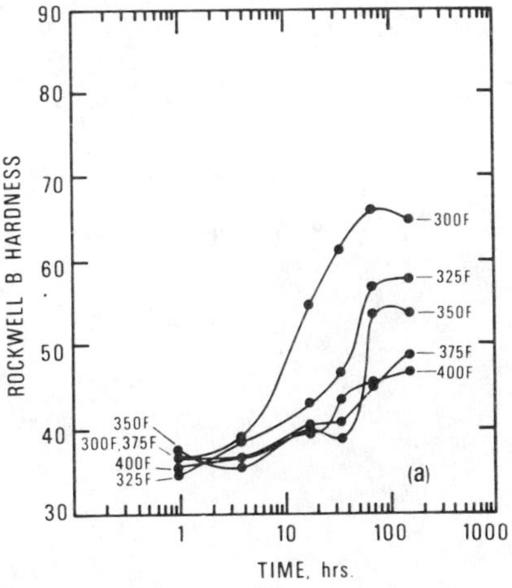

Figure 1. Isothermal hardness curves illustrating the range in behavior encountered. (a) alloy 2 and (b) alloy 7.

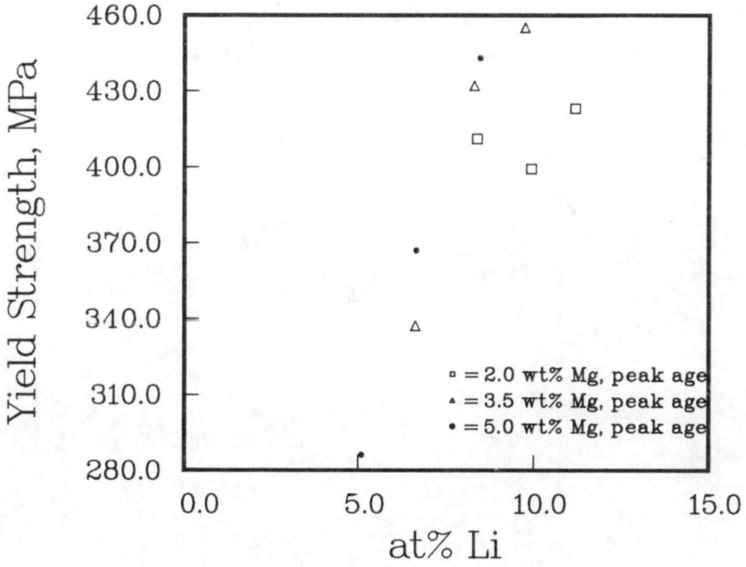

Figure 2. The effect of lithium content on the peak yield strength for the different Mg levels.

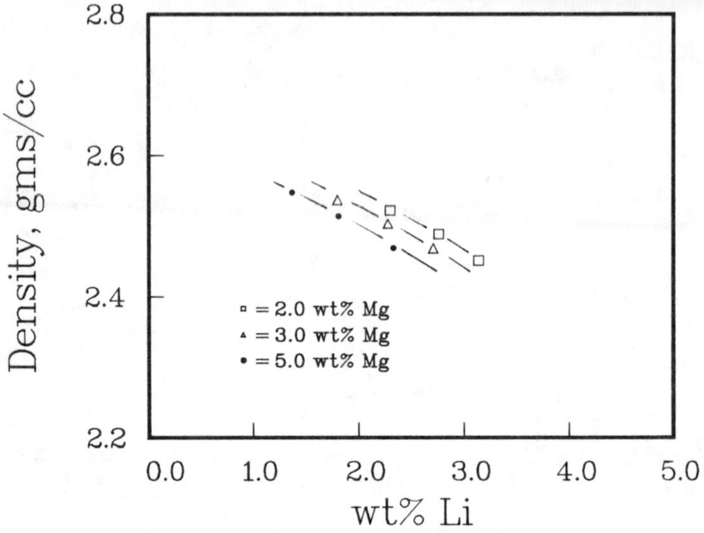

Figure 3. Variation in density as a function of Li content for the different Mg levels.

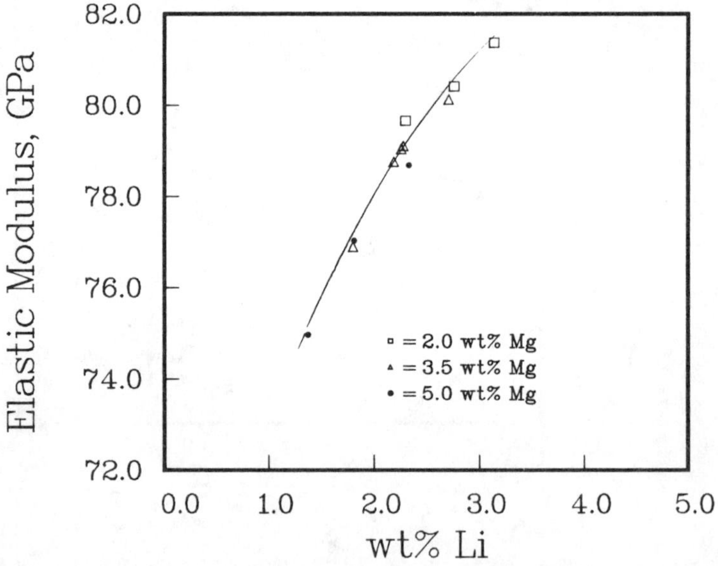

Figure 4. Variation in elastic modulus as a function of Li content for the different Mg levels.

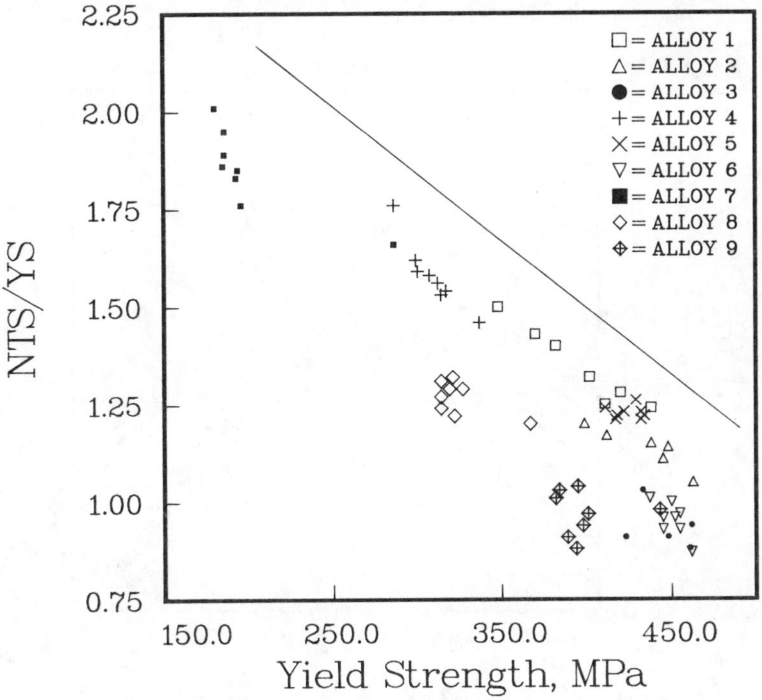

Figure 5. Relationship between NTS/YS and yield strength for the 9 experimental alloys. The solid line represents the trend line for 7075.

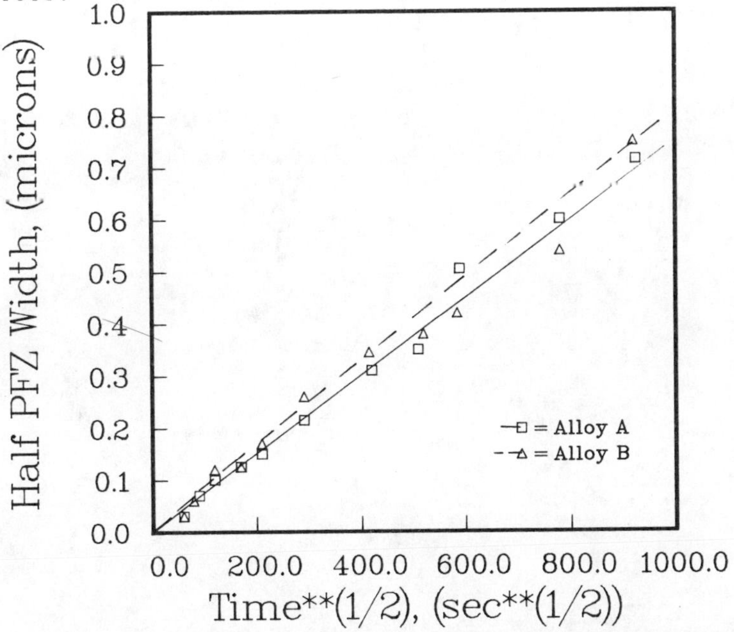

Figure 6. PFZ half-width as a function of isothermal aging time for alloys A and B.

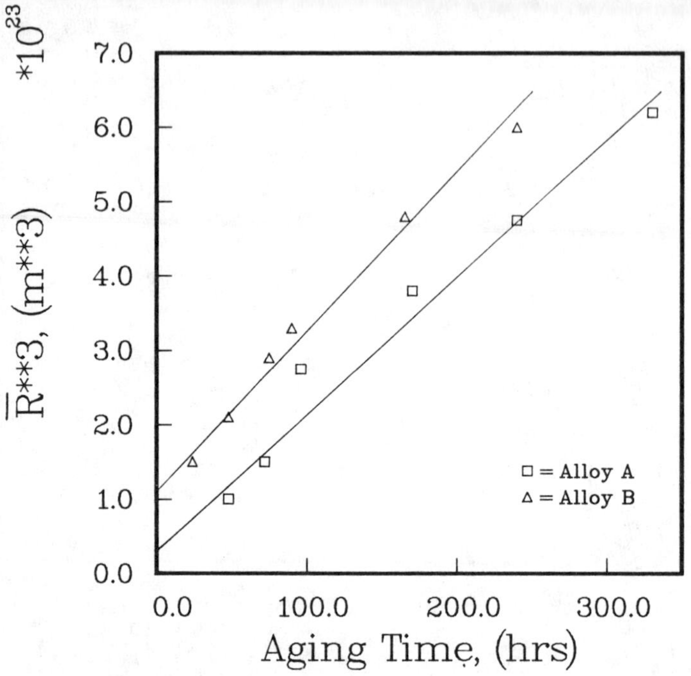

Figure 7. The coarsening behavior of δ' during isothermal aging for alloys A and B.

1st International SAMPE Metals Conference
August 18-20, 1987

EFFECTS OF ALLOY COMPOSITION ON THE AGING RESPONSE OF AL-LI-CU-MG ALLOYS

J.W. Bohlen and K.M. Bresnahan
Northrop Corporation, Aircraft Division
Hawthorne, CA 90250-3277

A.P. Divecha and S.D. Karmarkar
Naval Surface Weapons Center, White Oak Laboratories
Silver Springs, MD 20910-5000

1. ABSTRACT

Northrop has investigated the characteristics of 17 experimental Al-Li alloys covering a broad range of compositions. The alloys had lithium contents ranging from 1.6 to 3.2, copper from 0 to 2.6, and magnesium from 0.5 to 3.5 weight percent with minor Zr, Ge, Y and Ti additions. Densities ranged from 2.52 to 2.60 gm/cm^3 (0.090 to 0.094 lb/in^3) Properties were initially determined for eight alloys in the as-cast condition. Most of these as-cast alloys exhibited modest strengths and low ductilities.

Most of the ductilities decreased to unacceptable levels upon artificial aging. Since the alloys exhibited low potentials as castings, but might be acceptable as wrought alloys, a thermomechanical treatment was selected to improve the alloy properties. Small ingots of 10 alloys were successfully hot upset and rolled in a sequence with a warm roll pass, several hot roll passes, and a final cold roll pass to make small plate samples that were 13 mm (0.5 in) thick. The aging responses determined for each of the alloys were very similar. These wrought alloys developed acceptable tensile properties. Tensile property comparisons indicated very small variations in strength and ductility for the range of compositions examined. Yield strength exhibited some sensitivity to composition, with higher Cu:Li ratios showing higher yield strengths.

2. INTRODUCTION

During the period prior to the registration of commercial alloys, alloy producers, government agencies, and aerospace companies investigated several different alloy systems and ranges of composition. The exploratory work described was focused on composition variations within the Al-Li-Cu-Mg system with Zr, Ti, and/or Y grain refiners, and on the investigation of the effects of Ge additions. Except for the alloy designated NOR 73, all the alloys listed in Table 1 were supersaturated compositions in two-phase regions of the equilibrium diagram at the solution heat treatment temperatures.(1)

The NOR 70 and NOR 90 series alloys listed in Table 1 were cast as small frustums (truncated cones) in segmented brass molds in the glove box facility at the Naval Surface Weapons Center (NSWC), White Oak Laboratories. NSWC supplied the NOR 70 series alloys in 1983 and the NOR 90 series alloys in 1984. The frustums were machined into cylinders to remove surface porosity. Ends were cropped to remove centerline pipe and mold bottom impurities. The NOR 70 series was first evaluated for potential as casting alloys. Only one alloy, NOR 73, which had unintentionally low solute content, developed a cast microstructure amenable to dissolution of segregation via homogenization treatments. As shown in Table 2, the tensile properties of the NOR 70 series alloys evaluated for the T4 and T6 conditions indicated very little promise as casting alloys. The alloys required thermo-mechanical treatment to break up second phase networks, such as those shown in Figure 1, and aid in the precipitation of strengthening precipitates.

3. EXPERIMENTAL

Differential scanning calorimetry was used to determine incipient melting and solid state reaction temperatures during heating of ten as-cast ingots. All alloys were homogenized below their incipient melting points and maintained below these temperatures throughout the additional processing.

Subsequently, 15 alloys from the NOR 70 and NOR 90 series were started through thermomechanical processing (TMP). The processing sequence shown in Figure 2 was used to work the grain structure and introduce a dislocation substructure. Ten alloys survived the TMP to provide small ellipsoidal plates that were 13mm (0.5 in.) thick. All the alloys with > 2.5 Mg content broke apart during the first or second rolling pass. All the successfully processed alloys were water

TABLE 1. CHEMICAL COMPOSITION OF THE NOR 70 AND 90 SERIES ALLOYS

Alloy	Alloying Elements (Weight-Percent)*								
	Li	Cu	Mg	Zr	Ge	Ti	Y	Fe	Si	Al
NOR 72	3.18	0.01	3.29	0.08	-	-	-	N	N	Balance
NOR 72A	3.12	2.55	3.51	0.093	-	0.001	-	0.003	N	Balance
NOR 73	1.56	1.51	0.98	0.13	-	-	-	N	N	Balance
NOR 73C	2.68	2.01	2.62	0.08	-	-	-	0.014	N	Balance
NOR 74	2.89	1.61	1.30	0.09	-	-	-	N	N	Balance
NOR 74H	3.06	1.57	1.36	0.16	-	-	-	0.03	N	Balance
NOR 75	2.02	3.08	0.95	0.20	-	-	-	N	N	Balance
NOR 76A	2.94	1.85	2.49	-	-	-	0.03	N	N	Balance
NOR 77A	2.90	1.85	2.53	0.10	-	-	0.12	N	N	Balance
NOR 78	2.70	1.80	2.57	0.09	-	0.06	-	N	N	Balance
NOR 91	2.49	2.60	0.55	0.14	0.006	-	-	0.03	0.06	Balance
NOR 92	2.44	2.60	0.57	0.11	0.54	-	-	0.02	0.09	Balance
NOR 93	2.32	2.61	0.52	0.13	-	0.09	-	0.03	0.07	Balance
NOR 94	3.01	2.64	0.63	0.18	-	0.01	-	0.02	0.08	Balance
NOR 95	3.07	2.02	0.62	0.14	-	-	-	0.02	0.08	Balance
NOR 96	3.17	2.00	0.52	0.17	0.55	-	-	0.02	0.08	Balance
NOR 97	3.07	1.55	0.55	0.13	-	-	-	0.03	0.09	Balance

*-N indicates analysis not performed.

quenched following the eighth hot-reduction pass. Each received a final cold reduction pass of 2.2 to 3.2 percent of thickness to achieve uniform plate thickness without risking severe cracking.

Age hardening curves were generated for each of the ten surviving alloys. A typical aging response curve is shown in Figure 3. Most alloys exhibited a single broad peak or a broad double peak response. Aging times and temperatures were chosen to provide the peak hardness. NOR 73 and NOR 97 specimens were also aged for longer times to evaluate the properties at the second peak.

4. RESULTS

Tensile properties are summarized in Tables 3 and 4. Alloy NOR 73, which had the best potential as a casting alloy, showed improvement

TABLE 2. TENSILE PROPERTIES OF NOR 70 SERIES ALLOYS AS CASTINGS

Alloy (Composition)	Temper	TYS MPa (ksi)	UTS MPa (ksi)	elong (%)
NOR 72A (Al-3.1Li-2.6Cu-3.5Mg-0.09Zr)	T4	*	228(33)	OSG#
NOR 73 (Al-1.6Li-1.5Cu-1Mg-0.13Zr)	T4	221(32)	338(49)	24
	TX6	248(36)	351(51)	16
NOR 73C (Al-2.7Li-2Cu-2.6Mg-0.08Zr)	T4	*	241(35)	1
NOR 74 (Al-2.9Li-1.6Cu-1.3Mg-0.09Zr)	T4	241(35)	296(43)	4
	TX6	352(51)	365(53)	1
NOR 75 (Al-2Li-3.1Cu-1Mg-0.17Zr)	T4	310(45)	359(52)	1.5
	TX6	*	414(60)	<1
NOR 76A (Al-2.9Li-1.9Cu-2.5Mg-0.03Y)	T4	*	269(39)	1
NOR 77A (Al-2.9Li-1.9Cu-2.5Mg-0.10Zr-0.12Y)	T4	*	283(41)	1
NOR 78 (Al-2.7Li-1.8Cu-2.6Mg-0.09Zr-0.06Ti)	T4	276(40)	290(42)	1
<u>Reference Alloys</u>(2)				
A357 (Al-7Si-0.5Mg)	T6	290(42)	360(52)	8
A201 (Al-4.6Cu-0.7Ag-0.35Mn-0.35Mg-0.25Ti)	T4	215(31)	365(53)	20
	T6	435(63)	485(70)	7

*Ductility was inadequate to define YS
#Outside gage length

in yield strength following TMP. However, the NOR 73-T6 properties were significantly lower than those for wrought aluminum alloys such as 2024.

All the alloys in Table 4 exhibited acceptable strength relative to commercial alloys. Certain alloys, such as NOR 74 exhibited low ductility which may have been influenced by porosity.

The data from Table 4 is plotted in Figure 4 to show the relationship between tensile properties and Cu:Li ratio. For the Al-Li-Cu-Mg-Zr system examined, only the tensile yield strength (TYS) showed a good correlation by increasing in proportion to the Cu:Li ratio. Neither the ultimate tensile strength nor the elongation showed significant variations with the Cu:Li ratio.

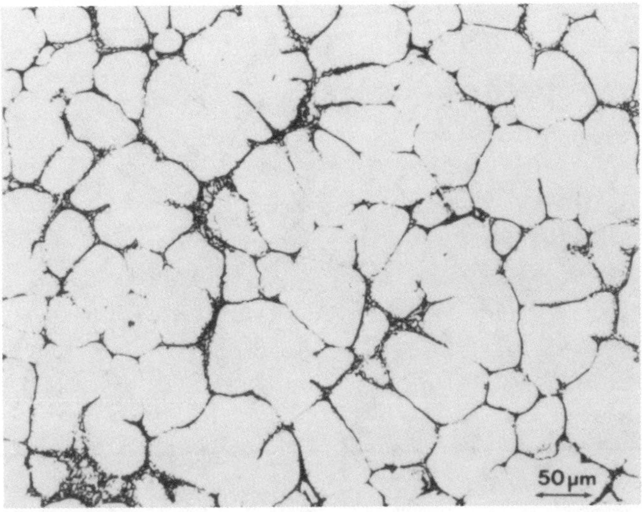

Figure 1. Networks of Second Phase in NOR 78 Alloy.

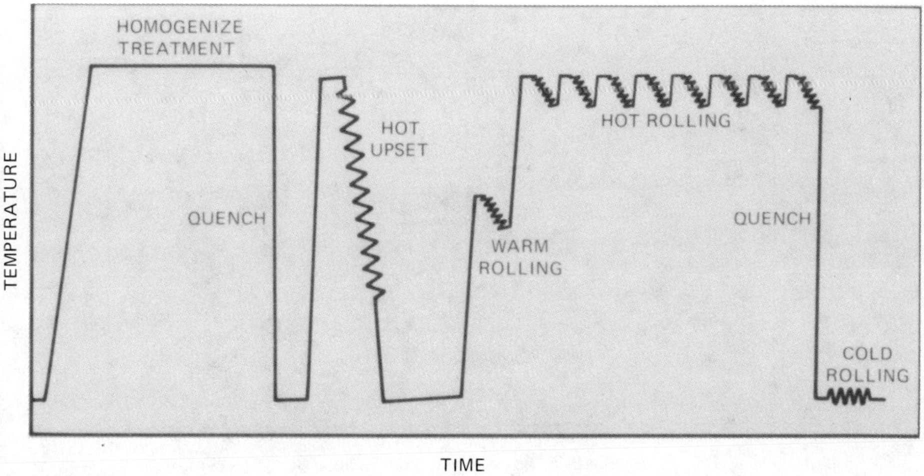

Figure 2. Processing Sequence for Working Alloy Structure

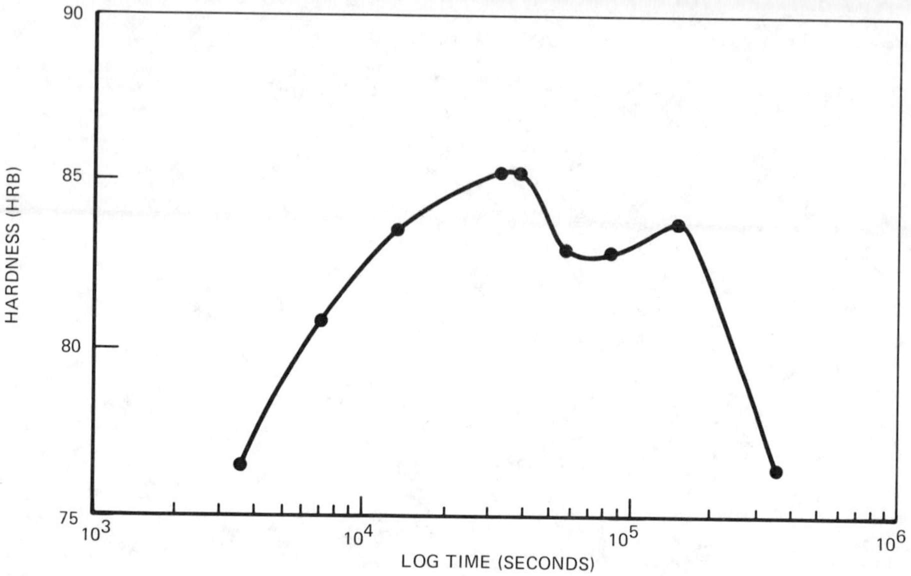

Figure 3. Typical Aging Response at 190C (375F) Exhibited by NOR 92

TABLE 3. TENSILE PROPERTIES OF NOR 73
(Al-1.6Li-1.5Cu-1Mg-0.13Zr)

Alloy-Temper	TYS MPa (ksi)	UTS MPa (ksi)	elong (%)
NOR 73-T4	221 (32)	338 (49)	24
NOR 73-T6X	248 (36)	352 (51)	16
NOR 73-T8*	310 (45)	359 (52)	16
Reference Alloys(2)			
A201-T4	215 (31)	365 (53)	20
A201-T6	435 (63)	485 (70)	7
A357-T62	290 (42)	360 (52)	8
2024-T4	325 (47)	470 (68)	20
2024-T361	395 (57)	495 (72)	13

X - Experimental temper aged at 190C (375F)
* - Experimental temper aged at 190C (375F) following thermomechanical treatment

TABLE 4. LONGITUDINAL TENSILE PROPERTIES OF T8 TEMPER ALLOYS FOLLOWING TMP

Alloy (Composition)	TYS MPa (ksi)	UTS MPa (ksi)	elong (%)	RA (%)
NOR 74 (Al-2.9Li-1.6Cu-1.3Mg-0.09Zr)	448(65)	527(76)	4.5	7.3
NOR 74H (Al-3.1Li-1.6Cu-1.4Mg-0.16Zr)	423(61)	533(77)	6.5	8.6
NOR 75 (Al-3.1Cu-2.0Li-0.95Mg-0.20Zr)	496(72)	534(78)	7	19
NOR 91 (Al-2.5Li-2.6Cu-0.55Mg-0.54Ge-0.14Zr)	492(71)	541(79)	6.5	13.5
NOR 92 (Al-2.4Li-2.6Cu-0.57Mg-0.11Zr-0.09Ti)	477(69)	523(76)	5	9.1
NOR 93 (Al-2.3Li-2.6Cu-0.52Mg-0.13Zr)	474(69)	521(76)	6	12.4
NOR 94 (Al-3.0Li-2.6Cu-0.63Mg-0.18Zr)	449(65)	512(74)	5.5	5.0
NOR 95 (Al-3.1Li-2.0Cu-0.62Mg-0.14Zr)	441(64)	524(76)	6	8.7
NOR 96 (Al-3.2Li-2.0Cu-0.52Mg-0.55Ge-0.17Zr)	450(65)	525(76)	5	8.8
NOR 97	421(61)	527(76)	6.5	9.0
(Al-3.0Li-1.6Cu-0.55Mg-0.13Zr)	449(65)*	528(77)*	5.5*	10.0*
<u>Reference Alloys</u>(3,4)				
8090-T651 Plate (L-orientation)	440(64)	490(71)	5.5	-
8090-T8 Sheet (L-orientation)	420(61)	490(71)	5.5	-
(Al-2.45Li-1.3Cu-0.95Mg-0.10Zr)				
8091-T651 Plate (L-orientation)	510(74)	550(80)	5.0	-
8091-T8 Sheet (L-orientation)	450(65)	510(74)	5.5	-
(Al-2.6Li-1.9Cu-0.85Mg-0.12Zr)				
2090-T8E41 Plate (L-orientation)	535(78)	565(82)	11.0	-
2090-T6 Plate (L-orientation)**	441(64)	517(75)	2.7	6.5
(Al-2.2Li-2.7Cu-0.12Zr)				

*Properties from aging time established by the second peak.
**Test data.

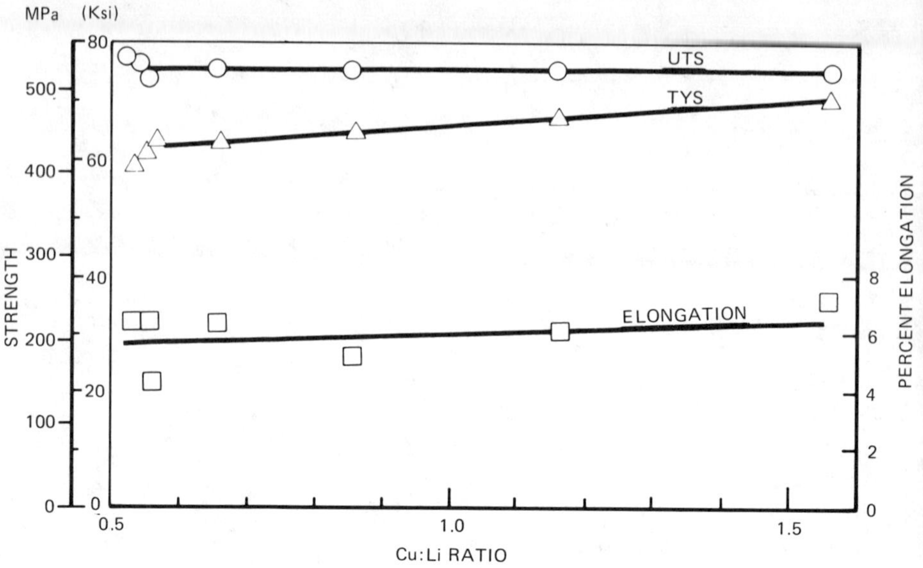

Figure 4. Relationship of Tensile Properties to Copper:Lithium Ratio in Al-Li Alloys

Analysis of other composition effects, such as Cu:Mg, Li:Mg, total amount of grain refiners showed poor correlation with tensile properties. The results from the study on germanium content were mixed. Germanium seemed to provide a small strength increase in NOR 91 relative to NOR 93, but showed little effect on strength in the NOR 96 to NOR 95 comparison with higher Li content and lower Cu:Li ratio. The addition of titanium to NOR 92 did not improve strength and may have reduced ductility relative to NOR 93.

5. CONCLUSIONS

a. For slightly supersaturated solutions in the Al-Li-Cu-Mg system, the aging response and tensile strength following a small amount of cold work did not vary significantly with composition. However, the yield strength increased slightly with increasing Cu:Li ratio without reducing ductility.

b. The addition of Ge to slightly supersaturated Al-Li-Cu-Mg systems produced a small positive effect on strength with a Cu:Li ratio of about 1.0, but little effect with a Cu:Li ratio of about 0.6 and/or a higher lithium content.

c. Al-Li-Cu-Mg-Zr alloys with about 2.5 Mg were difficult to warm or hot roll without cracking.

6. ACKNOWLEDGEMENTS

The authors acknowledge the assistance of Mr. Bruce Roberts at Kaiser Aluminum and Chemical Company, Center for Technology in thermomechanical processing.

7. REFERENCES

1. L.F. Mondolfo, Aluminum Alloys: Structure and Properties, 1976, p. 496, p. 498, and p. 555.

2. Metals Handbook Ninth Edition, Volume 2, Properties and Selection: Nonferrous Alloys and Pure Metals, American Society for Metals, 1979, p. 76, p. 153, and p. 168.

3. LITAL, Aluminium-Lithium Alloys, Current Data, British Alcan Alumin-ium plc brochure, 1986, pp. 2-4.

4. S.J. Cieslak, et al., "Alcoa Alithalite Alloy 2090 Technical Information," 17th Internatioinal SAMPE Technical Conference, 1985, p. 2 and p. 17.

8. BIOGRAPHIES

James W. Bohlen is a SeniorTechnical Specialist in the Materials and Processes Research and Development Organization of the Northrop Corporation Aircraft Division. Mr. Bohlen has an MS in Material Science from Cornell University and a BS in Metallurgical Engineering from the University of Illinois. Mr. Bohlen has been involved in Al-Li research at Northrop since 1980.

Keith M. Bresnahan is an Engineering Specialist in the Materials and Processes Research and Development Organization of the Northrop Corporation Aircraft Division. Mr. Bresnahan has an MS in Metallurgy and a BS in Chemical Engineering from the Georgia Institute of Technology. Mr. Bresnahan has been involved in Al-Li research since 1982.

A.P. Divecha, received his degree in Chemistry from the University of Bombay (1958) and his Masters in Metallurgy from University of Utah (1962). From 1976 onwards, he has been employed at the Naval Surface Weapons Center (NSWC) as a group leader in the Metallic Materials Branch. He is responsible for planning and directing activities in metal matrix composites, ceramic matrix composites, superconductors and aluminum-lithium alloys. He is author and co-author of more than 30 technical papers and reports composites, superconductors and aluminum-lithium alloys. He is author and co-author of more than 30 technical papers and reports and holds 8 patents relating to composite materials.

S. D. Karmarkar, received his BS in Mechanical Engineering from the University of Ranchi, Birla Institute of Technology, Ranchi, India in 1963 and his MS in Metallurgy from Deutsch Technische Schule, Dusseldorf, West Germany in 1967. Presently at NSWC, he is working as a Metallurgist and is responsible for the development of Al-Li alloys including TMP, melting, casting, deformation processing, mold and die design and overall research and development tasks. His other duties involve development of metal matrix and ceramic matrix composites and superconductor processing. He is the author and co-author of 32 technical papers and reports.

1st International SAMPE Metals Conference
August 18-20, 1987

ON CORROSION-FATIGUE CRACK GROWTH BEHAVIOR OF ALUMINUM-LITHIUM ALLOY 2090 IN SALT WATER

G. R. Yoder, P. S. Pao, M. A. Imam and L. A. Cooley
Naval Research Laboratory
Washington, DC 20375-5000

Abstract

Work has been undertaken to develop an understanding of corrosion-fatigue behavior of aluminum-lithium alloy 2090 in the T8E41 condition, for a salt-water environment. This alloy, which exhibits significantly superior strength-to-density ratio and elastic modulus when compared to 7075-T651 aluminum alloy, has been examined in the product form of 12.7-mm (0.50 in.) plate. In particular, precracked WOL-type fracture mechanics specimens of this material have been immersed in a flowing solution of 3.5 pct NaCl aqueous solution. Corrosion-fatigue crack growth rates have been determined over a broad spectrum of stress-intensity range levels. The influence of cyclic frequency has been examined at levels of 0.1 Hz and 5.0 Hz for a sinusoidal waveform and stress ratio of R = 0.10. The effect of crack-tip orientation, LT versus TL, has also been determined. Crack growth measurements have been made using the compliance based clip-gage technique, with evidence of crack closure duly recorded. Corrosion-fatigue sensitivity in the salt-water environment, as referenced to ambient air, is rationalized on a micromechanistic basis, including effects reflected in crack closure. The relation between cracking resistance and microstructure will be examined in light of evidence from textural analysis as well as scanning and transmission electron microscopy.

1. INTRODUCTION

Aluminum-lithium alloys have attracted a great deal of

interest as candidate materials for next-generation aerospace applications, owing primarily to superior specific strength and elastic modulus as compared to traditional aluminum alloys. Though a number of studies have also reported superior fatigue crack growth resistance in air (1-5), much less is known concerning the corrosion-fatigue sensitivity of these alloys in more aggressive, marine environments (6). In the present study, work has been undertaken to develop an understanding of the corrosion-fatigue crack growth behavior of Al-Li alloy 2090 in 3.5% NaCl aqueous solution.

2. MATERIAL AND PROCEDURE

The 2090-T8E41 alloy was in the form of 12.7-mm (0.50 in.) plate from Alcoa, with the following nominal composition (in wt pct): 1.9-2.6 Li, 2.4-3.0 Cu, 0.08-0.15 Zr, 0.12 Fe(max), 0.10 Si(max), 0.25 Mg(max), 0.05 Mn(max), 0.15 Ti(max), and Al (bal). Uniaxial tensile data, obtained from 6.35-mm (0.250 in.) diameter specimens, are given in Table I.

Microstructures (light micrographs) for the primary planes of the plate, L, T and S (cf. ASTM-E616), are given in isometric form in Fig. 1, with a thin-foil transmission electron micrograph in Fig. 2 showing δ', θ' and T1 precipitates as well as a relatively large Fe/Cu-rich constituent particle.

Fatigue crack growth rates (da/dN) have been determined in accord with ASTM E647 from WOL-type compact specimens (7), of width W = 64.8 mm (2.55 in.) and thickness B = 12.7 mm (0.50 in.). Specimens were cyclically loaded with a stress ratio of R =

Table I: Mechanical Properties

Orient.	Yield Strength, MPa (ksi)	Ultimate Tensile Strength, MPa (ksi)	El. (%)	Red. Area (%)	Young's Modulus, GPa (10^3ksi)
Long.	517 (75.0)	554 (80.4)	8	12	78.6 (11.4)
Trans.	518 (75.1)	551 (79.9)	6	17	78.9 (11.5)

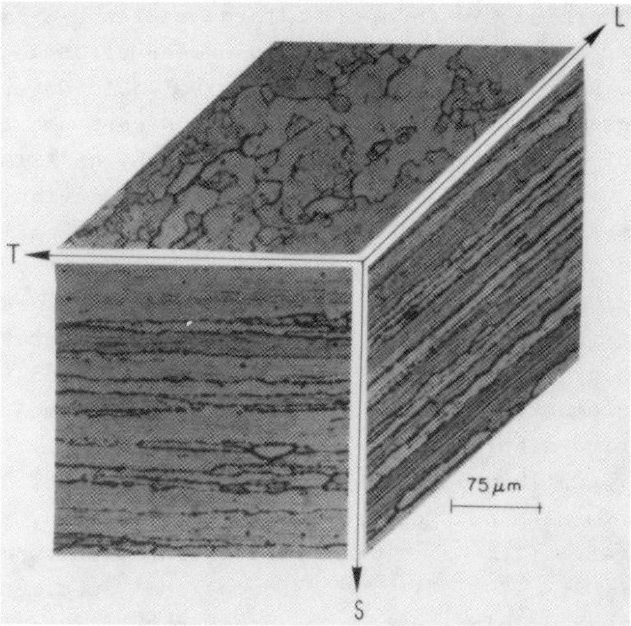

Fig. 1 Microstructure in the L, T, and S planes (etched with Keller's reagent).

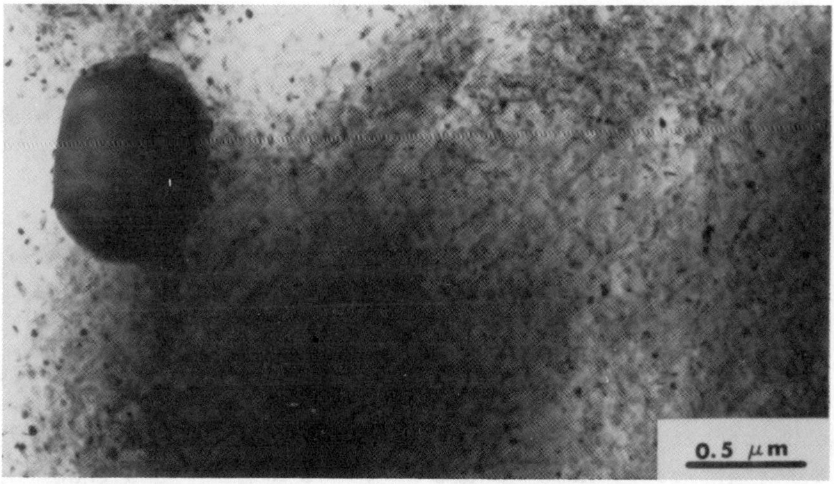

Fig. 2 Thin-foil transmission electron micrograph illustrating fine δ′, θ′ and T1 precipitates and relatively large Fe/Cu-rich constituent particle.

0.10, a haversine waveform, and with frequencies of 0.10 Hz and 5.0 Hz in 3.5 pct NaCl aqueous solution (flowing); in air, only the latter frequency was examined. Resistance to fatigue crack growth was determined over a substantial spectrum of stress-intensity range (ΔK), with crack growth determined from measurements of crack-mouth-opening-displacement (CMOD)[8]. Calibration equations for ΔK and CMOD are given in ref. (7).

Though it was originally planned to measure da/dN for both the LT and TL crack orientations (cf. ASTM-E616) from nonsidegrooved specimens, sidegrooves were soon adopted following early experience with crack-path deviations from the Mode I plane. These sidegrooves were cut to a 5-pct depth per side, with a vee-notch geometry of 45-degree included angle and 0.25-mm (0.010 in.) root radius.

3. RESULTS AND DISCUSSION

The propensity of this material for out-of-plane cracking is so great that even with sidegrooved specimens, crack paths typically deviated from the intended Mode I plane at large angles --- sometimes in excess of 35 degrees. Thus reliable determinations of da/dN were not feasible in most cases, as ASTM E647 limits such deviations to 5 degrees (NRL experience also raises reliability questions for significant deviations). For the TL orientation, this problem persisted in all specimens examined. Consequently, the only fatigue crack growth rate data reported in this study are for the LT orientation, where for an occasional specimen, the crack path did trace the Mode I plane to permit the limited da/dN data presented in Fig. 3.

These data indicate that corrosion-fatigue crack growth rates in 3.5 pct NaCl aqueous solution are elevated approximately 6-fold relative to region II fatigue crack growth rates in ambient air. However, inasmuch as growth rates in air are about 18 times less than counterpart growth rates for Al-7075-T651 alloy, the corrosion-fatigue crack growth rates for Al-2090-T8E41 in salt water are still 3-fold less than for Al-7075-T651 in air. The effect of cyclic frequency on corrosion fatigue in the Al-Li alloy appears to be

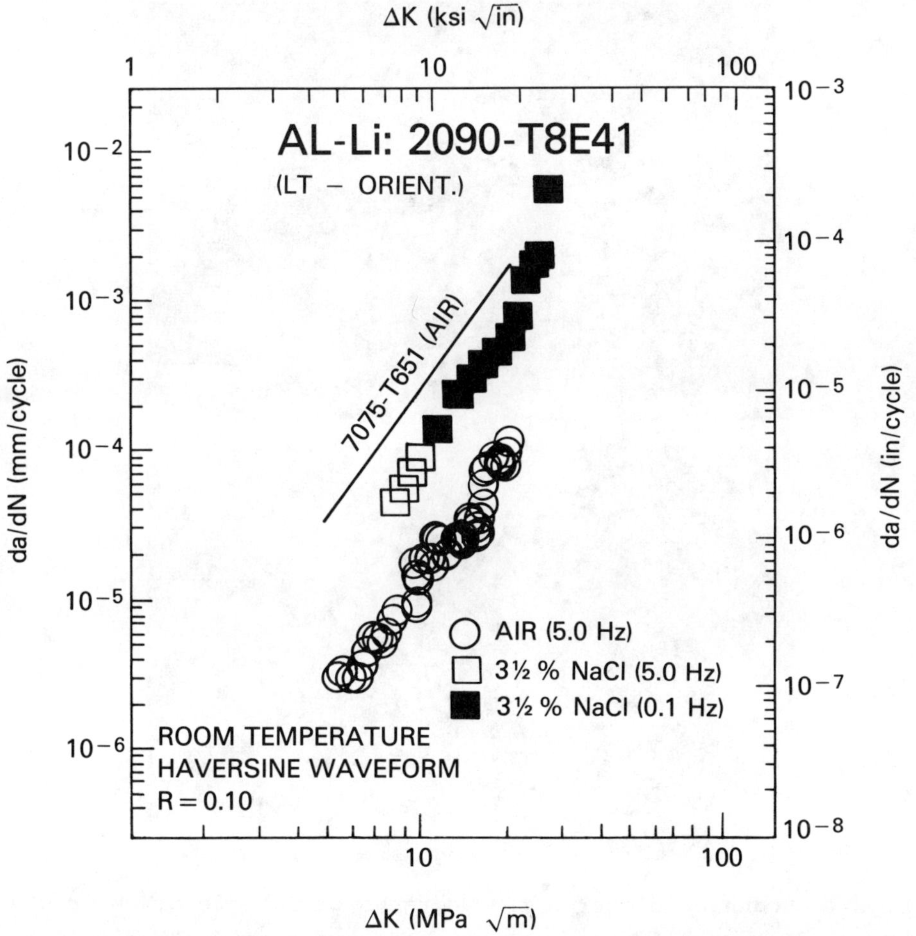

Fig. 3 Fatigue crack growth rates in salt water and air, for LT crack orientation.

minimal, though the ΔK regimes examined at 0.10 Hz and 5.0 Hz were very limited, owing to the crack-path deviation problem.

It is significant to observe, however, that the 6-fold increase in corrosion-fatigue crack growth rates of the Al-Li alloy in salt water relative to those in air is large compared to the 2-fold increase typically observed in the 7075-T651 alloy(9). This apparently greater sensitivity to salt water is likely a consequence of relative fatigue fracture surface roughness levels. The fatigue fracture surface of the Al-Li alloy in air, cf.

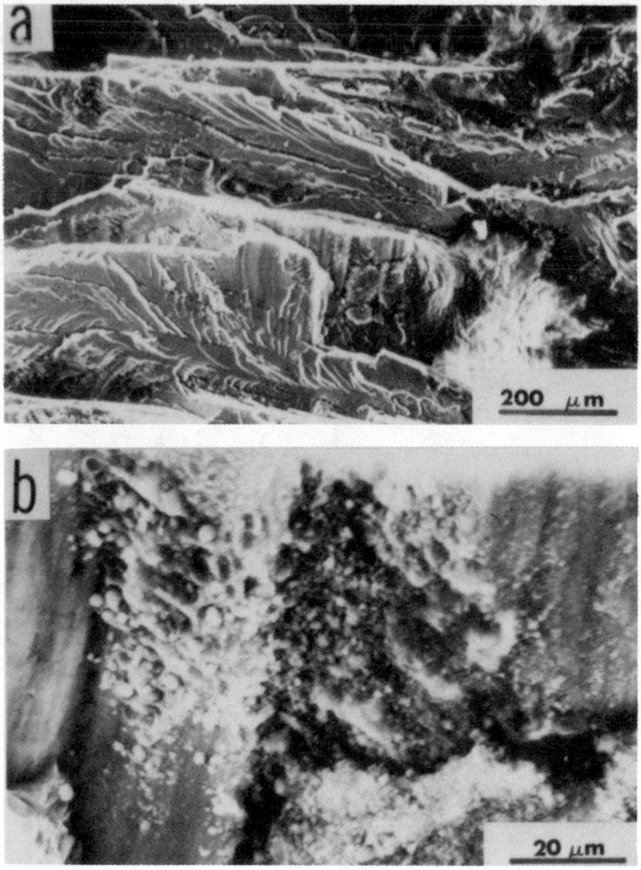

Fig. 4 Scanning electron fractographs of fatigue crack growth in air: a) overview, b) enlargement of fretted oxide debris.

Fig. 4, exhibits an extraordinary tortuosity, with considerable oxide debris likely attributable to fretting ---giving rise to a macroscopically blackish appearance. On the other hand, the counterpart fracture surface from salt-water corrosion fatigue, cf. Fig. 5, exhibits a flatter fractographic morphology, with corrosive debris giving rise to a "mudcake" appearance. Consistent with the flatter morphology of the corrosion-fatigue fracture surface, the apparent crack closure levels were significantly less than for fatigue crack growth in air; e.g. at $\Delta K = 11$ MPa$\sqrt{m}$ (10 ksi$\sqrt{in.}$), the apparent closure load in salt water is 27 pct of maximum load,

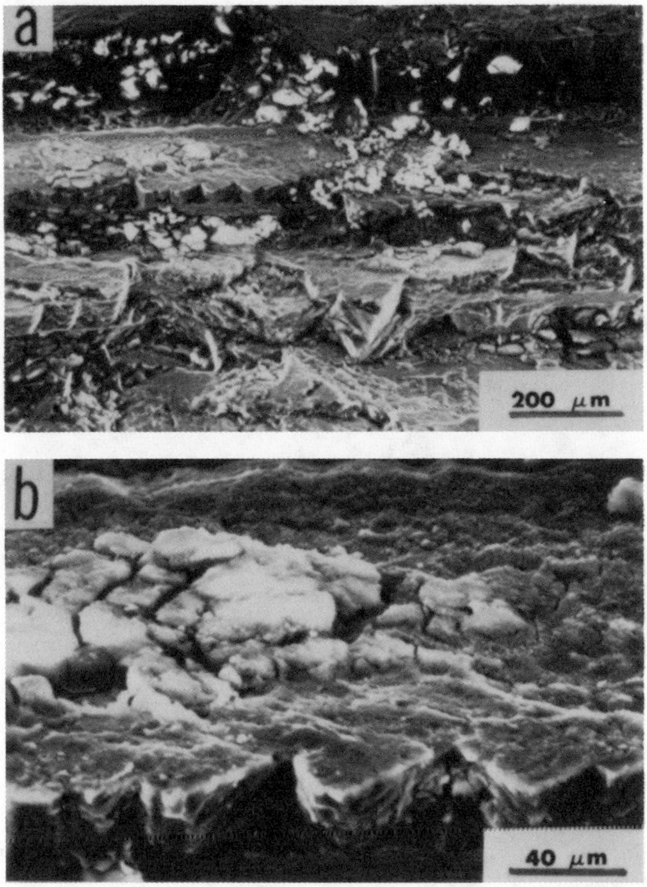

Fig. 5 Scanning electron fractographs of corrosion-fatigue crack growth in salt water: a) overview, b) enlargement of corrosive debris "mudcake" pattern.

whereas in air it is 70 pct . By contrast, the fatigue fracture surface in air for other, conventional aluminum alloys appears much flatter.

Finally, it is appropriate to further consider the unusual propensity of the Al-Li alloy for crack path deviation, if fatigue crack growth behavior in this material is to be truly understood. An example of this out-of-plane cracking tendency is illustrated in Fig. 6 for a sidegrooved specimen of TL orientation, where twin but symmetric crack-path deviations can be observed at roughly 35 degrees from the Mode I plane. A comprehensive study has

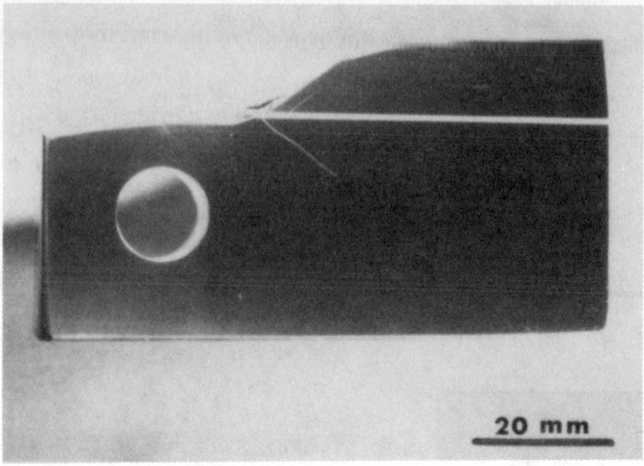

Fig. 6 Out-of-plane cracking in sidegrooved WOL-type fracture mechanics specimen of TL orientation.

been initiated to examine whether texture of the Al-Li plate material might be responsible for such behavior. Preliminary evidence from this work reveals a sharp contrast between texture in the plate surface and midthickness locations, as illustrated in Fig. 7. Tentatively, the texture in the near-surface regions has been identified as a recrystallization texture (10), (001) [$\bar{1}$10], and at the midthickness as a normal rolling texture, (110) [$\bar{1}$12] --- with a transition or combination of both textures at the thickness quarter-point, as illustrated. Work is in progress to document the microstructural variation associated with this textural change.It is further noted that this textural change contrasts with prior work with 8090 Al-Li alloy, in the form of 25.4-mm (1.00 in.) plate from Alcan International(11). In that work, the textural type remained the same for different through-thickness locations, but with increased intensity as the midthickness location was approached.

4. CONCLUSIONS

For the conditions examined in this study of Al-Li alloy 2090-T8E41, the results indicate that:

1. Corrosion-fatigue crack growth rates in 3.5 pct NaCl aqueous solution are elevated approximately 6-fold relative to region II fatigue crack

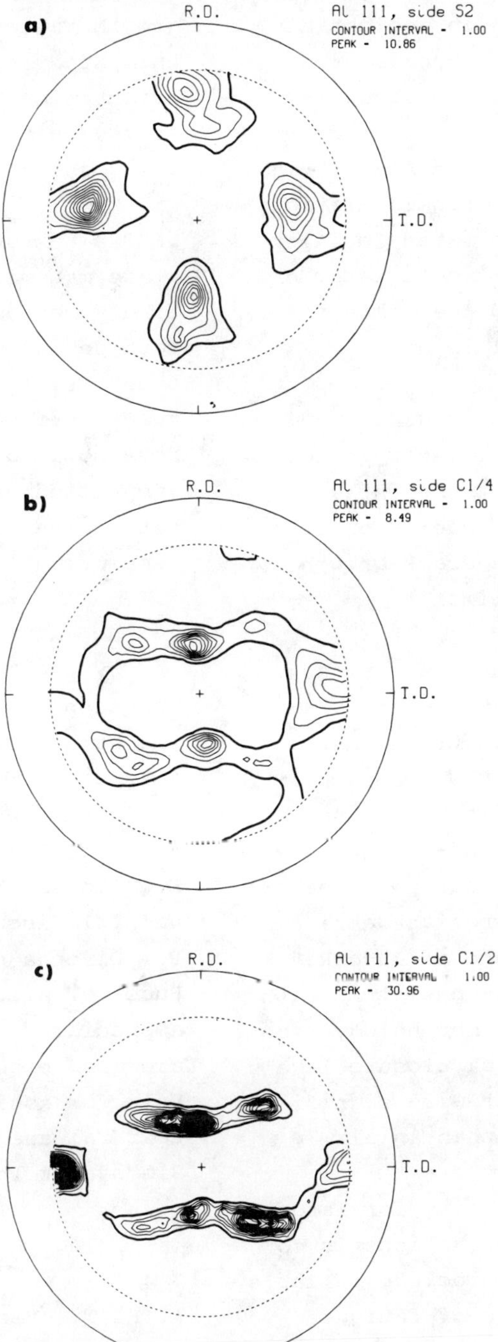

Fig. 7 Contrast in (111) pole figures for: a) surface and c) midthickness locations in plate, with b) transitional texture at the quarterpoint.

growth rates in ambient air. However, inasmuch as growth rates in air are about 18 times less than counterpart growth rates for Al-7075-T651 alloy, the corrosion-fatigue crack growth rates for Al-2090-T8E41 in salt water are still 3-fold less than for Al-7075-T651 in air.

2. For the salt-water environment, preliminary evidence suggests that the influence of cyclic frequency on corrosion-fatigue crack growth rates is minimal.

3. Though the fracture surface from fatigue in air exhibits an extraordinary tortuosity, the counterpart fracture surface from salt-water corrosion-fatigue exhibits a flatter morphology, with corrosive debris giving rise to a "mudcake" appearance. Consistent with this flatter morphology, the apparent crack closure levels in salt water were significantly less than in air.

4. The propensity for large crack-path deviations is so great as to preclude reliable determination of fatigue crack growth rates for the TL orientation, even with side-grooved specimens. Consequently, the only crack growth rate data reported in this study are for the LT orientation, where the crack stayed in the Mode I plane in an occasional specimen.

5. The texture of this Al-Li plate material, which is most likely responsible for crack-path deviation, exhibits a substantial difference between surface and midthickness positions, tentatively identified as recrystallization and rolling textures, respectively.

5. ACKNOWLEDGMENTS

The support of this work by the Naval Air Systems Command is gratefully acknowledged, as is the encouragement of J. F. Collins, W. Koegel and L. E. Sloter. A number of very helpful discussions with A. P. Divecha of NSWC and R. J. Bucci of Alcoa are also much appreciated. Special thanks are also extended to C. L. Vold for assistance with textural analysis and to P. Slebodnick for metallography.

6. REFERENCES

1. K. T. Venkateswara Rao, W. Yu and R. O. Ritchie, "Fatigue Crack Propagation in Aluminum-Lithium Alloy 2090: Part I. Long Crack

Behavior," submitted to Met. Trans. A.

2. S. J. Harris, B. Noble and K. Dinsdale, Fatigue '84, C. J. Beevers, ed., 1, EMAS, Ltd., West Midlands, U. K., pp.361-370, 1984.

3. K. V. Jata and E. A. Starke, Met. Trans. A, 17A, pp. 1011-1026, 1986.

4. A. K. Vasudevan, P. E. Bretz, A. C. Miller, and S.Suresh, Mater. Sci. Eng., 64, pp. 113-122, 1984.

5. E. J. Coyne, T. H. Sanders and E. A. Starke, Aluminum-Lithium Alloys, T. H. Sanders, Jr. and E. A. Starke, Jr., eds., TMS-AIME, Warrendale, PA, pp. 293-305, 1981.

6. P. S. Pao, K. K. Sankaran, and J.E. O'Neal, Aluminum-Lithium Alloys, T. H. Sanders, Jr. and E. A. Starke, Jr., eds., TMS-AIME, Warrendale, PA, pp. 307-323, 1981.

7. A. Saxena and S. J. Hudak, Jr., Inter. Journ. Fracture, 14, pp.453-468, 1978.

8. G. R. Yoder, L. A. Cooley and T. W. Crooker, Fatigue Crack Growth Measurement and Data Analysis, S. J. Hudak, Jr. and R. J. Bucci, eds., ASTM STP 738, ASTM, Philadelphia, PA, pp. 85-102, 1981.

9. P. S. Pao, Ming Gao, and R. P. Wei, Scripta Met., 19, pp. 265-270, 1985.

10. J. E. Hatch, ed., Aluminum: Properties and Physical Metallurgy, ASM, Metals Park, OH , pp. 105-133, 1984.

11. M. Peters, J. Eschweiler and K. Welpmann, Scripta Met., 20, pp.259-264, 1986.

7. BIOGRAPHIES

George Yoder, Peter Pao and Laurence Cooley are with the Fatigue Criteria Section, and Ashraf Imam is with the Physical Metallurgy Branch of the Naval Research Laboratory. Dr. Yoder, a metallurgist, has degrees from Syracuse University and the University of Michigan. Dr. Pao, a materials research engineer, has degrees from North Carolina State University and National Taiwan University. Dr. Imam, a metallurgist, has degrees from George Washington University, Carnegie-Mellon University and the Regional Institute of Technology (India). Mr. Cooley, a metallurgist, has B.S. and M.S. degrees from the University of Missouri-Rolla.

1st International SAMPE Metals Conference
August 18-20, 1987

ELECTRON BEAM PROCESSING, A KEY TO METAL CLEANLINESS

Robert Bakish
Bakish Materials Corporation
P.O. Box 148, Englewood, NJ 07631

Abstract

The electron beam as a heat source and its role in melting and refining leading to the production of super clean alloys is presented. This is followed by a discussion of the effect of inclusions on the properties of alloys and the need for super clean alloys. The electron beam melting test for evaluation of alloy cleanliness follows. The paper concludes by reference to the entities responsible for the development and support of this effort to produce super clean alloys.

I INTRODUCTION

The electron beam melting and refining process was patented[1)] in 1907 in the U.S., but not until the late 50's under the pioneering effort of H. Smith and his associates of Temescal did it become a process to be reconed with. In the 60's and 70's we saw this technology expand and in fact become the backbone for the processing of many reactive and most refractory metals and alloys. The 80's with major advances in vacuum, computer and electron gun technology brought us the real big hearth furnaces, furnaces exceeding 2 megawatts of power. We had a truly viable economic option to produce super clean metals, and we are in the process of deriving the benefits from it.

The writer referred briefly to these historic milestones only to show that we are dealing not with a new technology, but only with the latest phase of a proven industrial technology, a technology with outstanding capabilities. He would like to add that this is a technology which is far from being truly understood, much less appreciated by those in the metallurgical industry who stand to derive the greatest economic benefits from it.

The electron beam, the heat source here is extremely flexible and capable of most precise controls. Fig. 1 shows a cross section of a typical electron gun. The recent major advances in computer technology have given its users extreme capabilities for management of its applications. The state of the art electron guns have outstanding reliability, minimal maintenance and operational simplicity. In melting installations around the world they are in service 24 hours a day and up to 7 days a week, exceeding by a notable margin the performance of their predecessors.

To day Leybold-Heraeus and The Institute M. von Ardenne share the world electron gun market. In the U.S. the von Ardenne electron gun installed melting and refining capacity exceeds that of Leybold-Heraeus by a factor of more than three. The von Ardenne Institute is the only producer of 1200KW. All its high power guns feature gas focussing offering stability as melt chamber pressure exceed the 10^{-4}Torr range.

II ELECTRON BEAM MELTING TODAY

Electron beam melting and refining today is a two mode technology. The first, drip melting, is the mainstay of the refractory metals industry. The second electron beam cold hearth refining (EBCHR) is the key to to days exciting area of growth, the production of super clean metals and alloys. Fig. 2 shows a schematic of the two modes.

Electron beam cold hearth refining makes possible the removal of both lighter and heavier than the metal or alloy processed inclusions. The lighter inclusion are floated on the surface for removal, while the heavier ones sink and are removed in the skull. This capability has been enhanced by a recent break through which makes possible real time composition monitoring and control.

Electron beam processing has thus become a most powerful technology for the production of a wide range of clean materials.

Super clean metal production and results obtained by EBCHR depend on a number of parameters including:

1. The size of, i.e. depth and width of the hearth.
2. The molten metal pool depth.
3. The melting rate.
4. The metal stream flow velocity and the average resistance time in the hearth.
5. The presence and/or absence of pool turbulence or agitation.
6. The presence or absence of skim dams and weirs and their design.
7. The power input controls and distribution on the pool surface.

Fig. 3 shows mechanism proposed by Jarrett[2)] which is operational in reducing defects as function of the melting rate, while the effect of superheat is shown in Fig. 4.

Let us look at the A. Johnson Metals Corporation furnace, a 2 megawatt installations powered by four 600KW von Ardenne guns, Fig. 5. To this writer this is the best operated electron beam melting facility in the U.S. It was designed, built and its operation is still supervised by H. Harker. He is one of the pioneers of electron beam melting and refining technology, having entered the field in the late 50's with a Berkeley physics degree. His latest contribution has been his intimate involvement in the development, the support and the placing in service of real time composition monitoring. Fig. 6 shows a view of the top of the electron guns, while Fig. 7 is a look into the furnace showing the hearth and the water cooled circular mold. In Fig. 8 the hearth and mold are shown during the casting of a cylindrical ingot, one of a number of possible shapes. Fig. 9 shows a large ingot.

III BUT WHY CLEAN METALS?

The best way to appreciate the need for super clean metals and alloys is to note the manner in which defects and specifically inclusions affect the properties of these materials. As much as this is a paper on the electron beam processing no fracture mechanics analysis used to demonstrate this dependency will be presented. Instead reference will be made to the results of diverse studies in the hope of making the point. Consider data reviewed by Tien and Schwarzkopf[3)] by referring to Figs. 10 and 11. Here one can see the dependence of ductility on inclusion size and inclusion content per grain. From the same study Fig. 12 shows the low cycle fatigue dependence on inclusions while Figs 13 and 14 show fracture toughness dependence on inclusion diameter and inclusions per grain. These are the materials which are habitually fabricated into high pressure turbine discs, blades and vanes. They see high stresses, elevated temperatures and corrosive environments.

In addition to the fracture mechanics derived justification for cleaner metals needs, recent foundry work[4)] with EBCHR super alloys has demonstrated conclusively the virtues of this process. It has shown that such processing permits the recycling of here to useless super alloy revert. This product can now be used to replace both 100% virgin material or 50% virgin 50% recycled material in critical applications. The so processed revert is producing castings which have equivalent or superior properties of virgin metal castings. Also there is neither penalty on the mechanical nor on the chemical properties of the recycled alloys evaluated. This work has lead to major improvements in this foundry economics. Leybold-Heraeus for example has recently introduced Barstick furnace[5)] which could conceivably serve this market.

IV ELECTRON BEAM BUTTON REMELT TEST

Progress related to super clean metals has been intimately tied up to the development of the capability to monitor alloy cleanliness. The inability of existing metallographic and non destructive techniques to detect, much less to quantitatively characterize defects at the very low ppm range mandated the development of new techniques. The electron beam remelt test emerged as the leading candidate for inclusion content evaluation and monitoring. To fulfill this mission the electron beam button test,as it is now referred to, had to meet several requirements[6)]:

1. Provide adequate sensitivity to detect and qualify the nonmetallic particles of interest.

2. Provide accurate reliable and reproducible data.
3. Provide the necessary information on a timely basis.
4. Be cost effective (quality payoffs exceed costs.)

One can refer to three phases in the development of the test. The first phase was initiated in 1978 by E.E. Brown[7] and his associates[8,9]. It was followed in the period from 1981 to 1983 by the entering of other investigators[10,11,12] who introduced changes and modifications. We are now in the third phase which began in 1984, and which should be best referred to as the expansion and automation phase. A phase in which we also have the formation and involvement of the ASTM subcommittee E04-09. We now have an even larger group of companies active in this field.

Those known to the writer are listed in Table 1. Automated button production equipment has been discussed both by manufacturers[13,14] and users[15.16].

For those of you who are not familiar with it, the substance of the test is the production of a button under precisely controlled and reproducible conditions. The electron beam drip melts specially configured sample from the alloy, the cleanliness of which is to be determined, while maintaining the material in the water cooled semi spherical mold in the molten state. This makes possible the flotation to the surface and coalescence of essentially all light non metallics. For the materials in question we are essentially referring to oxide and nitrides of the reactive metals alloy additions. The button, including the ceramic content, is subsequently analyzed. Depending on the specific needs of the user this analyses can vary from simple to complex,and TableII[6] compares the methods used to assess the particle content of the button which essentially reflects the particle content of electron beam button rafts of the alloys tested.

TableIIIshows a comparison of the cleanliness levels of materials processed by various competing, if you will, melting processes.

To day test buttons are produced by one of the two fully programable commercial button melting systems. Fig. 15 shows the Qualitron produced by Consarc and powered by a 37.5 KW von Ardenne gun. Fig. 16 shows the Leybold-Heraeus ES1/07/30B Electron Beam Button Furnace. While these machines differ in details, they both have the capability to produce test buttons with full computer control.

V THE REAL PLAYERS AND THE DRIVING FORCES BEHIND THE EFFORT

The effort for clean metals was initiated and supported by the Air Force with Pratt & Whitney initially and then G.E. as the principal participants for the jet engine manufacturers. While there are other interested parties, to this writer at least, the initiating trio still carries the ball.

On the metal producer's side Special

Metals Corporation which operates a 60KW furnace, Viking Metallurgical which operates a 2.4 megawatt furnace, A. Johnson Metals Corporation which operates a 2.0 megawatt furnace and of more recent date Degussa Metals (formerly Electro Metals) which operates a 250KW furnace, have been the principal participants in these programs. Coincidentally all these producers' furnaces are powered by von Ardenne guns.

The beneficiaries from this pioneering effort have been a growing number of super alloys, several Titanium alloys and CP Titanium.

To this writer electron beam processing alone and/or in conjunction with VIM VAR or VADER is firmly established in the supply chain for super clean materials for the aerospace, energy and automotive industries. Though ESR, Plasma and Inducto Slag are potential processors of proven quality metals. EBCHR's virtues and recent advances will assure it an ever increasing segment in the production of super clean materials.

REFERENCES

(1) M. v. Pirani, U.S. Patent No. 848, 600, March 26, 1907.

(2) Jarrett, R.N., "Removal of Defects From Titanium with Electron Beam Cold Hearth Refining," Electron Beam Melting and Refining - State of The Art 1986, R. Bakish, (Ed.), Englewood, NJ, 1986, pp. 332-358.

(3) Tien, J.K., Schwarzkopf, E.A., "Cause and Effect of the Penetration of Superalloy Castings into High Temperature Engines Industry," pp. 8-18, of ref. 2.

(4) Mosher, J.R., "Electron Beam Cold Hearth refined Superalloys, An Investment Foundry Experience," pp. 277-285, of ref. 2.

(5) Romberg, et al., "Electron Beam Melting and Refining of Superalloys for Ingor and Barstick Production," pp. 260-276, of ref. 2.

(6) Sutton, W.H., "Electron Beam Remelt, Substance, Scope and Future as a Quality Control Tool," pp. 297-317, of ref. 2.

(7) Brown, E. E., et al., "The Influence of VIM Crucible Composition, Vacuum Arc Remelting, and Electroslag Remelting on the Nonmetallic Inclusion Content of MERL-76," Superalloys, 1980, J.K. Tien et al. (Eds.) ASM, Metals Park, OH. 1980, pp. 159-168.

(8) Brown, E.E., Hatala, R.W., "Electron Beam Refining of Nickel-Base Superalloys," Electron Beam Melting State of the Art 1985, Part II, R. Bakish, (Ed.), Englewood, NJ, pp. 103-117.

(9) Walsh, J.M., Gumz, K.P., "Classification of Nonmetallic Inclusions in Electron Beam Remelted Test Specimens," pp. 118-122, of ref. 8.

(10) Shamblen, C.E., Culp, S.L., Lober, R.W., "Superalloy Cleanliness Evaluation Using the EB Button Melt Test," Electron Beam Melting - State of the Art 1983, R. Bakish (Ed.), Englewood, NJ, pp. 61-94.

(11) C. d'A Hunt, "Evaluation of the Quality of Superalloys by the Preparation of Electron-Beam Drip Melted

Buttons," Electro-Metals Corp., Vallejo, CA, Feb. 1982.
(13) Hainz II, L.C., Bakish, R., "Evolution of the Qualitron - The Second Generation," Electron Beam Melting and Refining - State of the Art 1985, Part I, R. Bakish (Ed.), Englewood, NJ, pp. 120-131.
(14) ES 1/07/30B EB Button Melting Furnace, Bulletin 61-150.1/2, Leybold-Heraeus, Hanau, Germany.
(15) pp. 61-94, of ref. 10.
(16) Shamblen, C.E., "History of the Electron Beam Remelt Cleanliness Test at the General Electric Company Aircraft Engine Group," pp. 160-167, of ref. 13.

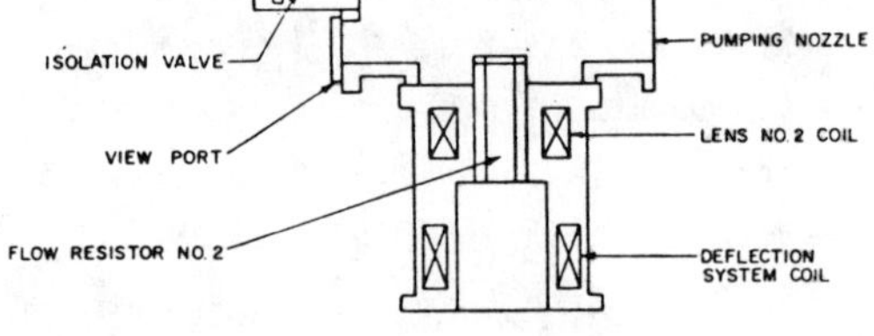

Figure 1.
Cross Section of a Typical High Power Gun.
(Courtesy: S. Schiller, Institute M. von Ardenne.)

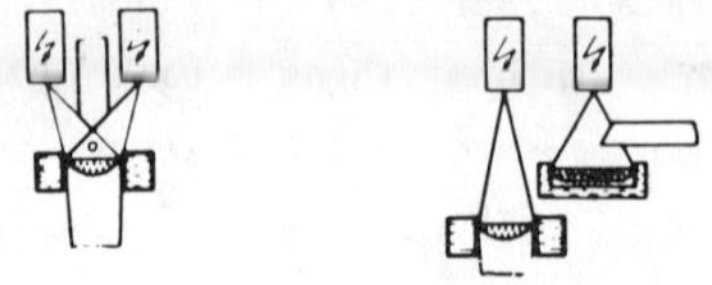

Figure 2.
Schematic of Drip Melting (right) and EBCHR (left).

Figure 3.

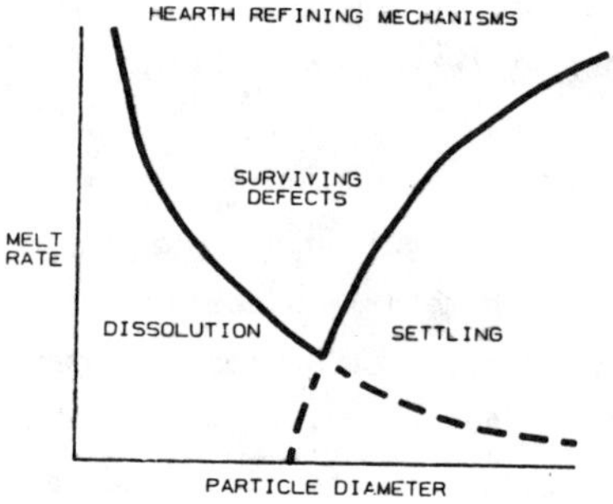

Mechanism map of dissolution and sedimentation for defect removal in hearth refining as a function of melt rate and initial defect size.

Figure 4.

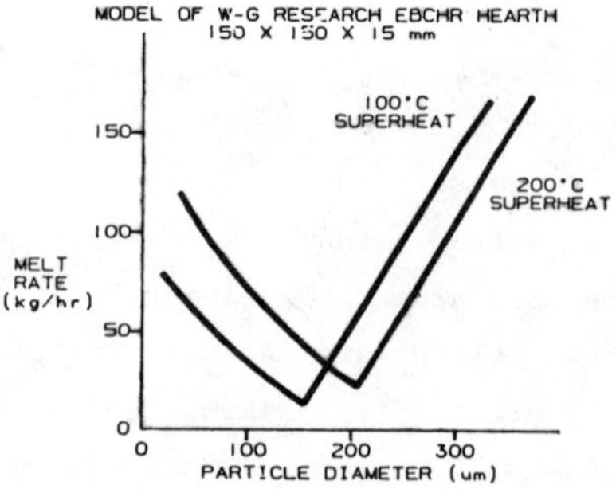

Dissolution and sedimentation curves for the experimental furnace as a function of superheat

Figure 5. Overall View of the A. Johnson Metal's 2 Megawatt Furnace.
(Courtesy: H. Harker, A. Johnson Metals Corporation.)

Figure 6. Top View of the Furnace Shown in Figure 5, Showing the 600KW Electron Gun.
(Courtesy: H. Harker, A. Johnson Metals Corporation.)

Figure 7. View into Furnace (Fig.4) Showing the Hearth and Water Cooled Cylindrical Mold.
(Courtesy: H. Harker, A. Johnson Metals Corporation.)

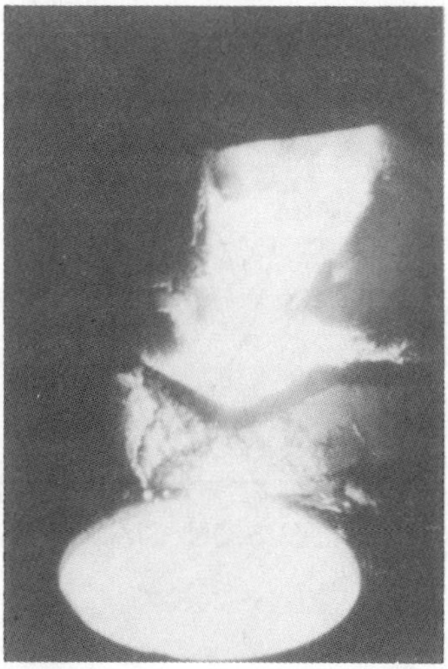

Figure 8. Hearth and Mold During Casting of Cylindrical Ingot.
(Courtesy: H. Harker, A. Johnson Metals Corporation.)

Figure 9. Large Rectangular Ingot, 12" x 45" x 10'.
(Courtesy: A. Johnson Metals Corporation.)

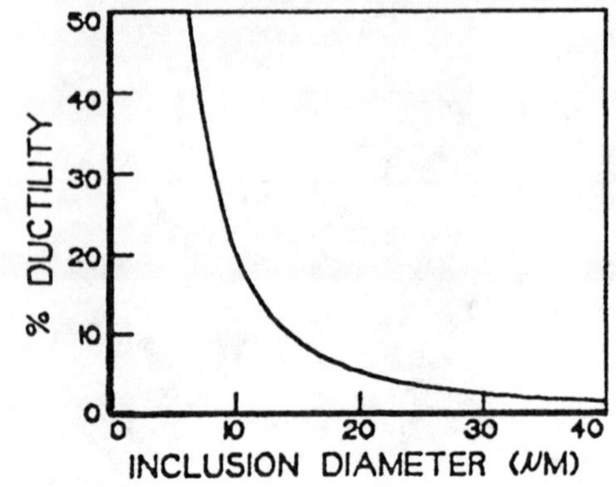

Figure 10.
Percent Ductility vs Inclusion Diameter.(3)

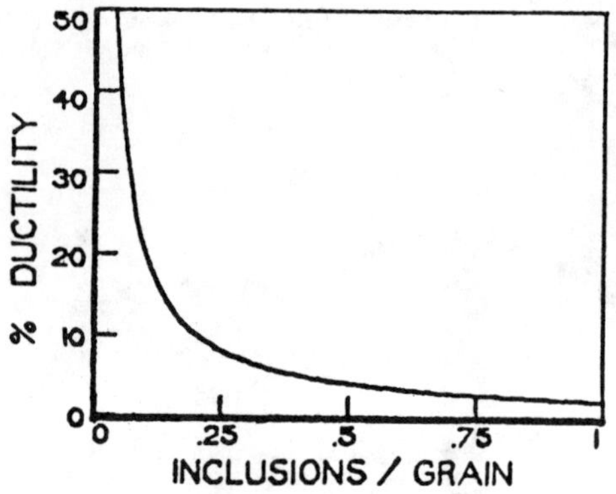

Figure 11.
Percent Ductility vs Inclusion per Grain.(3)

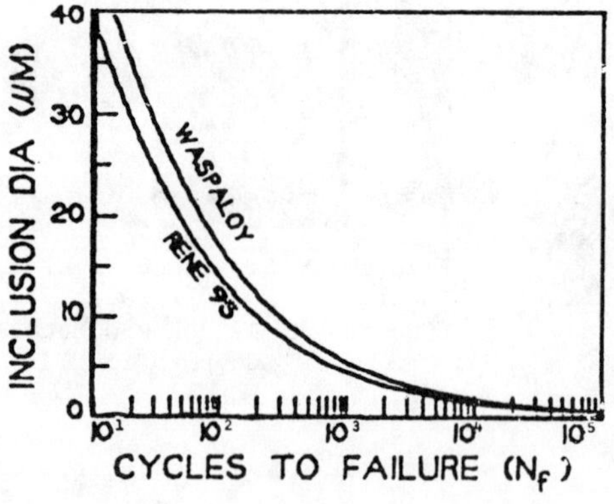

Figure 12.
Inclusion Diameter vs Fatigue Lifetimes.

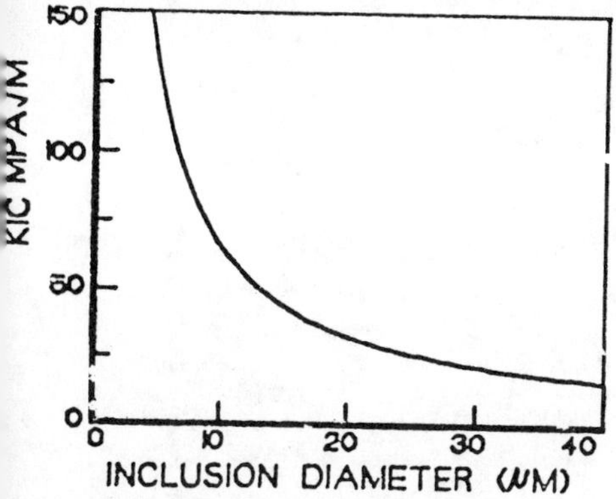

Figure 13. Fatigue Fracture Toughness Dependence on Inclusion Diameter For Rene 95.

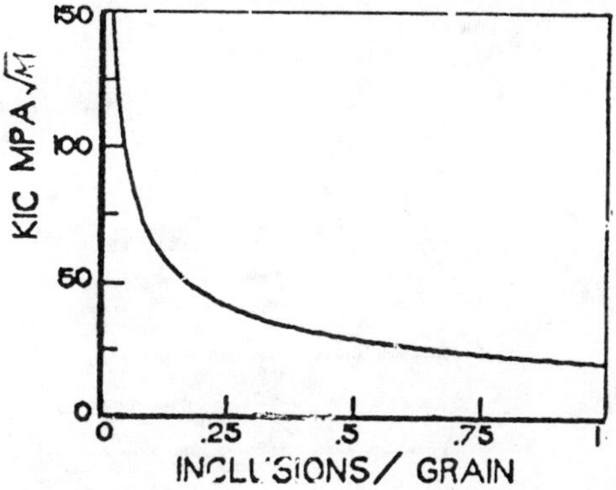

Figure 14. Fatigue Fracture Toughness Dependence on Inclusion Content For Rene 95.

TABLE 1

ELECTRON BEAM REMELT CLEANLINESS TEST FACILITIES

	Status by Years - See Code Below				
Company	1979-80	1981-2	1983-4	1985-6	1987
Pratt & Whitney	1,3	1,3	1,3	1,3	1,3
Special Metals Corp.	-	2,3	2,3	2,3	2,3
General Electric Co.	-	1,3,4	4	4	4
GNB Corp.	-	2	2	2	2
Electro Metals Corp. (now Degussa)	-	2	-	-	-
Howmet - Dover	-	-	2	2,4	2,4
Wyman-Gordon Co.	-	-	2,4	2,4	2,4
Teledyne Allvac	-	-	-	2	2
Carpenter Tech. Corp.	-	-	-	2,4	2,4
Univ. of British Columbia	2,3	2,4	2,4	2,4	2,4
Lawrence Livermore	-	-	-	-	2
British National Laboratory	-	-	-	-	2,4
IMPHY	-	-	-	-	2

Code

1. EB Melt Charge in Crucible
2. EB Drip Melt, Also Infers Capability for 1
3. Electrolytic Dissolution - Acid Digestion + SEM-EDS-IA of Extracted Particles
4. SEM-EDS-IA of Particles on Button Surface

Method of Analysis	Analytical Procedure (typical magnification,X)	Additional Options	Particle Quantity	Particle Composition	Size Dist.	Typical Analysis Time(h)	Rel. Cost
1. Visual	Visual comparison of rafts(1X)	----	5-50^{+} ppm,Q	----	----	1/6	1
2. Photo/Optical	Planimeter meas. of oxide rafts (2-10X)	----	5-50^{+} ppm,SQt	----	----	1/2	3-5
2A. Photo/Optical	Planimeter meas. of oxide rafts (2-10X)	SEM(10-500X)	1-50^{+} ppm,SQt	SQt	----	1/2-1	10-20
3. Wet Chemical	Electrolytic Dissolution & Acid Digestion	----	3-50^{+} ppm,Qt[a]	----	----	17	60-100
3A. Wet Chemical	Electrolytic Dissolution & Acid Digestion	SEM/IA (20-200X) EPMA/IA (20-200X)	<1-10,Qt[b]	Qt[a]	Qt[a]	17-19	80-120
4. Electron/ Optical	SEM/IA Direct Analysis of raft. (20-200X)	----	<1-10,QT[c]	Qt[a]	Qt[a]	1/2-2	20-50

*Q_l = qualitative data
Q_t = quantitative data
SQ_t = semiquantitative data

(a) - direct determination
(b) - calc. values based on calibration curve
(c) - calc. based on particle density & volume

Table 11
Comparison of Methods Used to Assess Particle Content of EB Button Rafts.

Alloy/Process	Largest Oxide Particle μm (mils)	Oxide Quantity ppm	Major Oxide Signature (EDS)	Analytical Method
R-95 (Ref. 5)				
VIM	264 (10.4)	2.3	Al	SEM/EDS/IA[1]
VIM/VAR	163 (6.4)	2.8	Al	SEM/EDS/IA[1]
VIM/PM/EXT[a]	147 (5.8)	2.9	Al	SEM/EDS/IA[1]
VIM/ESR	102 (4.0)	0.6	Al	SEM/EDS/IA[1]
VIM/EBCHR[b]	158 (6.2)	0.5	Al	SEM/EDS/IA[1]
VIM/EBCHR[c]	23 (0.9)	0.2	Al	SEM/EDS/IA[1]
IN-718 Ref. 5)				
VIM	666 (26.2)	2.9	Al,Si,Ca	SEM/EDS/IA[1]
VIM/UAR	281 (11.1)	1.4	Al,Ca	SEM/EDS/IA[1]
VIM/ESR	68 (2.7)	0.6	Al,Si,Ca	SEM/EDS/IA[1]
VIM/ECBHR[c]	37 (1.5)	0.1	Al,Si,Ca	SEM/EDS/IA[1]
MERL 76 (Ref. 11)				
VIM	> 508 (>20)	8.9	Hf,Al[29]	Wet Chem. EMPA/IA[2]
VIM (Filter)	-- --	2.0	--	Wet Chem. EMPA/IA
VIM/EBCHR[b]	180 7.1	0.25	--	Wet. Chem. EMPA/IA
VIM/EBCHR[c]	132 5.2	0.14	--	Wet. Chem. EMPA/IA
VIM/EBCHR[d]	81 3.2	0.03	--	Wet. Chem. EMPA/IA

(a) VIM/PM/Extruded - powder process, extruded billet
(b) No barrier dam; use of EB beam sweep to retain floating particles in hearth
(c) One skimmer dam
(d) Two skimmer dams

(1) EB-drip melt process
(2) EB-double melt process

Table III
Comparison of Cleanliness Levels of Various Melting Processes.

Figure 15. "Qualitron" Button Melter.
(Courtesy: L. Hainz, Consarc.)

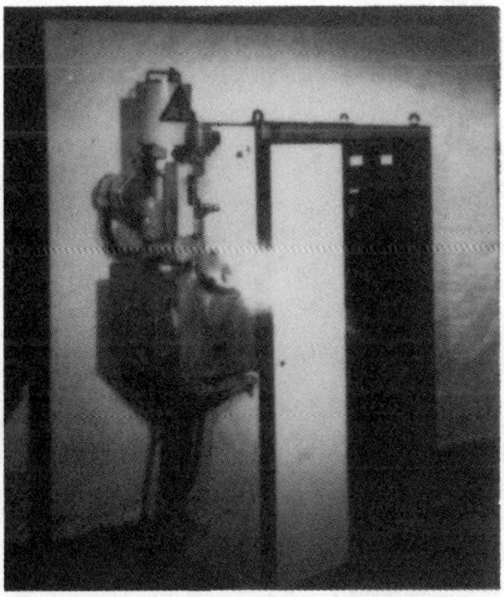

Figure 16. Electron Beam Melter ES1/07/30B.
(Courtesy: R. Schumann, Leybold-Heraeus Technologies, Inc.)

BIOGRAPHY

R. Bakish, President, Bakish Materials Corporation, is a metallurgist with a B.S. from Columbia and M.Eng. and D.Eng. from Yale. He has extensive industrial and academic experience and has been involved with electron beams since 1959. He authored and co-authored over 90 publications including co-authorship, contributions to and the editing of 22 volumes on electron beam technology. He is the chairman of the Electron Beam Melting and Refining and the Laser vs the Electron Beam - State of the Art Conferences and is on the Stirring Committee of the International Symposium on Electron, Ion and Photon Beam Technology. He is member of AIME, ASM, TSM, NACE, AVS,SVC and ASTM.

1st International SAMPE Metals Conference
August 18-20, 1987

ADVANCEMENTS IN THE PROCESSING OF SUPERALLOYS

G. E. Maurer

Special Metals Corporation
New Hartford, NY 13413

ABSTRACT

The ability to extend the performance of superalloys and future high temperature alloys depends on the ability to produce and guarantee ultra-clean and microstructurally uniform material. This presentation will highlight recent competitive advancements in the primary and secondary melting of superalloys.

The cleanliness of superalloy billet starts with raw material selection and vacuum induction melting. A discussion on parameters affecting oxide/nitride formation and removal will include refractories, filters, stirring and pouring methods.

Subsequent remelting techniques of VIM electrodes in the future will be chosen from many alternatives. Vacuum arc remelting (VAR), the present standard, improves the cleanliness of the VIM electrode but also has limitations, such as macrosegregation defects. Electroslag remelting (ESR) offers attractive additional cleaning potential, but positive segregation may limit the ingot size that can be melted. Electron beam cold hearth refining (EBCHR) and plasma cold hearth refining (PCHR) have demonstrated the production of very clean ingots under reduced melt rates and special pouring conditions. Difficulties associated with EB melting include chemistry and macrostructure control in the final product. Plasma melting is potentially less expensive, but more difficult to control. A new process, VADER, offers the opportunity to produce uniform fine-grain ingots from ultra-clean electrodes of various macrostructures.

A little known, but commercially available technique, vacuum centrifugal casting, will be shown to have produced the cleanest metal available for supercritical parts.

The vacuum remelting process or combination of processes that will eventually be developed for the superalloy industry must not only fulfill the "ultra-clean" criteria but must also be cost competitive with present processing. A semi-quantitative comparison of the various melting alternatives will be discussed.

1st International SAMPE Metals Conference
August 18-20, 1987

THE SEARCH FOR ECONOMICAL METHODS TO PRODUCE CLEANER SPECIALTY STEELS

Ronald E. Bailey, William W. Timmons and Anthony J. Polito
Jessop Steel Company

Abstract

It is widely accepted that cleaner specialty steel is better specialty steel. How clean must specialty steel be? Will the customer pay for the extra cost to produce cleaner specialty steel? For approximately thirty years vacuum induction melting (VIM), vacuum arc remelting (VAR), electroslag remelting (ESR) and electron beam melting (EB) have been used to produce clean specialty steels primarily for aerospace applications where quality requirements have dominated price concerns and fortunately this perspective continues in the aerospace industry today. In contrast, in many other industries the desire for cleaner specialty steel has grown but the interest in paying for VIM, VAR, ESR or EB has not. Therefore, efforts have been directed towards producing cleaner specialty steel with electric arc furnace (EAF) melting and argon oxygen decarburization (AOD) refining with special attention to deoxidation, ladle treatment and teeming practices. The reasons cleaner specialty steel is desirable will be discussed for certain applications and the processing equipment and techniques to achieve cleaner specialty steel will be described.

1. INTRODUCTION

Jet engine builders have demanded clean steel since the industry was in its infancy. In fact, the demands of jet engine builders were instrumental in the commercialization of both vacuum induction melting (VIM) and vacuum arc melting (VAR). For many years the combination of these two processes produced the cleanest specialty steels and nickel base alloys available. In recent years it has been demonstrated that VIM plus electroslag remelting (ESR) produces cleaner alloys than VIM + VAR and that VIM + electron beam melting (EB) produces even cleaner metal.[1-5] Moreover, triple melting (VIM + ESR + VAR or VIM + EB + EB) produces the cleanest metal.[1-3] The commercialization and technological advancements associated with these processes have resulted in the production of safe, reliable jet engines. However, these melting processes add significantly to the costs of producing metal. For jet engines the additional costs are easily justified by the saving of human lives and expensive aircraft.

On the other hand, there are manufacturing processes where clean specialty steels are needed but the additional cost of the traditional clean steel processes cannot be economically justified or the customer is simply unwilling to pay for the extra cost. These needs will represent the focus of this paper.

2. DISCUSSION AND RESULTS

Jessop Steel Company of Washington, Pennsylvania is a producer of specialty steel plate. Stainless steel, tool steel, nickel base alloys, high manganese non-magnetic steel and armor plate are the main product lines of Jessop Steel Company.

Why must we strive to make cleaner specialty steel? It is generally recognized that cleaner specialty steel is better because it allows a) better polished surface quality, b) better electrical discharge machining (EDM) performance c) better formability, d) better fatigue life, e) better impact resistance and f) better pitting resistance in certain corrosive media.

The pitting resistance of stainless steels in chloride environments can be affected by inclusion chemistry, shape and size. Sulfide inclusions, in particular, are susceptible to chemical attack and pit initiation in certain corrosive media.[6] Cleaner stainless steel would be less prone to such pitting corrosion.

Jessop Steel Company maintains a substantial inventory of tool steel plate. These plates are hot rolled, annealed and machine ground to produce a decarburization free surface. Various configurations can be cut economically from such plates. Tool steel plates, particularly S7, are frequently used to make molds which are used to produce plastic parts. This application prohibits the presence of large, nonmetallic inclusions in the steel; otherwise, when the mold is machined and polished, large inclusions will permit pits to form on the steel and the pits will lead to blemishes on the plastic parts. Despite the high machining costs to make plastic molds, very little steel for such applications is refined by ESR or VAR.

The major advantage of metal armor plate is its ability to withstand multiple hits due to the inherent toughness of metal. However, excessive amounts of inclusions will render the material susceptible to cracking during forming, heat treatment or ballistics testing. Cleaner steel as provided by ESR or VAR has provided better ballistics.[7-10] However, very little armor plate is refined by ESR or VAR.

Electrical discharge machining is a

common method of economically producing simple and complex parts from tool steel. The process works well provided the equipment is properly maintained, properly set up and the steel does not contain elongated, nonmetallic inclusions. The electrical discharge will not cut through the refractory inclusions which will cause the equipment to malfunction.

A critical application for 347 stainless steel is a tank that is used to contain a corrosive and explosive material on fighter planes. The tank head is formed by spinning; consequently, elongated inclusions in the steel will permit separations to form and result in surface defects. After expensive surface problems were encountered, the customer accepted our recommendation to use ESR 347. Customers involved with aerospace applications are more willing to accept the additional cost of ESR to achieve a distinctly higher quality level to solve a problem.

Cleaner specialty steels contain lower levels of oxides and sulfides. Next let us consider the methods of achieving lower levels of oxides and sulfides. Specialty steels are normally melted in an electric arc furnace and refined in an AOD vessel. The electric arc furnace is an efficient melter while the AOD is an efficient refiner. Metal is refined in the vessel by introducing argon and oxygen through tuyeres at the bottom of the vessel. AOD refining is one of the steps in producing cleaner specialty steel. The advantages of AOD refining in producing cleaner specialty steel are as follows:

1) Blowing of argon and oxygen mixes molten metal and lime based slag and lowers the sulfur content.
2) Easily oxidizable elements form oxides which are absorbed by the slag.
3) Argon purging before tapping the molten metal from the vessel promotes the flotation of large inclusions.
4) Low levels of carbon and nitrogen (.010% maximum each) are attainable if desired.
5) Reduced but not low levels of oxygen are achieved.

The lower sulfur levels in AOD refined tool steels, for example, result in better sulfide ratings as shown in Figure 1. The top row of microstructures display H13 tool steel without AOD refining, while the bottom row exhibits H13 tool steel with AOD refining. The AOD refined steel is clearly cleaner. Also, AOD refining has let to better silicate ratings as illustrated in Figure 2. The top row of microstructures shows D2 tool steel without AOD refining, while the bottom row characterizes D2 with AOD refining. Again the AOD refined steel is certainly cleaner. The silicates depicted in even the AOD refined material can cause problems in plastic molds and illustrate a problem area that we have addressed.

An important part of the refining operation is deoxidation. Common deoxidants in steelmaking are silicon, aluminum, calcium, magnesium, titanium and rare earth elements. The deoxidant is usually added either a) to the vessel just prior to tapping into the ladle or b) to the ladle before or during tapping. The physical chemistry of the deoxidant and the deoxidation products must be considered in determing the favored deoxidation process for a given alloy system.

The next step is tapping the metal from the vessel into the teeming ladle. The teeming ladles are preheated before use. The ladle refractory can be a source of inclusions. Therefore, use of the highest quality refractories that the economics will permit is recommended. As an example, fireclay brick contains liquid phases at steelmaking temperatures and will act as a source of inclusions.[11] In contrast, high alumina and dolomitic bricks do not contain liquid phases at steelmaking temperatures and do not wear as fast as fireclay bricks.[11] When Jessop switched from fireclay to high alumina ladle brick, the ladle lining life increased by a factor of 4 to 5 times which also means that fewer brick fragments reach the molds. High alumina brick costs approximately twice as much as fireclay brick.

The physical chemistry properties of the common deoxidants are compared with iron and nickel in Table 1. Calcium, magnesium, aluminum, titanium, lanthanum and cerium are all strong deoxidizers while silicon is a moderately strong deoxidizer. Calcium and magnesium have not only low density but also high vapor pressure. The low density means that the elements tend to float rather than sink into the molten metal. Also, a portion that enters the molten metal will tend to evaporate due to the high vapor pressure. These tendencies are combated by a) using higher density alloys or blends that contain the desired elements b) using special addition methods such as injection and c) adding a larger portion of the desired element than the aim retention level. The retention of calcium and magnesium may be only 10 - 30% of the amount added. An auxiliary benefit of calcium and magnesium is that retained amounts of these elements will counteract the harmful effects of residual sulfur. In addition, titanium will react with carbon to form titanium carbide. Moreover, aluminum and titanium have a strong affinity for nitrogen which may be an advantage or disadvantage depending on the metallurgy of the alloy involved. For many stainless steels nitrogen is a beneficial alloying element which should not be tied up with aluminum or titanium. Also, aluminum can lead to defects that occur during polishing that are called alumina tails. The rare earth elements lanthanum and cerium are the most powerful deoxidants listed and they also have a strong desulfurizing potential. Note that they are dense relative to iron and nickel.

The densities of deoxidation products must be considered to understand how well they may be expected to float to the surface. The densities of the most common deoxidation products of the deoxidants are presented in Table 2. The densities of SiO_2, CaS, MgS, CaO, MgO, Al_2O_3 and TiO_2 are relatively low compared with specialty steels. On the other hand, the densities of La_2S_3, Ce_2S_3, La_2O_2S, Ce_2O_2S, La_2O_3 and Ce_2O_3 are relatively dense compared with specialty steels. This is a disadvantage of using rare earth elements as deoxidants because the deoxidation products of lanthanum and cerium are dense they will not float readily. However, lanthanum and cerium are suitable additives for deoxidation and desulfurization in vacuum melting because the oxygen level is already low and vacuum arc remelting or electroslag remelting will follow and reduce the frequency of such inclusions.[12] In air melting, unfortunately lanthanum and cerium will promote dirty steel because the deoxidation products probably will not float to the surface.

An expression that describes the movement of spherical particles in fluids is Stokes Law.[13,14] The velocity at which particles smaller than 100 μ in diameter will float to the melt surface can be approximated with Stokes Law as follows:

$$V = K\,\frac{2}{9}\,\frac{\Delta D g d^2}{M} \qquad (1)$$

V = velocity of the floating particle (cm/sec)

ΔD = density difference between the floating particle and the liquid metal (g/m^3)

d = particle diameter (cm)

M = viscosity of liquid metal (P)

g = acceleration due to gravity (980 cm/sec^2)

k = a constant

Stokes Law indicates that inclusion flotation is favored by a) low inclusion density, b) large inclusion diameter and c) low liquid metal viscosity, i.e., high liquid metal temperature. A deficiency of Stokes Law is that it does not account for frictional forces at the inclusion/melt interface for solid nonwetting inclusions or the adsorption of surface active elements in the case of liquid inclusions.[14] Nevertheless, it is a useful tool particularly when used with pertinent experimental data.

The time for inclusion flotation can be calculated as follows:

$$t_p = \frac{H}{V} \qquad (2)$$

where t_p = theoretical time for particle to travel distance H.

The time for flotation would be decreased by a) particle coalescence to form larger inclusions, b) argon rinsing to push the inclusions upward and c) an upward current in the molten metal via induction stirring.

The effect of titanium deoxidation as compared with silicon deoxidation can be evaluated by comparing inclusion ratings of top poured stainless steels 304 and 321. The nominal compositions of 304 and 321 are similar with the exception of the titanium present in 321 as shown in Table 3. Titanium acts as the main deoxidant in 321 while silicon serves as the deoxidant in 304. The inclusions ratings of 304 and 321 presented in Table 4 show that the microcleanliness of 321 is generally better than that of 304 due to the titanium addition.

Silicon is a specified element in virtually all specialty steels. The presence of silicon and the absence of stronger deoxidants promotes the formation of silicate inclusions in an air melting environment. Silicate inclusions are plastic at hot working temperatures; consequently, they exist as elongated particles in the wrought product. Investigators have reported that silicates float more slowly in molten steel than aluminates;[15-19] apparently the interfacial energy of silicates in the molten metal is lower than that of aluminates. Moreover, investigators report that silicate inclusions generally are larger than aluminates;[18,19] our evaluation also suggests that silicate inclusions typically are larger than aluminate inclusions. Calcium addition by injection tends to compensate for calcium's low density and high vapor pressure.[20-24] Calcium will promote the formation of calcium rich silicates or aluminates which unfortunately float more slowly than silica or alumina.[24] However, calcium rich inclusions do not elongate as much during hot rolling so they are more spherically shaped and are less likely to be detrimental to mechanical properties or to EDM than are stringer or elongated inclusions.[23] Calcium should not be added until the sulfur and oxygen levels have been reduced to low levels and preferably only after argon rinsing; otherwise, the calcium effectiveness will be diminished.[23]

Another critical step in production of clean specialty steel is the teeming operation. When the metal falls from the teeming ladle into the molds during top pouring, the molten metal is exposed to the atmosphere not only as part of the stream but even more so as molten metal impacts the bottom of the mold and subsequently the molten metal pool as it rises in the mold. Such exposure of molten metal to the atmosphere is substantially reduced during bottom pouring. As depicted in Figure 3, bottom pouring consists of pouring molten metal from the teeming ladle down a refractory lined, cast iron pipe called a trumpet. The molten metal travels through hollow brick in the base plate and gently fills the mold from the bottom while a mold powder floats on the molten surface as the molten metal rises in the mold. Bottom pouring favors cleaner specialty steel due to reduced exposure to air and therefore less reoxidation because a) the metal stream is largely protected by the trumpet b) there is less molten metal turbulence in the mold, c) the mold powder covers the molten metal as it rises in the mold and d) argon shrouding is easily applied. There has been speculation that bottom pouring behaves as a filtering system as inclusions attach themselves to the bottom pouring refractory.

Jessop AOD refines and bottom pours Alloy 600, a titanium and aluminum bearing nickel base alloy. Metallographic evaluation reveals that the material satisfies the microcleanliness requirement of MIL-N-23228C type II or type III which specifies ESR or VAR as shown in Table 5. However, the consistency and reliability of bottom poured metal does not equal that of ESR metal because ESR product is not only less likely to possess occasional large inclusions but also is more homogeneous and contains lower sulfur levels.

Argon rinsing in the teeming ladle and bottom pouring have been evaluated separately on tool steel plates as reported in Table 6. The rating levels are depicted by the photomicrographs in Figures 4 and 5. Compared with the standard, top poured, non-argon rinsed product a typical field of argon rinsed product contained a higher percentage of small, globular inclusions and was virtually free of stringer inclusions. Relative to the standard product a typical field of bottom poured product contained a lower percentage of the smallest and largest levels of globular inclusions but a much higher percentage of the small, stringer inclusions. Argon rinsing provided a clear improvement in microcleanliness while bottom pouring seemingly converted the small, globular inclusions into small, stringer inclusions. It is surprising that the microcleanliness of the bottom poured tool steel was not better; the results indicate that our bottom pouring techniques for tool steel must be optimized.

Ultrasonic testing is another tool which can be used to assess the effectiveness of methods to produce cleaner steel; this technique detects macroinclusions and other imperfections which can create problems in critical applications. Tool steel plates were ultrasonic tested using a 3/64" flat bottom hole as the reference standard and the recordable indications per plate were reduced by argon rinsing as illustrated in Table 7.

Another technique to achieve cleaner specialty steel is the use of ceramic filters.[25-27] These filters can be placed in the bottom pouring refractory system. The ceramic filters have the appearance of a sponge as revealed on Figure 6. The pore size of the ceramic filter must be adjusted to the cleanliness level of the incoming metal, that is, the more inclusions that are present in the incoming metal the larger the pore size; otherwise, the filter will become clogged. The inclusions tend to be trapped by the filter because the interfacial energy of the inclusion is lowered when attached to the filter. The use of ceramic filters today is more common in vacuum melted metal than air melted metal. Ceramic filters offer the potential of attaining improved levels of microcleanliness to satisfy many demanding but noncritical applications as illustrated in Figure 7.

If the economical methods of producing clean specialty steel such as careful deoxidation, argon rinsing, bottom pouring, argon shrouding, ceramic foam filtering do not product a sufficiently clean product, then one of the remelting techniques such as ESR, VAR or EB must be used. The beneficial effect of ESR on microcleanliness of stainless steel is depicted by the inclusion ratings in Table 8. The improvement in microcleanliness provided by ESR is most noticeable in the worst fields.

3. SUMMARY

The focus of this paper has been on relatively low cost methods that can be applied during the melting, refining, teeming and casting of specialty steels in an air environment. More certain and reliable methods of producing clean specialty steel include the secondary refining processes of electroslag refining (ESR), vacuum arc remelting (VAR) and electron beam (EB) melting. In each method metal from a prior melting source is introduced to the ESR, VAR, or EB unit, the metal is remelted, refined, inclusions are removed and the metal resolidifies. These processes have evolved to a high level of sophistication and produce the highest quality levels commercially available; however, they also involve a cost premium that some customers are unwilling to accept. Jessop Steel Company supplies ESR material[28] as its cleanest product but often despite demanding and costly applications customers will order commercial quality steel rather than pay a premium for the best quality. Consequently, we have explored low cost methods of attaining better microcleanliness levels and we propose the following approach:

1. Use high quality refractory materials.
2. Deoxidize with appropriate element or elements.
3. Control molten metal temperatures during tapping and teeming.
4. Employ argon to accelerate

inclusion flotation in the teeming ladle.

5. Use argon to shroud the molten metal stream and perhaps the molds during teeming.
6. Bottom pour to minimize exposure of molten metal to air.
7. Install ceramic filters in the bottom pouring refractory system.
8. If the above methods do not produce a sufficiently clean product, evaluate electroslag remelting, vacuum arc remelting or electron beam melting.

REFERENCES

1. Cordy, Kelley and Lherbier, "Chemistry and Structure Control in Remelted Superalloy Ingots", 1984, AIME, 1985.

2. Kennedy, Chien and Bougham, "Electroslag Remelted Waspoloy", Ibid.

3. Shamblen, Culp and Laper, "Superalloy Cleanliness Evaluation Using the EB Button Melt Test", Ibid.

4. Dominque, Sutton, and Yu, "Correlation of EB-Britton Raft Structures with Manufacturing Variables", Ibid.

5. Entrekin and Clarkson, "EB Refining Promises Super Clean Metal Production", Metal Prog., June, 1986, pp. 35-39.

6. Zhong, Jargelius, and Otterberg, "Influence of REM Additions on Pitting Resistance of Some Austenite Stainless Steels", Proc. of Stainless Steels '84, Inst. of Metals, London, 1985, pp. 158-165.

7. Hickey and Thomas, "Comparing a Split Heat of ESR/VAR 4340 Steel", Metal Prog., Oct. 1985, pp. 69-74.

8. Thomas and Anctil, "Characterization of Electroslag Remelted and Ladle Refined, Electric Furnace Melted 4340 Steel Armor", J. Heat Treating, Dec. 1986, pp. 317-325.

9. Hickey and Anctil, "Split Heat Mechanical Property Comparison of ESR and VAR 4340 Steel, J. Heat Treating, Dec. 1985, pp. 177.

10. Garrison, "The Design of Ultra High Strength Steels of Improved Toughness", CIT Eng. News, Spring, 1986, pp. 21-24.

11. Stubbles and Skrabec, Engineered Clean Steels, "Advanced Secondary Refining - The Quest for Clean Steels, ISS-AIME Advanced Tech. Symp., Birmingham, Ala., Sept. 9-11, 1986.

12. Bailey, Shiring, and Anderson, "Effects of a Rare Earth Addition on Unitemp 901", Superalloys Metallurgy and Manufacture, Proc. of the Third Intern. Conf., Claitor's Publishing, Baton Rouge, La., Sept. 1976, pp. 109-118.

13. Herbertson and Harris, "Inclusion Removal by EB Hearth Refining", Proc. of the Conf. EB Melting and Refining State of the Art 1985, Part 2, Bakish Materials Corp., Englewood, N.Y., pp. 99-102.

14. Ototani, Calcium Clean Steel, Springer - Verlag, New York, p. 49.

15. Plockinger and Rosegger, Stahl Eisen, Vol. 77, 1957, pp. 798-804.

16. Plockinger and Wahlster, Rodex Rund, Vol. 516, 1957, pp. 754-770.

17. Simon, "Study of Factors Influencing the Number and Nature of Inclusions Originating from Complex Deoxidation in Ferroalloys", Production and Application of Clean Steels, Iron and Steel Inst., London, pp. 50-55.

18. Pickering, "Effect of Processing Parameters on the Origin of Non-Metallic Inclusions", Ibid., pp. 75-91.

19. Ichinoe, Kajioka, and Kokubo, "Production of Clean Steel at Yawata Works", Ibid, pp. 137-166.

20. Weimer, "Calcium Injection by

Wire - A Whole New Technology", Iron Age, July 5, 1985, pp. 42 and 46.

21. Dwyer, "Ladle Met Update", 33 Metal Producing, April, 1985, pp. 55-61.

22. Fruehan, "Physical Chemistry and Process Dynamics of Reactions in Ladle Metallurgy", J. of Metals, March, 1985, pp. 50-54.

23. Turkdogan, "Clean Steel Technology - Inclusion Control to Improve Steel Castability, Processing and Performance in Service", Advanced Secondary Refining - The Quest for Clean Steels, ISS - AIME Advanced Tech. Symp., Birmingham, Ala., Sept. 9-11, 1986.

24. Lindon and Billington, J. Iron and Steel Inst., Vol. 207, 1969, p. 340.

25. Aubrey, Brockmeyer and Mauhar, "Ceramic Foam - An Effective Filter for Molten Steel", Proc. Fifth Intern. Iron and Steel Congress, April, 1986, pp. 977-991.

26. Aubrey, Wieser, Dulta and Ilhan, "Cast Steel Quality Improvement by Filtration with Ceramic Foam Filters, "AFS Trans., 85-22, pp. 177-182.

27. Luyckx and Komanecky, "A New Approach to Steel Filtration: Vertical Inclusion Collector Baffles in the Tundish", Electric Furnace Conf. Proc. 1986, AIME, Warrendale, Pa., p. 81.

28. Timmons and Bailey, "Jessop's Experience in Producing Nickel Base Alloys and Stainless Steels by a 12" x 42" ESR Ingot Practice", Vac. Met. Conf. 1986, to be published.

TABLE 1
COMPARISON OF PHYSICAL PROPERTIES
OF
COMMON DEOXIDANTS

ELEMENT	RELATIVE VAPOR PRESSURE	DENSITY (g/cm^3, lb/in^3)	MELTING POINT (C,F)
Ca	VERY HIGH	1.52 (.055)	840 (1542)
Mg	EXTREMELY HIGH	1.74 (.063)	649 (1200)
Si	LOW	2.33 (.084)	1410 (2570)
Al	LOW	2.68 (.097)	660 (1220)
Ti	LOW	4.48 (.162)	1660 (3020)
La	LOW	6.12 (.221)	920 (1688)
Ce	LOW	6.64 (.240)	797 (1468)
Fe	VERY LOW	7.89 (.285)	1549 (2802)
Ni	VERY LOW	8.91 (.322)	1455 (2651)

TABLE 2
DENSITIES OF VARIOUS DEOXIDATION PRODUCTS
RELATIVE TO SPECIALTY STEELS

DEOXIDATION PRODUCT	DENSITY (g/cm^3, lb/in^3)
SiO_2	2.63 (.095)
CaS	2.16 (.078)
MgS	2.80 (.101)
CaO	3.35 (.121)
MgO	3.57 (.129)
Al_2O_3	3.96 (.143)
TiO_2	4.24 (.153)
La_2S_3	4.90 (.177)
Ce_2S_3	5.15 (.186)
La_2O_2S	5.76 (.208)
Ce_2O_2S	5.98 (.216)
La_2O_3	6.48 (.234)
Ce_2O_3	6.84 (.247)
Tool Steel	7.83 (.283)
Stainless Steel	7.89 (.285)
Nickel Base	8.30 (.300)

TABLE 3
NOMINAL COMPOSITIONS OF 304 AND 321

	C	Mn	P	S	Si	Ni	Cr	N	Ti
304	.05	1.70	.035 max.	.030 max.	.50	8.20	18.25	.09	--
321	.04	1.70	.035 max.	.030 max.	.60	9.70	17.50	.02	.50

TABLE 4
INCLUSION RATINGS OF AOD 304 AND 321
PER
ASTM E45 METHOD D

	A		B		C		D	
	T	H	T	H	T	H	T	H
304	1	0	0	0	2-1/2.	0	1/2	0
321	0	0	1/2	0	0	0	1/2	1/.

TABLE 5
INCLUSION RATINGS OF BOTTOM POURED ALLOY 600
PER ASTM E45 COMPARED WITH NAVY NUCLEAR REQUIREMENTS

	A		B		C		D	
Mil-N-23228C	1	1/2	1-1/2	1	1	1/2	1-1/2	1
Jessop Bottom Poured Alloy 600	1	0	1	0	1/2	0	0	1/2

TABLE 6
PERCENTAGE OF TYPICAL FIELDS PER INCLUSION LEVEL
IN
ARGON RINSED AND BOTTOM POURED TOOL STEEL PLATES

	Globular				Stringer			
	1	2	3	4	1	2	3	4
Standard	24.2	42.5	12.1	3.0	15.2	0	0	3.0
Argon Rinsed	37.8	56.8	2.7	2.7	0	0	0	0
Bottom Poured	9.1	45.4	18.2	0	27.3	0	0	0

TABLE 7
ULTRASONIC TEST RESULTS OF TOOL STEEL PLATES

	AOD Refined and Top Poured	AOD Refined Argon Rinsed and Top Poured
Recordable Point Indications Per Plate	2.5	1.7
Sample Size	98 plates	44 plates

TABLE 8
316L AND 347 INCLUSION RATINGS PER ASTM E45

AOD - Method A
ESR - Method D

	A		B		C		D	
AOD	T	H	T	H	T	H	T	H
Typical Field	1	0	0	0	1	0	0	0
Worst Field	1	0	1	0	2	3	2	0
ESR								
Typical Field	0	0	0	0	0	0	1	0
Worst Field	0	0	1	0	0	0	1	1

NON AOD 100X

NON AOD 500X

AOD 100X

AOD 500X

FIGURE 1 - AOD VS. NON-AOD REFINED
H-13 TOOL STEEL

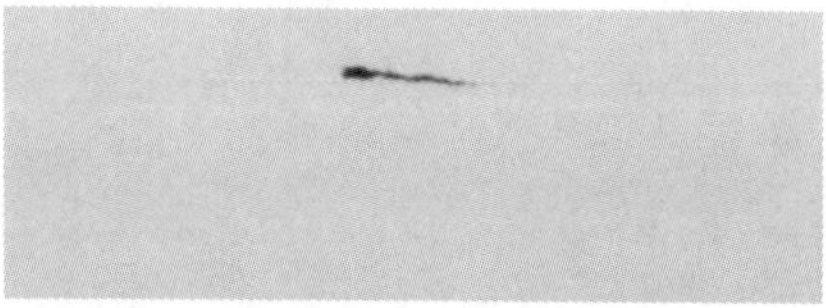

NON AOD 100X

NON AOD 500X

AOD 100X

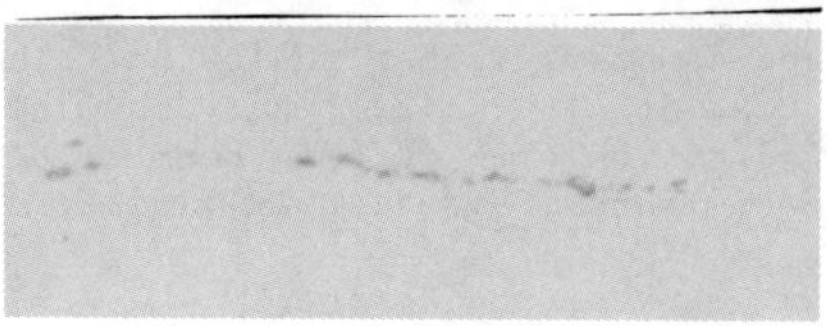

AOD 500X

FIGURE 2 - AOD VS. NON AOD REFINED
D2 TOOL STEEL

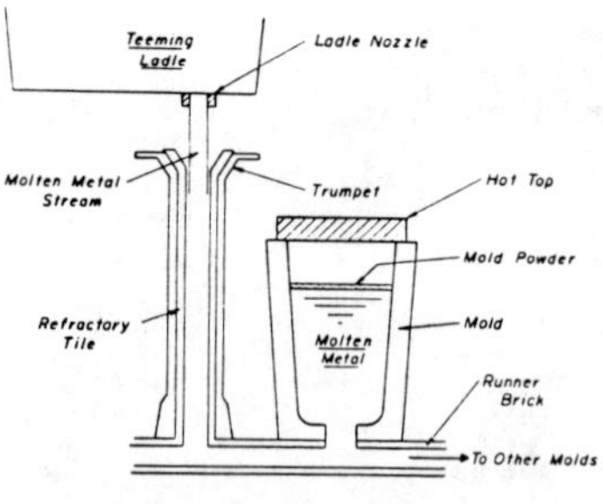

Conceptual Diagram of Bottom Pouring Set Up

FIGURE 3

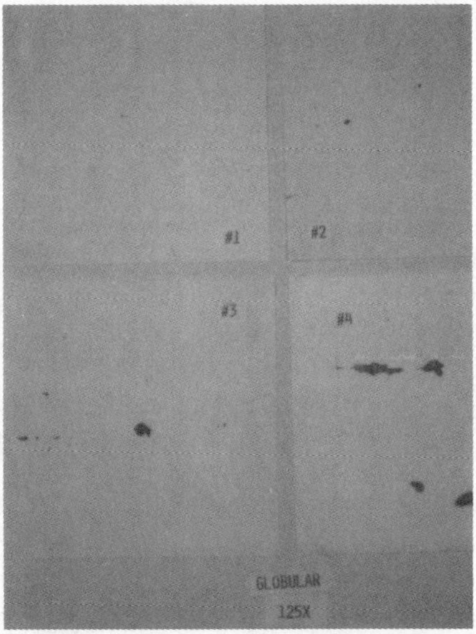

FIGURE 4: GLOBULAR INCLUSION RATING CHART (INCLUSION PHOTOS AT 125X)

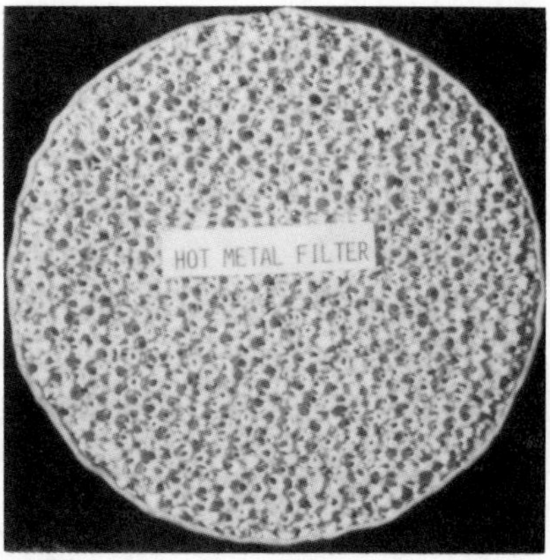

FIGURE 6: CERAMIC HOT METAL FILTER

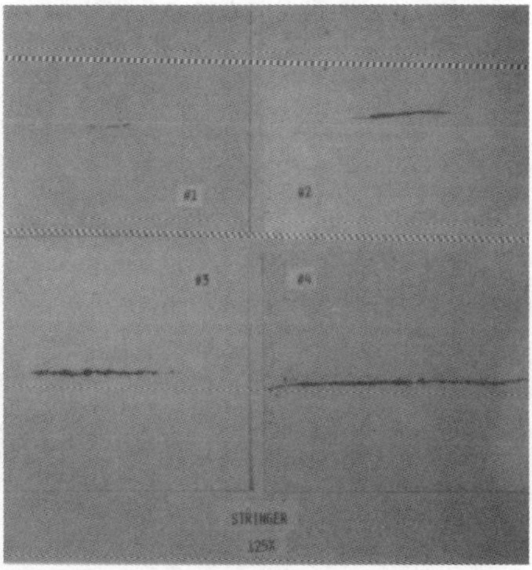

FIGURE 5: STRINGER INCLUSION RATING CHART (INCLUSION PHOTOS AT 125X)

FIGURE 7

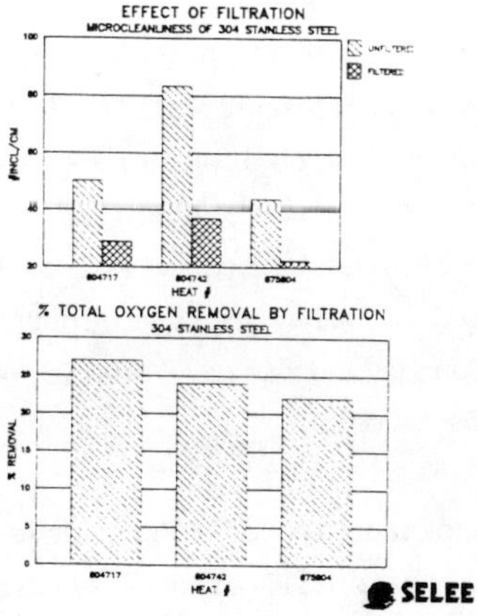

1st International SAMPE Metals Conference
August 18-20, 1987

MAGNESIUM CASTING DEVELOPMENTS AND OTHER FOUNDRY TECHNOLOGY

Alan H. Braun
Director of Engineering
Fansteel/Wellman Dynamics

ABSTRACT

Recent developments in magnesium casting technology have resulted in two new alloys with salt water corrosion resistance equal to aluminum alloys. Another new alloy offers high mechanical properties at elevated temperatures.

Techniques are being evaluated to produce thinner wall castings (0.90") with lower processing costs. Methods are also being evaluated to reduce NDT costs. SPC is being employed in the foundry to improve quality and reduce costs.

A discussion of future development programs will also be covered.

1. INTRODUCTION

Magnesium and magnesium alloys offer some very good advantages as a structural engineering material. Magnesiums primary asset is its strength to weight ratio. Others included is fine machinability, weldability and castability. It also has good dampening characteristics and is plentiful through out the world.

Magnesium is not without problems in its use as an engineering material. Some of these stem from improper use or lack of knowledge. Two major complaints are magnesium is a fire hazard and that it corrodes too rapidly. The magnesium industry has addressed these problems and feels we have come up with solutions.

2. ALLOY DEVELOPMENTS

Due to the energy crisis in the late 1970's, edicts were put in place out of Washington, D.C. requiring higher miles per gallon on future automobiles. One of the best ways to accomplish this goal is with lighter weight cars. Looking at the structural materials in helping to accomplish this goal magnesium was quickly noticed. The problem was, although magnesium could be readily cast or fabricated, exposure to road elements such as dirt, stones and salt would cause unacceptable corrosion rates.

It has been long known but never thoroughly explored that heavy element contamination levels in AZ type alloys cause significant increases in the salt water corrosion rate of magnesium(1). Work done by several primary producers in the die casting area reduced this corrosion rate by a factor of 100 making it comparable to aluminum alloys. After observing this work a development program was put together between Dow Chemical and Wellman Dynamics to see if the results were applicable to sand castings. The findings were positive and a new alloy revision of AZ91 was registered with ASTM. The new designation is AZ91E for sand castings and D for die castings. Mechanical properties were evaluated and found to be identical to the current AZ91C. Techniques were developed for melt control and casting processing to insure maintenance of the low corrosion levels. As well as an ASTM specification being approved an AMS 4446 has also been issued. This specification covers the Fe-Mn ratios required to maintain the low corrosion rate(2). It also includes a periodic testing criteria, in accordance with ASTM B117, of less than 50 mils per year corrosion rate to assure foundry conformance.

The Effect of Iron: Manganese Ratio on the Salt Fog Corrosion Performance of AZ91 Sand Cast Panels

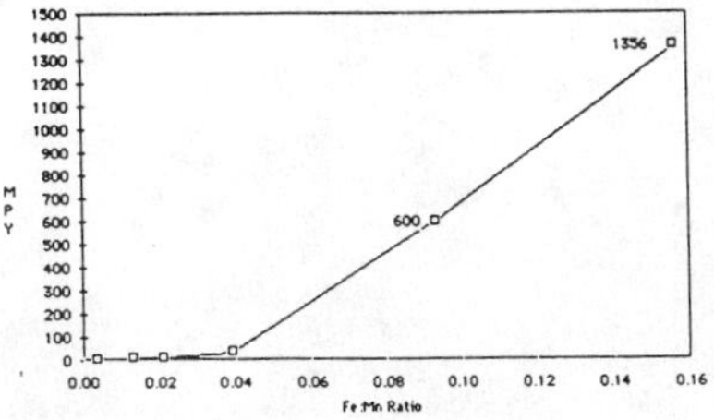

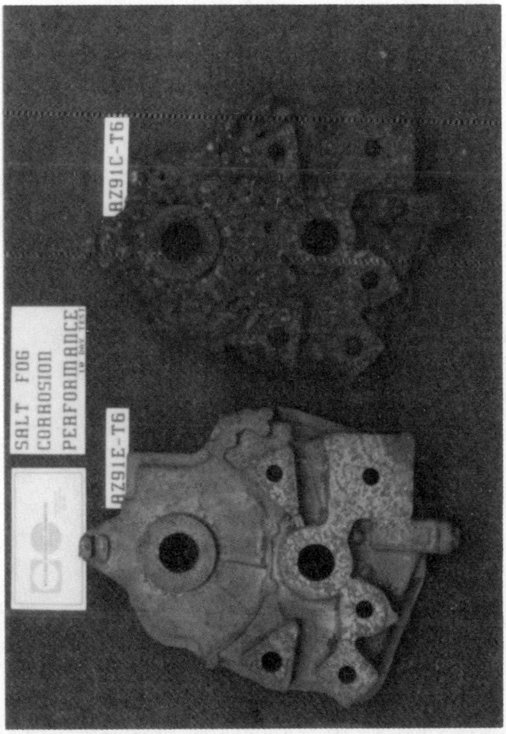

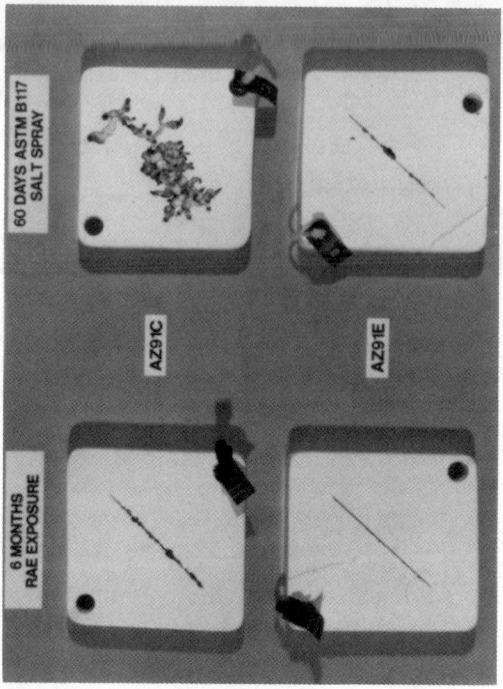

Work has also been done on AZ92 alloy and the same low corrosion rates have been obtained. An AMS for this alloy has not yet been requested.

EQ21A is another alloy development, this one accomplished by MEL. This alloy is a variant of QE22 which is a silver containing alloy along with rare earths. The silver content is about 2.5% in QE22. It is a good high strength alloy. When silver prices began climbing toward $40 per oz. the alloy fell into some disfavor because of its cost. The new alloy reduces the silver content by about 1%. As a result there is about a 1 KSI loss in both tensile and yield strength at room temperature but no loss in strength at the 400°F level. At current silver prices this alloy may begin to gain favor. An AMS is currently out for review for the material.

TYPICAL PROPERTIES AT 480°F
separately cast bars

alloy	yield	tensile	elong
QE22A-T6	22.5	25.0	19.0%
QH21A-T6	24.3	27.3	17.6%

A new alloy, WE54 containing about 5% yttrium has also been developed by MEL. This alloy offers high strength at room temperature similar to QE22 but maintains good strength up to 580°F. The alloy was developed primarily for elevated temperature use (3). Corrosion tests revealed that the alloy is similar to AZ91E in corrosion rate which was an added bonus. The one drawback is the currently expensive cost of yttrium which puts the alloy at over $10.00 per pound for ingot.

Effect of Temperature on Tensile Properties of WE54 v Aluminium Casting Alloys A356 and RR350

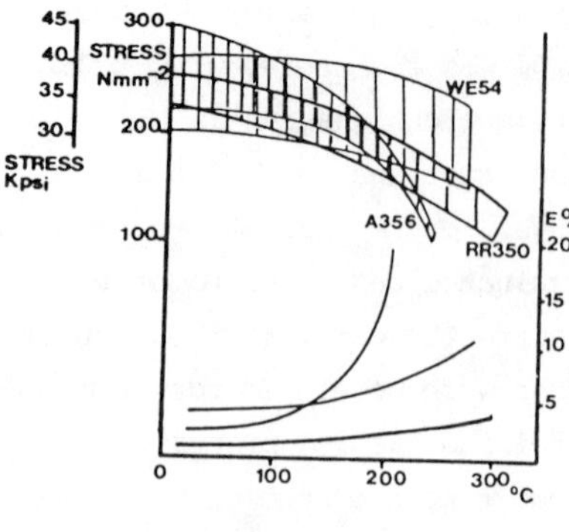

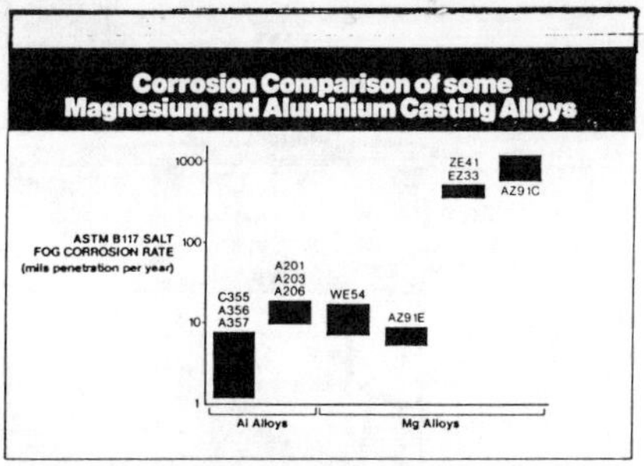

We feel these developments offer answers to the corrosion problems of the parent material. Unfortunately these developments do not change the electro potential of magnesiums when used with dissimilar metals. Galvanic potential is still there but can be eliminated by proper design, assembly and protection schemes. In design we must eliminate water traps particularly where there is use of dissimilar metal. The proper protection scheme should be used for the type of environmental exposure that is expected. A surface sealing procedure should be used. Wet assembly is required for inserts, studs and bolts. Sharp edges should be broken. Following these techniques will eliminate the corrosion problems with magnesium.

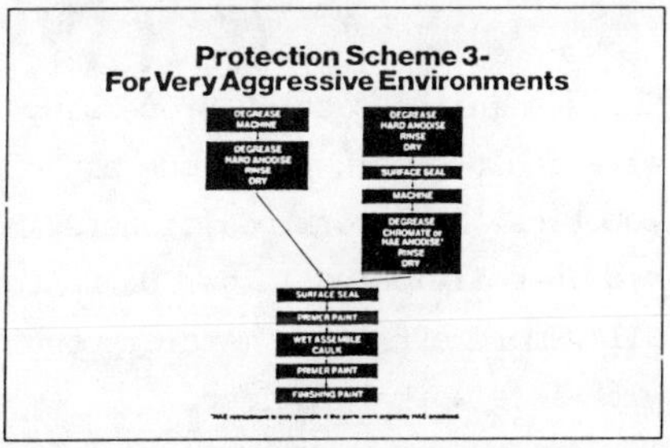

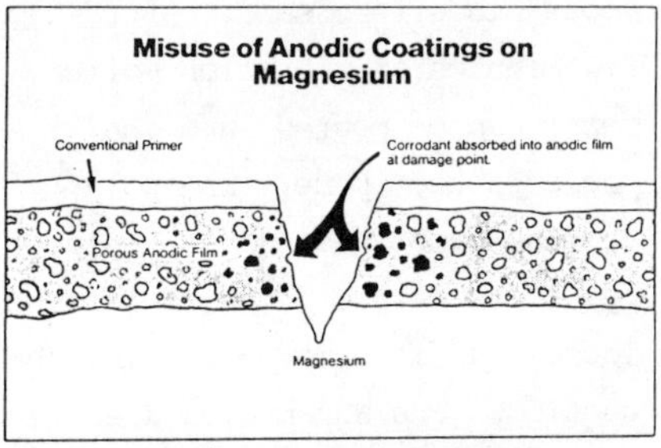

With regard to the fire hazard, this is more myth than fact. True, fines and chips will burn if exposed to flame. Solid pieces of metal will have to be melted before you can get them to burn. Once any material is molten it is not much good as a structural material.

NEW METHODS

Several new methods and technologies are going on in the casting business. One is the use of different types of sands to create better surface finishes, greater dimensional accuracy and induced thermal gradients. This results in sounder castings with higher mechanical properties.

The lost foam process is being applied more widely to high production less complex castings. In this process a die of the mold is made and filled with expandable polystyrene. A ceramic coating is then applied to the polystyrene. A base of loose sand is placed in a flask; the coated polystyrene model is then placed in the flask and the remainder filled with loose binderless sand. After a few minutes of vibration molten metal can be poured into the pre-attached sprue. The polystyrene vaporizes and is replaced by molten metal. When cooled the loose sand is removed leaving the casting. The advantages are re-use of the sand, no parting lines on the casting and no cores required for some complexity of shape.

Another technology that is now being used is low pressure sand casting. The metal in the furnace is pressurized, a tube runs from the lower part of the melt into a mold setting on top of the furnace. The advantages here are a slow orderly fill of metal that does not produce oxides. The overpressure at the end of the cycle can then be used to enhance feeding. Gating ratios can be reduced because of this, thus increasing yields and reducing cleaning and fettling costs. Consistency can also be improved due to computerized control of the pressure system.

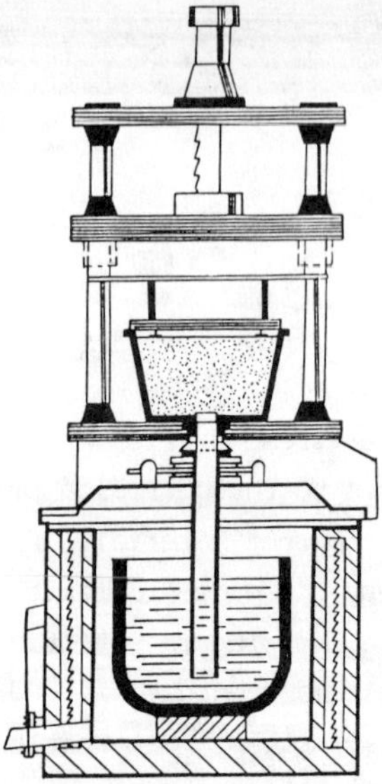

In the melting area the use of SF_6 (Sulfur hexafluoride) as a cover gas in melting and processing magnesium has also been employed in most foundries. This significantly reduces the use of flux minimizing the potential for flux in the casting. It also gives a cleaner melt and improves all over cleanliness in the melt room.

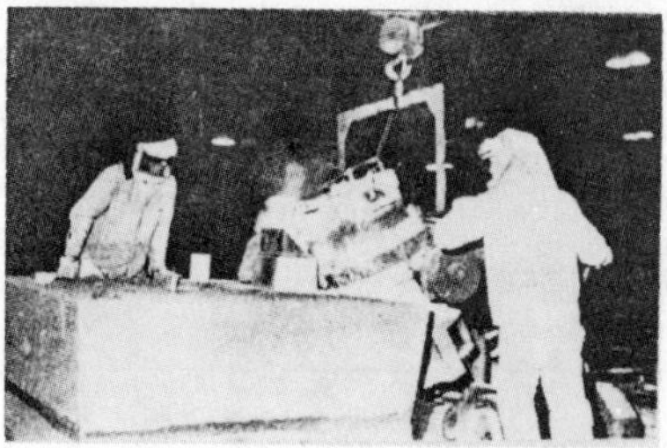

Changes in the casting processing area include work positions and robotics. The work positioners aid in positioning larger parts to allow more efficient metal removal with less worker fatigue.

Robotics present some difficulty to the jobbing shop foundry because there are so few long runs. This makes automation less able to produce pay backs. We are looking at systems that allow a simple tape change to re-program a unit to change jobs more rapidly. Due to some of the tolerance variations in castings, machine vision, though currently expensive, may have to be employed.

In the NDT area real time imaging offers significant savings. In this technology, X-Rays are passed through the part and according to the degree of X-Ray absorption excite fluoroscopic screens or electronic optics. These then produce a positive image on a screen which can be viewed by an inspector. With mechanical and electronic devices the part can be rotated, the image enhanced, enlarged and contrast increased. The images can then be stored on video tape. The main benefit is reduced cost because there is no radiographic film to buy or process. Capital equipment costs are high for this process.

SPC is another method coming into its own in the foundry industry. It is being used to control sand into the plant, core making, molding and cleaning operations to improve quality and decrease rework costs.

THE FUTURE

The future for the magnesium foundry will have to offer metal matrix composites. Magnesium has the unique characteristic of readily wetting most fibers. This is required to increase the modulus of the material. Two fiber approaches appear feasible for casting operations. The more difficult of these is the continuous fibers which are laid up in a preform and infiltrated with magnesium alloys. A great deal of problems are encountered here in getting complete wetting because both the fiber and the magnesium oxidize quickly. Discontinuous or chopped fibers offer quicker opportunities for success even if the properties are not as high as the continuous fiber approach. Here the fiber can be wetted by immersion into the melt. This produces good wetting and using special casting techniques can maintain reasonably uniform volume % fiber distribution. Early test results indicate 25 to 50% strength increases. Wear resistance is also significantly enhanced and may be able to replace race insert surfaces in some cases.

In summary a lot of new development work is going on and significant changes enhancing the use of magnesium and its alloys is in the not too distant future. Progressive managements must continue to support these efforts if the future materials requirements are to be met.

REFERENCES

1. Reicheck, K.N., Clark, K.J., Hills, J.E.. "Controlling the Salt Water Corrosion Performance of Magnesium AZ91 Alloy", SAE Technical Paper #580417, March, 1985.

2. Clark, K.J., "AZ91E Magnesium Sand Casting Alloy, The Standard for Excellent Corrosion Performance", June, 1986 IMA World Conference.

3.Unsworth, W., "Meeting The High Temperature Aerospace Challenge", June 1986 IMA World Conference.

1st International SAMPE Metals Conference
August 18-20, 1987

A NEW HIGH STRENGTH CORROSION RESISTANT MAGNESIUM CASTING ALLOY

William Unsworth
Magnesium Elektron, Ltd.
Swinton, Manchester, England

Abstract

A recently developed magnesium alloy containing yttrium, neodymium and zirconium in the fully heat treated condition provides high ambient temperature strength with excellent retention of properties up to 600°F. The alloy has good castability is weldable and should be suitable for premium quality aerospace castings. In addition the alloy offers excellent corrosion resistance which, under salt fog conditions, approaches that of some aluminum base casting alloys.

1. INTRODUCTION

The unique grain refining action of zirconium lead to the development of a diverse range of alloys the earliest of which were based on the Mg-Zn-Re-Zr and Mg-Zn-Th-Zr systems (e.g. EZ33, ZE41, HZ32 and ZH62). These alloys are designed either to provide creep resistance at elevated temperatures (EZ33, HZ32) or good ambient temperature strength with retention of properties up to 300°F and have been in use in aerospace applications for many years.

A major advance arose from the discovery that alloys containing silver and a high neodymium fraction rare earth could be solution treated and artifically aged to give high tensile properties at room temperature[(1)] which were well maintained to around 400°F (e.g. QE22). Subsequent variants on the Mg-Nd-Ag-Zr system such as QH21 and EQ21 offer improved performance, or lower cost respectively compared with QE22, have recently been developed. The development of engines with increasing power and efficiency, and consequently higher temperatures, have generated a continuing requirement for increasing strength and elevated temperature strength. Reduction in overall aircraft weight also indicated a move towards higher strength alloys which could be used in lighter castings. Consequently, it is in the area of high strength and temperature stability that recent work has concentrated.

2. MAGNESIUM-YTTRIUM-NEODYMIUM SYSTEM

Yttrium has a high solubility in magnesium (12.5 wt%) and the progressive decrease in solubility with decreasing temperature should permit an age hardening response. Alloys based on Mg-Y-Zn-Zr(2) and Mg-Y-Nd-Zn-Zr(3) appear in patent literature.

These claim good elevated temperature properties but ambient temperature strength is moderate and the alloys do not respond to a full solution precipitation heat treatment. This is probably because a major proportion of the yttrium remains tied up as a stable Y-Zn compound preventing solution/precipitation reactions. However addition of yttrium to Mg-Nd-Ag-Zr alloys gave fully heat treatable alloys with both high ambient and elevated temperature strength although at least 4 wt% yttrium was necessary to obtain a full effect.(4) It therefore appeared that fully heat treatable alloys combining high strength with good temperature resistance could be developed in the Mg-Y-Nd system.

Additions of Yttrium and Neodymium result in a progressive increase in eutectic content of the as-cast alloys.

Yttrium has an appreciable grain refining effect but this is less marked than zirconium. In the presence of zirconium the eutectic has a more irregular appearance and is also lameller in contrast to other grain refined alloys (fig. 1a).

On solution treatment the eutectic is almost completely dissolved (Fig. 1b); the amount remaining depends more on the neodymium content than the yttrium content as would be expected.

On subsequent precipitation treatment only one hardness peak has been identified.(5) This develops slowly at low temperatures (350°F), attains peak hardness at around 400°F and is barely discernable at 650°F. The low temperature precipitate has an ordered hexagonal structure of the DO_{19} (Mg_3Cd) type which slowly transforms at higher temperature (400°F) to a base center orthorhombic structure. At temperatures above 575°F this precipitate transforms to a face center cubic structure which is tentatively assumed to be the equilibrium precipitate and appears to be a ternary compound. Composition of the intermediate precipitates, which are responsible for the good ageing response, have not been established.

3. DEVELOPMENT OF A PRACTICAL CASTING ALLOY

The effect of varying the yttrium and neodymium content on the yield and tensile strength of the fully heat treated quarternary alloys is shown in Fig. 2. Strength increases with increasing yttrium and neodymium content having the greatter effect on properties.

To achieve maximum strength with

acceptable ductility the alloy should contain around 6 wt% of yttrium and around 2 wt% neodymium. Yttrium is expensive, particularly when used as a major alloying addition and the use of cheaper, less pure, grades is commercially desirable.

Yttrium is widely distributed in association with other rare earth metals in ores such as xenotime, monazite, etc. and yttrium rich concentrates of varying content can be obtained as intermediate products during the processing of these ores.

An "yttrium mischmetal" derived from concentrates containing 75 wt% yttrium with the balance heavy rare earths, mainly Dysprosium, Erbium, Ytterbium and Gadoliniun, has proved to be an effective substitute for pure yttrium. Substitution does not effect the precipitation sequence and results in only a slight reduction in strength. The heavy rare earths added with the yttrium appear to restrict the solubility of neodymium and some reduction in the latter addition is necessary to achieve a satisfactory compromise of tensile strength and castability. This gives an alloy (WE54) with the following compositional range:

Yttrium 5 - 5.5 wt% (plus HRE elements)
Neodymium 1.5 - 2 wt%
Zirconium 0.3 - 0.6 wt%

Alloys within this compositional range meet the following properties (separately cast bars)

0.2% yield strength 27 kpsi ($185\ Nmm^{-2}$)
Tensile strength 37.5 kpsi ($255\ Nmm^{-2}$)

Elongation 2%

Optimum heat treatment for this alloy comprises solution treatment for 8 hours at 975^oF, hot water or polymer quench followed by artificial ageing for 16 hours, at 480^oF.

The conventional method of melting magnesium alloys relies on the use of fluxes, based on alkali and alkaline earth halides, to protect against oxidation.

Such techniques cannot be used on alloys containing appreciable amounts of yttrium since this reacts with $MgCl_2$, an essential constituent of these fluxes, resulting in unacceptable loss of yttrium. High yttrium losses are also encountered with fluxless techniques using CO_2/SF6 gas mixtures now being used successfully on other magnesium base alloys. Alloys can be produced satisfactorily in an inert atmosphere (i.e. Argon + 0.5% SF6) and the standard CO_2/SF_6 protection techniques can still be used to protect the melt and the pouring stream at the casting stage. A processing technique using these gas mixtures with simple closed crucibles has been developed and proven under practical foundry conditions (Fig. 3).

Table 1 TENSILE PROPERTIES OF WE54-T6

	Yield Strength KSI(Nmm^{-2})	Tensile Strength KSI(Nmm^{-2})	% Elongation
Max.	32.6(225)	45.7(315)	11
Min.	26.8(185)	37.0(225)	2
Typical	29.0(200)	40.6(280)	4

4. PROPERTIES OF WE54

Ambient temperature tensile properties obtained from vests on over 250 separately cast bars poured from 100 melts are summarized in Table 1.

These properties are little affected by short term exposure at temperatures up to 575°F. In this respect the alloy is superior to currently used high strength magnesium alloy QE22 and compares favourably with some of the high strength aluminium casting alloys (Fig. 4). On long term temperature exposure (10,000 hours at 500°F) the alloy is clearly superior to any currently used magnesium alloys (Fig. 5) and again compares favourable with RR350.

The alloy is more creep resistant at temperatures up to 500°F than the currently used high strength magnesium alloys and similar to that of the thorium containing alloys HK31 (Mg-3Th-Zr) and HZ32 (Mg-3Th-2.5 Zn-Zr) (Fig. 6). Stress to rupture in 1000 seconds at 500°F is 210 N/mm^2 which is almost double that of QE22.

Fatigue endurance limit on rotating/bending tests at 5 x 10^7 cycles is 97 N/mm^2 and of the same order as that of QE22 and A356. At 500°F the fatigue endurance value is 75 N/mm^2 which is significantly better than QE22 (50 N/mm^2).

A range of castings have been made under practical foundry conditions using sand gravity die and precision casting techniques. These have confirmed the good casting characteristics of the alloy and its ability to be welded using tungsten arc inert gas techniques. (Fig. 7) Testing of specimens cut from 25 different castings with finished weights ranging from 3.5 to 14 lbs gave ambient and elevated temperature properties similar to those obtained on separately cast test bars. Properties were not significantly affected by section thickness variations ranging from 0.2 to 1.8 inches. On the basis of these vests proposed minimum tensile properties in designated areas of castings to Mil-M-46062 have been derived and these are summarized in Table 2.

Table 2 TENSILE PROPERTIES IN DESIGNATED AREA'S OF WE54 CASTINGS TO MIL-M.46064 (PROPOSED)

Alloy Designation	Designated Area Class	Minimum Properties YS. KSI(Nmm^{-2})	Tensile KSI(Nmm^{-2})	%Elongation
WE54-T6	1	30.2(208)	40.8(281)	4.8
	2	28.9(199)	39.3(271)	3.0
	3[(b)]	27.6(190)	37.1(256)	1.5
	Undesignated (a)	24.4(168)	31.0(214)	0.5

a = 99% conformance at 95% confidence level

b = 90% conformance at 95% confidence level

5. CORROSION RESISTANCE

The question of corrosion resistance is invariably raised when magnesium alloys are being considered for aerospace applications and interest has been raised by the recent introduction of "high purity" AZ91.[(6)]

Zirconium containing alloys are naturally "high purity" in that Fe, Cu and Ni contents are invariably below 50 ppm with Cu and Ni contents typically in the range 10-20 ppm. Corrosion behavior under salt fog or immersion test conditions are more influenced by some of the major alloying additions, and in particular Zn and Ag, rather than impurity levels. In milder corrosive environments, i.e. industrial atomspheric exposure, there is little to choose between any of the magnesium alloys.

WE54 is unique among the zirconium refined alloys in that its corrosion resistance, under salt fog or immersion testing, is excellent approaching that of some currently used aluminium casting alloys (Fig. 8). No evidence of stress corrosion cracking has been observed with stress levels up to 100% of the yield strength.

6. CONCLUSIONS

WE54, combining high ambient temperature strength, excellent thermal stability and good corrosion resistance is particularly suited to aerospace/military applications e.g. gear boxes designed to operate at high temperatures.

Some limited improvement in properties by compositional and heat treatment modifications may be possible but it is believed that these will be relatively small. Opportunities for further alloy development appears greatest for alloys aimed at high temperature applications where some of the rare earth elements, currently available in commercial quantities, may be capable, given suitable heat treatment, of producing temperature stable precipitates.

Finally, with the increasing interest in fibre reinforcement for improving rigidity and elevated temperature performance, it would appear that the combination of properties obtainable with WE54, and Mg-Y-Nd-Zr alloys in general, would make them attractive as matrix materials.

7. REFERENCES

1) R.J.M. Payne and N. Bailey, J. Inst. Metals 1959-60, 88 417.
2) G.S. Foerster, J.B. Clark, U.S. Pat. 3,419,385 (1968)
3) N.M. Tokhova et.al. Brit. Pat. 1,378,281 (1973).
4) W. Unsworth, J.F. King, S.L. Bradshaw. Brit. Pat. 1,463,609 (1976)
5) H. Karimzadeh, P.h.D. Thesis, University of Manchester (1985)
6) Reichels, J.K. Clark and J.E. Hillis, SAE Meeting, Detroit, February 1985.

BIOGRAPHY

For 15 years, William Unsworth has been Manager, Research and Development, of Magnesium Elektron Limited Swinton, Manchester, England with which he has been associated all his working life. He is the author or coauthor of numerous papers dealing with his field of research, and is the inventor of some 40 patents covering all aspects of magnesium technology. He is a fellow of the Royal Chemical Society and a chartered chemist.

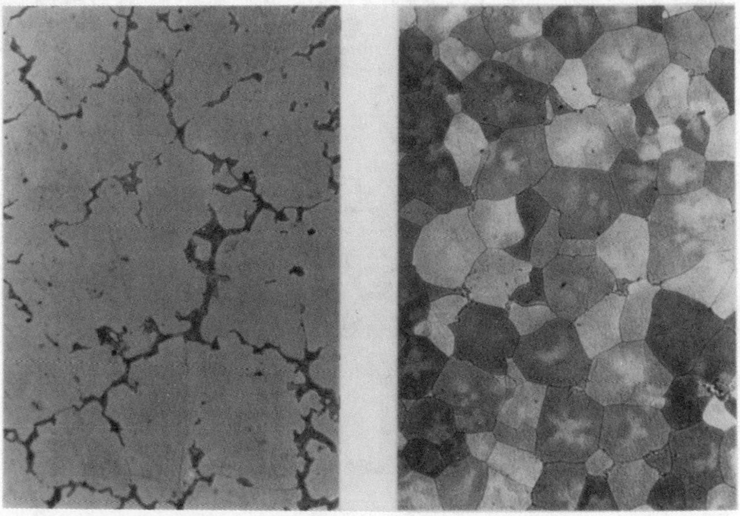

(a) As-cast (b) After full heat treatment

Figure 1. Microstructure of Mg-Y-Nd-Zr Alloy WE54
(Original Magnification X200)

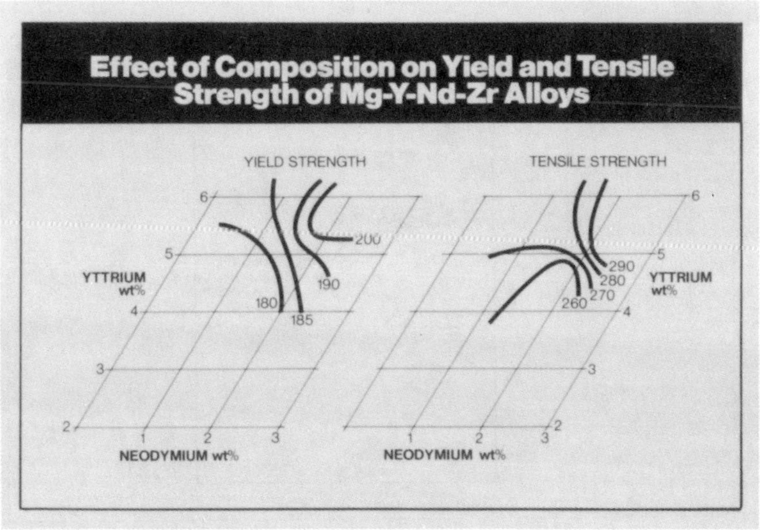

Figure 2.

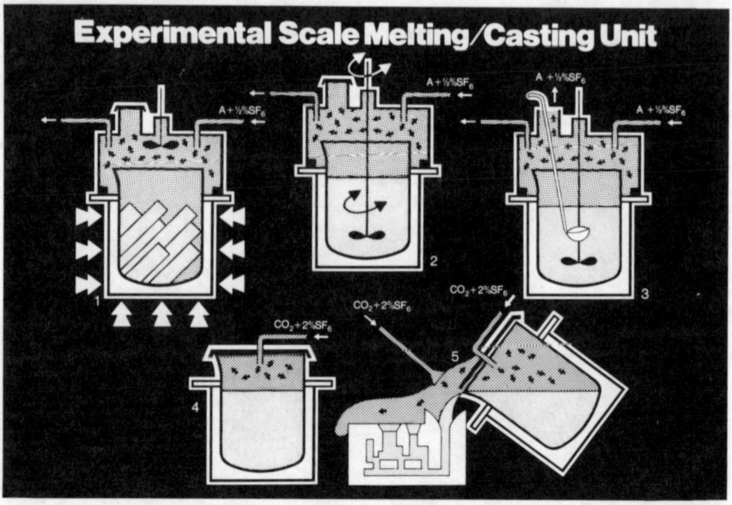

Figure 3.

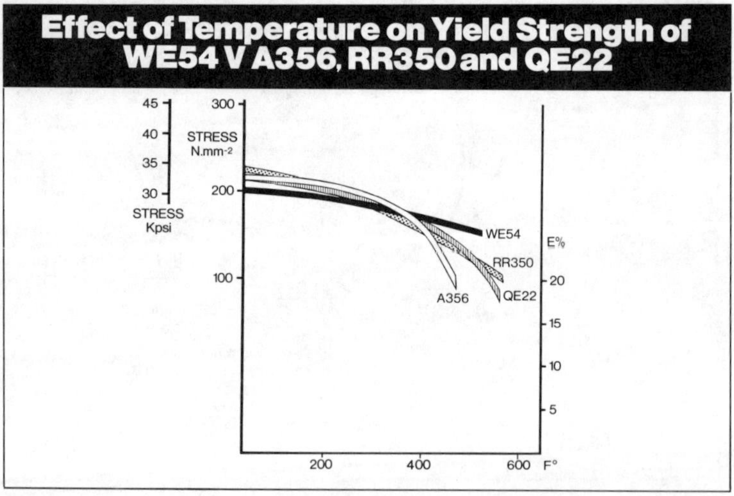

Figure 4.

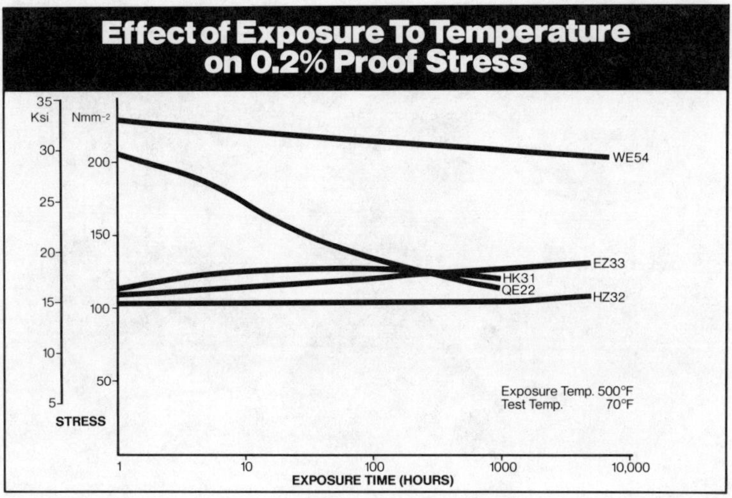

Figure 5.

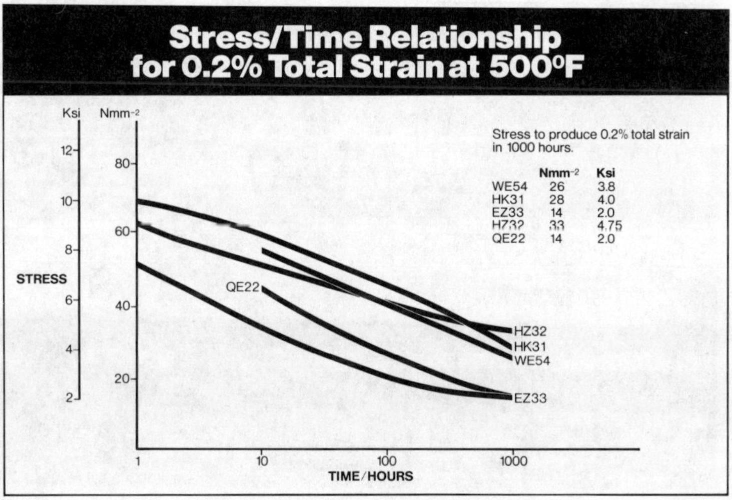

Figure 6.

Figure 7 Selection of WE54 Sand Castings

COMPARATIVE CORROSION RATES

Corrosion rates expressed in m.c.d. ($mg/cm^2/day$)

ALUMINIUM-BASE CASTING ALLOYS	ASTM B117 SALT FOG
A356-T6	0.03-0.07
A357-T6	0.07
C355-T6	0.14
KOI-T6	0.14
RR350-T6	0.21
ELEKTRON WE54-T6	0.1-0.2

ASTM B117 – 5% Salt Fog
– 35°C
– 20 days

Figure 8.

1st International SAMPE Metals Conference
August 18-20, 1987

PRECISION FORGING OF GEARS

J.R. Douglas and A.M. Sabroff
Corporate Manufacturing Services
Eaton Corporation
Willoughby Hills, Ohio 44094

ABSTRACT

Close tolerance forging of spiral bevel gears requiring only minimal machining, offers both manufacturing and performance advantages. The reduced stock on the flanks of the teeth reduces the material usage and machining cost. Also, gears forged with net or near-net teeth have been shown to have considerably improved fatigue life.

The forging of straight bevel gears with net teeth is a well-established process for certain applications such as truck axle differential gears and right-angle gear boxes. However, the forging of spiral bevel gears, even with near-net teeth, until recently, has been done only in a few isolated research programs.

In a recent U.S. Army Tank Automotive Command project, conducted jointly by Eaton Corporation and Battelle Columbus Division, the methodology for forging spiral bevel gears with near-net teeth was developed. It was also demonstrated that precision forging of spiral bevel gears is a practical production process.

The program first involved creating the spiral bevel tooth surface geometry by simulating the gear cutting process using computer techniques. Then the calculations for the forging die geometry were made, taking into account the thermal shrinkage and elastic distortion of the die. Finally, the geometry of the EDM electrodes are calculated for machining the forging die.

Forging trials were conducted and the forgings were inspected to demonstrate the accuracy of the design process. Spiral bevel gears up to 16-inch diameter with near-net forged teeth were produced in the pilot production trials.

1st International SAMPE Metals Conference
August 18-20, 1987

PERFORMANCE MECHANICS OF ARALL LAMINATES

J. L. Teply, S. M. Reed and R. W. Schultz
Alcoa Laboratories
Alcoa Center, PA 15069

ABSTRACT

ARALL, aramid aluminum laminate, is a family of new structural hybrid materials being commercialized by Alcoa. These materials consist of alternating layers of thin aluminum sheet bonded by adhesive impregnated with strong aramid fibers. All three constituents, aluminum, aramid fibers and resin, exhibit a nonlinear material response when subjected to tensile, compressive, or shear load. The tensile and shear stress-strain curves of the constituent materials and ARALL plies are presented. A nonlinear lamination theory is then applied to derive the overall stress-strain curves of ARALL laminates subjected to the same loading conditions; these calculated stress-strain curves are contrasted with experimental results.

Theoretical and experimental results are used to derive ARALL overall mechanical properties from the stress-strain curves. The effect of residual stresses on the mechanical response of ARALL laminate is discussed.

1. INTRODUCTION

ARALL Laminates are a family of new structural composite materials that show promise for use in weight sensitive, fatigue and fracture critical structures (1-5). The laminates consist of layers of thin aluminum alloy sheet bonded by adhesive impregnated with high strength aramid fibers. An ARALL-1 Laminate schematic representation is shown in Figure 1.

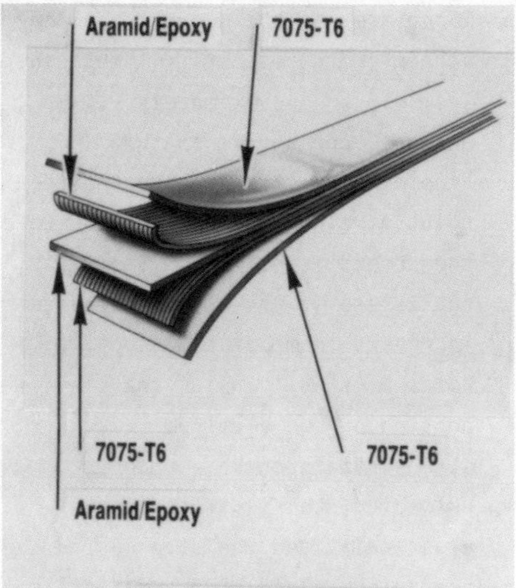

Schematic of ARALL-1
Figure 1

ARALL Laminates have outstanding fatigue properties as compared to monolithic aluminum materials. Under cyclic loading, cracks tend to grow very slowly or arrest. This is attributed to the load transfer from the metal to the unbroken fibers. The fibers which bridge the crack, restrain it from opening. Additional fatigue property enhancement of ARALL can be achieved by prestraining the composite layup after curing. This imparts a favorable compressive residual stress state in the metal that retards crack initiation.

In this paper, the performance mechanics in tension and shear of ARALL-1 and ARALL-2 Laminates are studied. ARALL-1 is a laminate fabricated from three 0.3-mm (0.012 inch) thick layers of 7075-T6 aluminum alloy and two 0-degree, 0.2-mm (0.0085 inch) thick layers of aramid/epoxy. The full designation for the described stacking sequence is 3/2 ARALL-1 Laminate. ARALL-1 Laminates are prestrained (prestretched) to 0.5 percent residual strain. ARALL-2 consists of four aramid/epoxy layers and three layers of 2024-T3 aluminum alloy. The layer thicknesses are the same as for the ARALL-1 Laminate. ARALL-2 Laminates are not stretched. The present study is primarily intended as the first stage in the verification of the two step integrated finite element model for fibrous composites developed in Alcoa Laboratories.

2. MANUFACTURING PROCESS AND PERFORMANCE MECHANICS

The yield stress, ultimate strength, elongation and other properties of ARALL Laminates are functions of the material properties of the constituents (fibers, resin and aluminum alloy) as well as the process variables such as curing temperature and stretching strains. The relation between material and process variables is illustrated in Figure 2. The selected example is a simplified one-dimensional tension model of a generic unidirectional ARALL Laminate with fibers running along the model's axis. The dashed lines in Figure 2 represent the stress-strain curves of the constituents: aramid/epoxy and aluminum layers. Line AGB represents the stress-strain curve of the aramid/epoxy system and line CDH characterizes the aluminum alloy. The solid line is the resultant curve of the ARALL Laminate. Points A and C represent the thermal residual stresses of aramid/epoxy and aluminum layers after the cooling of the laminate from the curing temperature (250°F) to room temperature (70°F). The cooling process, coupled with the mismatch in the coefficients of thermal expansion gives rise to a compressive residual stress in aramid/epoxy layers (point A) and a tensile stress in aluminum layers (point C). This residual stress state has a negative effect on the fatigue crack growth resistance of the laminate, and can be reversed by prestretching the laminate.

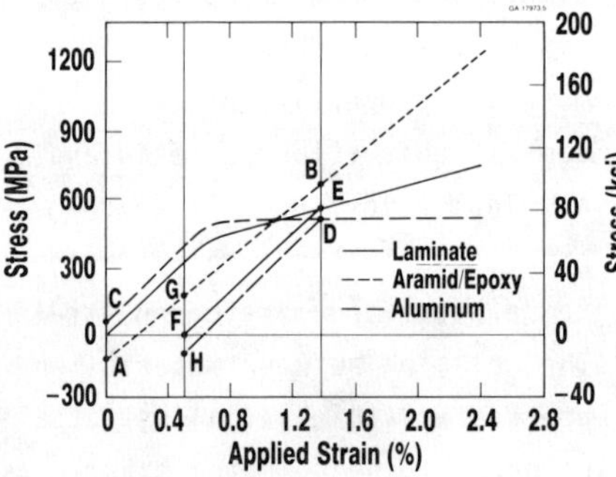

Laminate and Layer Stresses During Prestraining
Figure 2

For a thin unstretched ARALL sheet, strains are uniform through the thickness of the laminate. Therefore, as the overall applied strain increases during stretching from 0.0 to approximately 1.4 percent, the stress in the aramid/epoxy layers increases from point A, along the dashed line to the stress at point B. Similarly, the stress in the aluminum layers increases from point C to point D. Point D is well inside the plastic portion of the aluminum stress-strain curve. Upon unloading, the laminate behaves elastically and the stresses in the aramid/epoxy and aluminum decrease in accordance with the original Young's moduli of these layers. In Figure 2, the unloading is represented by the lines B-G and D-H. The laminate is also unloaded elastically from Point E, along the solid line, to zero stress at Point F. Now, there exists an overall residual strain of about 0.5 percent. Through this procedure the thermal residual stresses are reversed and the fatigue crack growth resistance increases.

It is obvious that processes and material variables influence the curves of Figure 2. For example, an increased yield stress for aluminum can make prestretching less effective, but improves the subsequent yield stress upon reloading. A higher strain hardening constant has similar effects. A more compliant aramid/epoxy layer (a lower volume fraction of fibers) may lead towards either increased or decreased effective stretching, depending on the magnitude of prestretching strain. By varying these material and process variables, an optimum laminate can be designed for given thermo-mechanical requirements.

Optimization for performance, for example tension-tension fatigue and fracture toughness, usually has been made with the aid of an expensive experimental program. A verified model for performance and process mechanics of ARALL Laminates can alleviate the expense of experimental optimization. The first stage in optimization, modeling and verification of the tension and shear responses of ARALL, is discussed in Sections 3 and 4.

3. TENSION

A two-step model was used to predict the tensile stress-strain response of 3/2 ARALL-1 Laminates. In the first step, the micromechanical model, a periodic hexagonal array (PHA) model [6,7], was utilized to calculate the overall stiffness moduli, coefficients of thermal expansion and tensile stress-strain curves of the aramid/epoxy composite layers. In the second step, results from the PHA analysis and material properties of 7075-T6 alloy were used in an elastic-plastic lamination code to predict overall tensile behavior of ARALL-1 Laminates.

Results of the PHA micromechanical analysis of the aramid/epoxy layers are summarized in Tables I and II and Figure 3. Table I contains mechanical properties of the individual constituents: aramid fibers and 3M AF163-epoxy resin.

TABLE I. MECHANICAL PROPERTIES OF ARAMID FIBERS AND 3M AF-163 EPOXY RESIN

	Aramid	Epoxy
Longitudinal Young's Modulus - GPa (Msi)	126.7 (18.4)	2.8 (0.4)
Transverse Young's Modulus - GPa (Msi)	7.0 (1.0)	2.8 (0.4)
Longitudinal Shear Modulus - GPa (Msi)	2.6 (0.38)	1.0 (0.14)
Transverse Shear Modulus - GPa (Msi)	3.0 (0.44)	1.0 (0.14)
Major Poisson's Ratio	0.33	0.38
Ultimate Strength MPa (ksi)	2758 (400)	---

TABLE II. BOUNDS ON OVERALL PROPERTIES OF ARAMID/EPOXY LAYERS

Fiber Volume Fraction 0.484

	Upper Bound	Lower Bound	Experiment
E_{xx} = GPa (Msi)	62.74 (9.099)	62.74 (9.099)	61.09 (8.860)
$E_{yy} = E_{zz}$ GPa (Msi)	4.61 (0.668)	4.51 (0.654)	4.13 (0.599)
ν_{xy} =	0.3554	0.3552	0.435
G_{xy} = GPa (Msi)	1.65 (0.239)	1.60 (0.232)	1.54 (0.224)
G_{yz} = GPa (Msi)	1.55 (0.225)	1.50 (0.218)	---
α_x = C (F)	---	-0.58E-6 (-0.32E-6)	-1.62E-6 (-0.9E-6)
α_y = C (F)	---	71.0E-6 (39.0E-6)	76.0E-6 (47E-6)

x-axis is the fiber direction

Table II contains overall elastic properties of the aramid/epoxy layers as predicted by PHA. Because of a lack of reliable experimental data on the nonlinear, viscoplastic behavior of the epoxy resin, aramid/epoxy layers are, in this paper, considered as an elastic material. This assumption agrees well with the tensile tests of 0 and 90 degree specimens.

A deviation from a linear stress-strain curve was observed for small angles of the off-axis tension tests as illustrated in Figure 4 (for detailed information see [8]). For elastic properties, the PHA model supplies upper and lower bounds on all five independent moduli of the unidirectionally reinforced aramid/epoxy system. In addition to these five moduli, the bounds on two independent coefficients of thermal expansion are listed in Table II. The fiber volume fraction, 0.484, was determined from the densities of the fibers, resin and composite. To further demonstrate the accuracy of PHA predictions, the calculated directional modulus E is compared with off-axis tension tests in Figure 3.

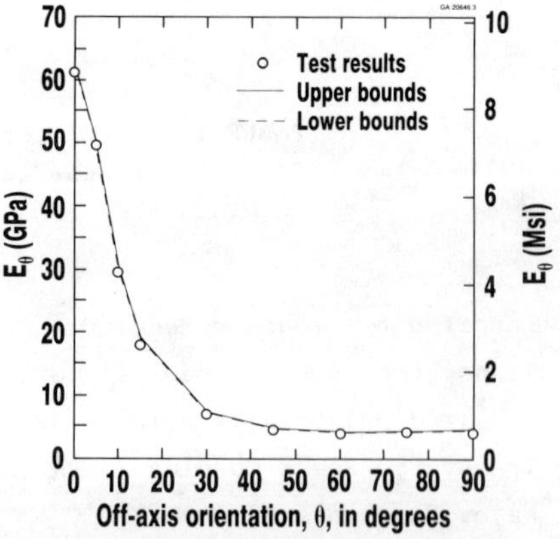

Elastic Modulus of Aramid/Epoxy Layers-Upper/Lower Bounds and Experimental Data
Figure 3

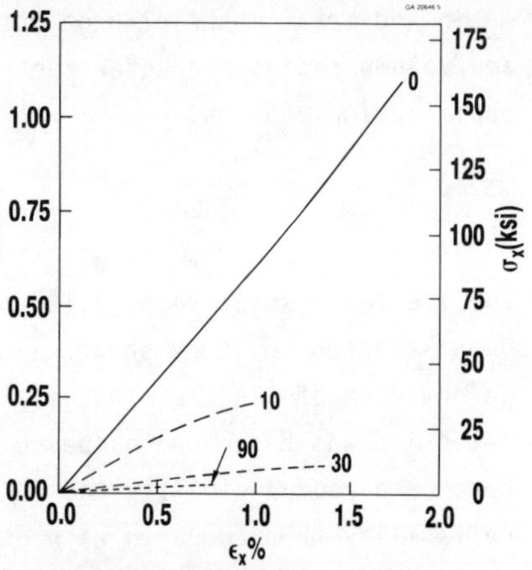

Aramid/Epoxy Off-Axis Tests
Figure 4

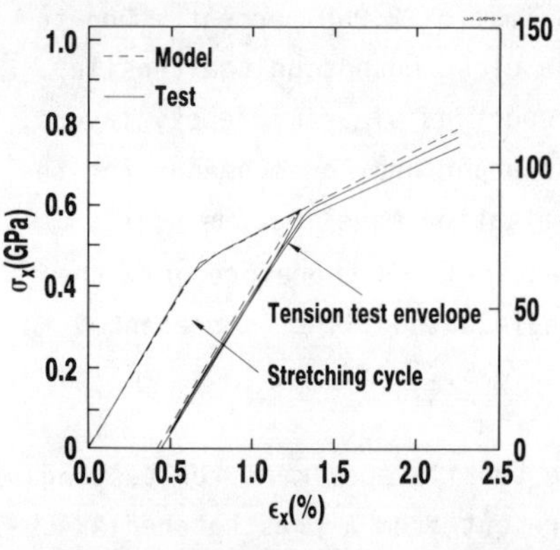

Stretching and Tensile Test of ARALL-1
Figure 5

A nonlinear elastic-plastic analysis of tension tests of ARALL-1 Laminates is shown in Figure 5. The solid line represents experimental data, the dashed line represents elastic-plastic lamination model. The stress-strain curves of the aramid/epoxy and aluminum layers are omitted in order to keep the figure simple. The aramid/epoxy stress-strain curve can be derived from the data in Table II. The aluminum stress-strain curve can be constructed with the aid of material properties of 7075-T6 alloy listed in Table III. Also, the linear kinematic hardening rule with Ziegler's modification is assumed for aluminum.

The nonlinear elastic-plastic lamination model is a finite element model which consists of eight-noded three-dimensional elements stacked to create the 3/2 ARALL-1 Laminate. Boundary conditions for stretching and bending of an infinite laminate are prescribed, and the membrane and bending response of the laminate can be derived for both plane stress and plane strain conditions. The plane stress condition was used in the modeling of the tension tests. The fabrication process, cooling cycle and 1.29 percent prestrain was modeled first.

After the release of the prestraining displacements, the tensile test was simulated. The maximum applied tensile strain was equal to the average experimental

TABLE III. MECHANICAL PROPERTIES OF 7075-T6 ALLOY

Young's Modulus GPa (Msi)	70 (10.14)
Poisson's Ratio	0.33
Yield Stress MPa (ksi)	517 (75)
Ultimate Stress MPa (ksi)	558 (81)
Elongation at Ultimate Stress	0.05

maximum of 2.268 percent. Due to the close bounds on the tensile properties of aramid/epoxy layers, the upper and lower bounds for the lamination model become nearly identical and therefore only the single dashed line is presented in Figure 5.

The tensile specimens for testing were cut from a prestretched ARALL-1 panel. Stress-strain measurements of the panel were recorded during the prestretching at Alcoa Laboratories. The stretching cycle and envelope of the ten tensile stress-strain curves were combined.

The comparison in Figure 5 also suggests that an entire manufacturing process; i.e., curing and stretching cycles, must be modeled in order to get a realistic tensile behavior of ARALL Laminates. ARALL-1 is a unidirectional laminate and one-dimensional models [3] will produce approximately the same results as the Alcoa integrated finite element model. However, to model future multidirectional ARALL Laminates, the use of the approaches illustrated in Figures 3 and 5 will be required. Also, for future unidirectional ARALL Laminates, the integrated two-step model provides an additional micromechanical dimension for optimization. This means that in addition to the stacking sequence and material properties of aluminum and aramid/epoxy layers, the fiber, resin and metal material properties and volume ratios can enter the optimization process.

4. SHEAR

The Iosipescu shear test [9,10] has been selected to study shear deformation of ARALL Laminates. The selection was based on Iosipescu test data reproducibility and reliability when compared with other shear test methods [11].

A two-step analysis similar to that for tension testing was applied to model the ARALL-2 Laminate shear response. As mentioned above, ARALL-2 is an unstretched laminate; therefore only thermal residual stresses are included in the model. The first step, the micromechanical analysis of the aramid/epoxy system is unchanged from the tension test model; i.e., the properties in Table II and results in Figure 4 are applied. The aluminum alloy is 2024-T3. Properties are listed in Table IV. As with the 7075-T6 alloy, the linear kinematic hardening rule with Ziegler's modification is assumed.

The finite element mesh for the Iosipescu test analysis is shown in Figure 6. The Iosipescu specimen has 90° notch and radius equal to 0.05 inch. The omitted elements in the center of the specimen mark the position of $\pm$ 45° rosette strain gage.

TABLE IV. MECHANICAL PROPERTIES OF 2024-T3 ALLOY

Young's Modulus GPa (Msi)	1.00 (10.42)
Poisson's Ratios	0.33
Yield Stress MPa (ksi)	345 (50)
Ultimate Stress MPa (ksi)	490 (71)
Elongation at Ultimate Stress (in/in)	0.17

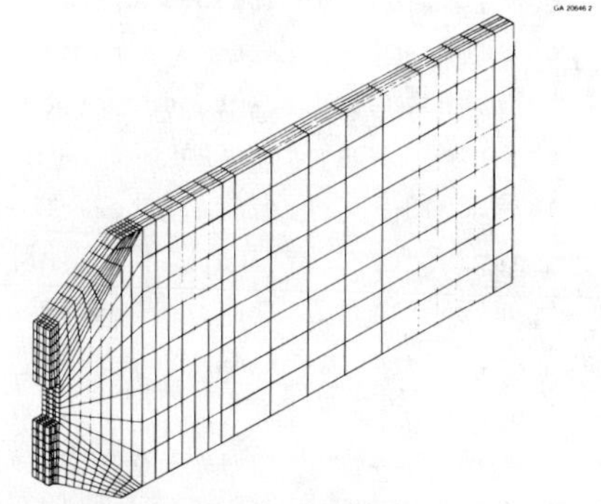

Finite Element Mesh of Iosipescu Specimen
Figure 6

Initially, the specimen was subjected to the cooling cycle of the laminate. The thermal residual shear stress in the top aluminum layer is shown in Figure 7. The application of the cooling cycle to the fabricated Iosipescu specimen tacitly assumes that machining does not effect the existing specimen residual stresses. This assumption is true if the machining is slow and careful. In addition to that, the notch radius has to be "large" enough not to cause stress concentrations in excess of the yield stress of the aluminum alloy. The finite element analysis has confirmed that for the specimen notch radius of 0.05 inch the thermal residual stresses stay well below the yield stress of 2024-T3 alloy (Figure 7).

ARALL-2 Laminates are not prestretched, hence, the thermal stresses are the initial stresses for the shear test model. Boundary displacements equivalent to those of the experimental test were used in the Iosipescu shear test model. The analysis results are compared with those from experiments in Figure 8. Both shear stresses, measured and modeled, are apparent shear stresses calculated from the ratio of the overall load to the area between the two notches. The experimental strain is measured from the strain gage placed in the center of the specimen. The strain from the finite element model is the strain averaged from the elements underneath the strain gage area.

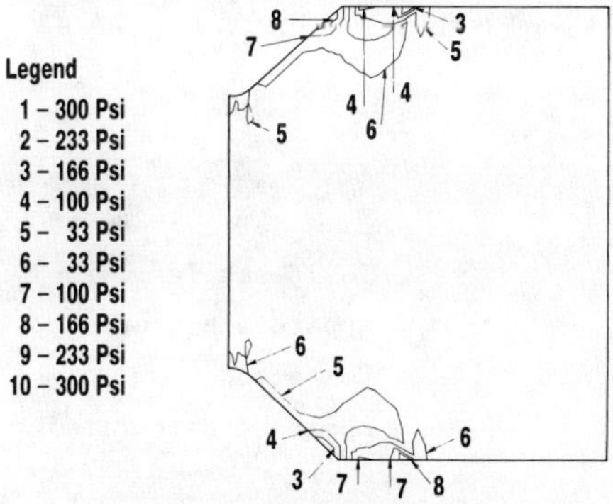

Thermal Shear Stresses in ARALL-2 Iosipescu Specimen
Figure 7

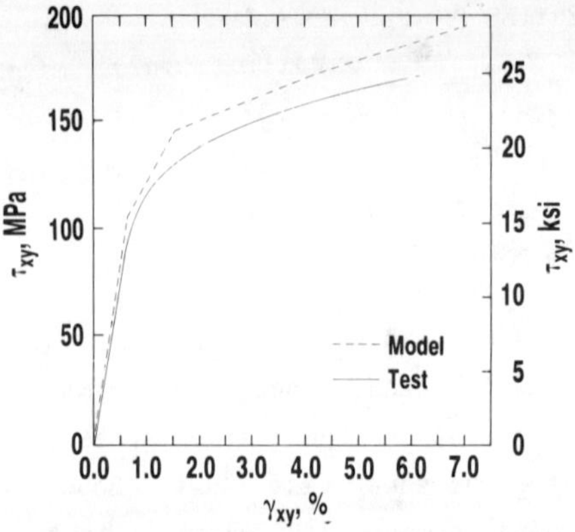

Iosipescu Shear Test of ARALL-2
Figure 8

This strain averaging is a different approach then in [9] and [10], and it appears to be more accurate for ARALL Laminates than the other two.

A close agreement between the apparent experimental and modeled shear stress strain curves has two implications. First, it indicates that the PHA model supplies a close estimate for the shear stress strain curves of the aramid/epoxy layers. Second, it suggests that the same averaging technique used for the shear stress-strain curves can be applied to calculate a correction factor between the actual and apparent shear modulus of ARALL Laminates.

According to the elastic-plastic analysis of the Iosipescu specimen, the correction factor, a ratio between the apparent and actual shear modulus, for ARALL-2 Laminates is 1.0613. Therefore, the actual shear modulus of ARALL-2 Laminates is 2.5695 Msi. The summary of shear modulus results is presented in Table V.

Table V also includes an estimate of the shear modulus from the two step lamination analysis described in Section 3. The estimate comparison has been provided in order to illustrate that the shear properties of ARALL Laminates can also be predicted and optimize based on the PHA and lamination model rather than on the complicated FE analysis of Iosipescu tests.

TABLE V

SHEAR MODULUS OF ARALL-2 LAMINATES

Iosipescu Test	
Apparent Measured	2.4211
Corrected	2.5695
Apparent Calculated	2.5639
Predicted from Lamination Model	
Upper Bound	2.6354
Lower Bound	2.6331
ARALL-2 Correction Factor	1.0613

5. CONCLUSIONS

Performance mechanics of tension and shear properties of ARALL-1 and ARALL-2 Laminates were modeled and analyzed. A good agreement between analytical and experimental results has been obtained for both. Although the micromechanical and lamination models were verified by experiments on unidirectional ARALL Laminates, the off-axis results for aramid/epoxy layers (see Figure 3) suggest that the same close predictions of tensile and shear behavior can be expected for crossplied ARALL Laminates.

Introduction and verification of the PHA micromechanical model into the ARALL Laminate models brings an additional depth to the laminate optimization. Various fiber/resin combinations and their effect on residual stresses can be studied with a minimum of expensive testing. The optimization of residual stresses is an important step in the optimization of fatigue and fracture toughness of ARALL Laminates.

REFERENCES

1. Marissen, R. and Vogelesang, L. B., "Development of a New Hybrid Material: ARALL," at Intercontinental SAMPE Meeting, Cannes, 1981 January.

2. Vogelesang, L. B., Marissen, R. and Schijve, J., "A New Fatigue Resistant Material: Aramid Reinforced Aluminum Laminate (ARALL)," at the 11th ICAF Symposium, Noordwijerhout, The Netherlands, 1981 May.

3. Vogelesang, L. B. and Gunnink, J. W., "ARALL, A Material for the Next Generation of Aircraft; A State-of-the-Art," Delft University Report LR-400, 1983 August.

4. Mueller, L. N., Prohaska, J. L. and Davis, J. W., "ARALL (Aramid Aluminum Laminates): Introduction of a New Composite Material," at AIAA Aerospace Eng. Conf. and Show, Los Angeles, 1985 February.

5. Schultz, R. W., Gregory, M. A. and Teply, J. L., "Lightweight Structural Hybrid Laminates," Proc. of the First OMAE Specialty Symp. on Offshore and Arctic Frontiers, ASME, 1986.

6. Teply, J. L. and Dvorak, G. J., "Evaluation of Bounds on Instantaneous Elastic-Plastic Properties of Fibrous Metal Matrix Composites," Proc. of the 1985 Preserve Vessel and Pijoring Conference, ASME, 1985.

7. Teply, J. L., "Periodic Hexagonal Array Models for Plasticity Analysis of Composite Materials," Ph.D. Dissertation, University of Utah, University Microfilm International, 1984.

8. Pindera, M., Gurdal, Z., Hidde, J. S. and Herakovich, C. T., "Mechanical and Thermal Characterization of Unidirectional Aramid/Epoxy," at Virginia Tech Center for Composite Materials and Structures, Report, CCMS-86-08, 1986.

9. Swanson, S. R., Messick, M. and Toombes, G. R., "Comparison of Torsion Tube and Iosipescu In-Plane Shear Test Results for a Carbon Fibre-Reinforced Epoxy Composite," J. of Comp. Materials, Vol. 16, July 1968, p. 220-224.

10. Walrath, D. E. and Adams, D. F., "Analysis of the Stress State in an Iosipescu Shear Test Specimen," UWME-DR-301-102-1, University of Wyoming Composite Materials Research Group Report to NASA-Langley Research Center, June 1983.

11. Lee, S and Munro, M., "Evaluation of In-Plane Shear Test Methods for Advanced Composite Materials by the Decision Analysis Technique," J. of Comp. Materials, Vol. 17, January 1986, p. 13-20.

1st International SAMPE Metals Conference
August 18-20, 1987

SYNERGISM OF ELECTROPLATING AND VACUUM PROCESSES

J. W. Dini
Lawrence Livermore National Laboratory
Livermore, California 94550

Abstract

Synergism can be obtained by combining electrodeposition and vacuum processes. In a number of cases, better coatings and improved adhesion can be obtained when the two processes are combined. Also, for some applications, parts that could not be fabricated by either method by itself, can be produced when the two processes are combined. A number of these examples are discussed in this paper.

1. INTRODUCTION

Although electroplating and vacuum processes (physical vapor deposition, plasma etching, sputter etching, etc.) are occasionally in competition with one another to satisfy coating requirements, they can also be highly complimentary processes. By combining the two processes, almost any plastic or metal can be coated with a much wider range of deposits than if either process is used alone. A combined vacuum deposition/ electroplating process offers the potential of producing a better coating than obtainable with either deposit by itself. Advantages include a significant cost reduction due to material savings (thinner deposits), reductions in energy usage, and lower expenditures associated with pollution control and disposal of wastes.

Electrodeposition methods feature applicability to more complex shapes, generally lower cost, better control of deposit properties including internal stress, and the ability to deposit very thick coatings. By comparison, vacuum-deposited coatings feature controllable close tolerance for thin functional coatings, a far wider choice of substrate and the ability to deposit coatings, particularly non-metallics and active metals, not obtainable by wet methods.

In this paper, four topics will be covered to help demonstrate the synergism that can be obtained with electroplating and vacuum processes. The topics to be covered include: plating on molybdenum, fabrication of sputter cathodes, electroforming parts with precision small inside diameter holes, and plating on plastics.

2. PLATING ON MOLYBDENUM

Molybdenum is a difficult metal to coat with an adherent electroplate because of its natural protective oxide. Many procedures have been reported in the literature for accomplishing this purpose and these were reviewed by Levy, et al.[1] in 1980. The processes typically include one or a combination of the following: 1- pickling in a variety of concentrated acids, 2- use of intermediate strikes such as chromium which has a thermal coefficient of expansion similar to molybdenum, 3- diffusion in a dry hydrogen atmosphere, and 4- magnetron ion plating. Ring shear tests[2] were used to quantitatively assess the effectiveness of various procedures for preparing molybdenum for plating. The results shown in Table 1 reveal that the best bonds were obtained with the magnetron ion plating process. Good results were also obtained with the processes which included hydrogen firing, (e.g., Bell Labs) and alkaline ferricyanide. It's important to note that the higher strengths obtained with the magnetron ion process (216 MPa) compared to the values of 100 to 125 MPa for the hydrogen firing process represent the difference in strength levels between copper and gold electrodeposits. Typically, copper deposits are about twice as strong as gold deposits and this difference is reflected in the data in Table 1 for those processes wherein failure occurred in the electrodeposit rather than at the substrate-plating interface. Results with the Martin-Marietta process were considerably inferior to all the rest.

By combining the magnetron ion deposition process with subsequent electroplating, parts with very complex geometries and large size (1.5 m diameter) were successfully coated and machined to very precise tolerances.[3]

2.1 Fabrication of Sputter Cathodes by Electroplating

Magnetron sputtering offers many advantages as a physical vapor deposition process. It can frequently produce thick and adherent coatings at relatively high rates, it is an inherently stable process permitting excellent rate control, and it allows the production of dense coatings on relatively cool substrates.

The primary disadvantage of magnetron sputtering involves the necessity of obtaining a sputtering target of the material to be deposited. This target must be fabricated into a well-defined shape to be useful for sputtering. With materials such as chromium, such fabrication can introduce impurities such as oxygen into the metal. In addition, conventional magnetron sputtering does not make full use of the target material and spent targets of expensive materials must be reclaimed in order to prevent costly waste.

Use of electroplating as an aid in fabricating sputtering targets alleviates some of these disadvantages. Benefits from electroplating include the deposition of material precisely where it is needed and high purity of the depositing species. At LLNL we've used electroplating to produce sputtering targets of Ag, Cr, Au, Fe, Co and Ni-Co.(4) The fabrication process consisted of depositing 20 to 40 mils of the coating of interest on a target substrate containing erosion grooves which are typical of those obtained during sputtering(5) (Fig. 1). This way the coating is applied precisely in the regions it is needed thus minimizing waste during subsequent sputtering.

Table 2 compares impurities in a commercially produced sputter target with those in an electroplated chromium target. The electroplated target is considerably purer even though special precautions were not taken during its fabrication. It should be pointed out that the electroplating solution used for this work was greater than five years old and had been exposed to a wide variety of metal parts. Much higher purity electroplated chromium deposits could be obtained if needed.(6)

2.2 Electroforming Parts With Precision Small Inside Diameter Holes

Electroforming is used to produce precision parts for measuring the shape of neutron emitting sources. These parts consist of holes (referred to as pinholes) varying in diameter from 1 mil to 10 mils in the center of a 1/4 inch diameter cylinder which is 1-1/2 inch long. The pinholes have very precise requirements, i.e., an internal diameter with a straightness of 0.001 inch, and parallel to the outer wall of the electroformed cylinder. The process for fabricating pinholes consists of electroplating a thick deposit on a wire or tube mandrel, removing the mandrel by chemical dissolution and then final machining. Further details on the electroforming process can be found in reference 7. Some pinhole designs require mandrels with a 2 mil center that increases in thickness to 24 mils over a 3-inch length (Figure 2). Mandrels with these dimensions cannot be purchased or fabricated by machining. However, by use of electron beam evaporation and a unique fixture to vary the source to substrate distance, acceptable mandrels can be fabricated (Figure 3). A copper wire, 2 mils in diameter is used as the starting material and copper is evaporated at a controlled rate to build up the final thickness to 24 mils (the diameter of a pinhole mandrel is increased by 2 mils per hour using a rod fed electron beam source). During deposition a movable shield is used to provide the desired gradient in coating thickness. The finished mandrel is then used for the electroforming process described above. Without the marriage of vacuum deposition and electroplating we do not know how it would be possible to produce these tapered pinholes.

3. PLATING ON PLASTICS

Chemical processes for plating on plastics are expensive, require rigorous chemical control, and present effluent treatment problems. A preplate process using vacuum techniques has been developed by Lindsay and LaSala[8] for ABS (acrylonitrile butadiene styrene) the most commonly used plastic substrate for electroplated decorative parts. This process offers the potential for reduced cost, is relatively simple, eliminates solution processing and has virtually no environmental problems.

The process consists of 1- an oxygen, radio frequency (RF) glow discharge plasma treatment to condition the surface, 2- a 1000 Å nickel deposit applied by sputtering or electron beam evaporation to establish a metal-plastic bond and 3- a 1000 Å copper deposit to prevent passivation of the nickel and enhance electrical conductivity for subsequent electrodeposition to final thickness. With the proper plasma voltage during the first step, adherence equal to that obtained by conventional chemical preplating (17.5 N/cm and above) is obtained (Figure 4). The plasma treatment changes the basic nature of the polymer surface by increasing the number of single and double bonds between carbon and oxygen atoms.[9]

Another process that works well on ABS and ABS alloys including polycarbonate varieties involves ozone etching followed by alkali conditioning.[10,11] The ozone gas "sees" and reacts uniformly with all exposed surfaces producing a very uniform etch. This process eliminates the possibility of Cr^{+6} contamination from the conventional wet etching process and this is critical because as little as 5 ppm Cr^{+6} poisons electroless nickel solutions causing complete decomposition. Another significant advantage of the ozone process is that the reaction is diffusion controlled and, therefore, self-terminating. By contrast, chromic acid etching never stops reacting. Vapor etching of plastics using sulfur trioxide is also gaining favor as a replacement for chromic acid etching prior to plating.[12] During this process, liquid sulfur trioxide is used and gasified on-site using a special generator system. At the end of the etch cycle, the gas in neutralized with ammonia before it is expelled from the chamber. The economics of the process are highly favorable since the cost for using vapor etching is 0.5 cents per square foot compared to approximately 4-5 cents per square foot for chromic acid etching.[13]

Plasma processing is becoming increasingly common in the fabrication of printed circuit boards, particularly for the removal of drill smear from multilayer boards.[14] Unlike chromic acid and sulfuric acid which are effective for epoxy resins only, plasma is the only currently available technology applicable to all commonly used substrates, including epoxy, polyimide, acrylic, and Teflon,[15] At present, plasma desmearing is favored for polyimide boards, but wet processing remains the method of choice for removing epoxy smear in quantity.[16] Though processing times are longer than those seen with aqueous desmearing, plasma desmearing is a very controllable process which produces a high energy substrate surface, and does not produce a significant waste disposal problem. The free radicals formed by the radio frequency applied to gases introduced to the plasma chamber at reduced pressures react with organic molecules to form innocously small amounts of volatiles. This finely dispersed layer of ash requires a post desmear treatment to preclude adhesion failures and voiding but it is easily removed by downstream rinses and other follow-on processes needed for electroless copper pretreatment.[17]

Sputtering processes have been developed for depositing the initial copper layer on printed wiring boards and flexible circuitry[18,19]. This approach allows through hole deposition without the use of electroless plating and permits the manufacturer to use standard unclad boards for the subtraction or semi-additive process. One system utilizes reactive plasma to desmear printed wiring boards and then sputter deposits a copper layer on the hole walls. Five vacuum chambers totaling 28 feet in length, are connected in series such that a conveyor moves printed circuit boards vertically from one chamber to the next. After removal from the final chamber, the boards are electroplated with 1 mil of copper. Poskanzer[20] predicts that in 15 years a process of this type will replace the presently used electroless copper systems.

5. SUMMARY

Synergism can be obtained by combining electrodeposition and vacuum processes. In a number of cases, better coatings and improved adhesion can be obtained when the two processes are combined. Also, for some applications, parts that could not be fabricated by either method by itself, can be produced when the two processes are combined. A number of these examples are discussed in this paper.

6. REFERENCES

1. Levy, D. J., Arnold C. R., and Ma, D. H., "Electroplating of Gold and Rhodium Directly On Molybdenum", American Electroplaters' Society, Symposium on Plating on Difficult to Plate Metals, Oct. 1980.

2. Dini, J. W. and Johnson, H. R., "Adhesion Testing of Deposit-Substrate Combinations", Adhesion Measurement of Thin Films, and Bulk Coatings, ASTM STP 640, K. L. Mittal, Ed., American Society for Testing

and Materials, 305 (1978).

3. Beat, T. G., Kelley, W. K. and Dini, J. W. "Plating on Molybdenum", Lawrence Livermore National Laboratory, UCRL-95940, Jan. 1987.

4. Beat, T. G., Kelley, W. K., Reiss, R. H., Leatherman, R. E., Sanders, D. M., "Fabrication of Sputter Cathodes by Electroplating", International Conference on Metallurgical Coatings, AVS, April 1985.

5. Gurvitch, M., J. Vac. Sci. Technol., A2 (4), 1550 (1984)

6. Esmore, L. H., Trans. Institute of Metal Finishing, 57, 57 (1979)

7. Kelley, W. K., Cowden, W. C. and Overturf III, G. E., "Electroforming Cylindrical Parts with Precision Small Inside Diameter Pinholes", AES Electroforming/ Deposition Forming Symposium, Jan. 1983.

8. Lindsay, J. H. and LaSala, J., "Plating and Surface Finishing", 72, 54 (July 1985).

9. Burkstrand, J. M., J. Vac. Sci. Technol. 15 (2), 223 (March/April, 1978).

10. Birkamier, A. A., et al., U.S. Patent 4, 442, 907.

11. Jobbins, J. M. and Sopchak, P., Metal Finishing 83, 15 (April 1985).

12. McCaskie, J. E., "Electroless Plating of Plastic Enclosures for EMI/RFI Shielding", Proceedings, AES Second Electroless Plating Symposium, Feb. 1984.

13. Berg, J. H., Electronics, 32, No. 10, 53 (Sept. 1986).

14. Rust, R. D., Rhodes R. J. and Parker, A. A., Solid State Technology 27, 270 (April 1984).

15. Spitz, S. L., Electronic Packaging & Production, October 1985, p. 34.

16. Murray, J., Circuits Manufacturing, May 1985, p. 44.

17. Murray, J., Circuits Manufacturing, Feb. 1985, p. 116.

18. Anderson, A., "Plating and Surface Finishing", 68, 44 (Dec. 1981).

19. Lee, R. A., "Eliminate Electroless Copper With Plasma Enhanced Deposition", Proceedings of the 29th Annual Meeting, Boston IPC, 1986.

20. Poskanzer, A. M., "Plating and Surface Finishing", 74, 20 (March 1987).

Work performed under the auspices of the U.S. Department of Energy by Lawrence Livermore National Laboratory under Contract W-7405-ENG-48.

TABLE 1

RING SHEAR DATA FOR VARIOUS PROCEDURES FOR
PREPARING MOLYBDENUM FOR PLATING

Process	Rod Diameter (cm)	Electrodeposit	Number of Tests	Ring Shear Strength (MPa)	Standard Deviation	Location of Failure
Bell Labs[1]	1.27	Gold	8	125	13	Gold Deposit
Alkaline Ferricyanide[2]	.64	Gold	5	79	15	Gold-Mo Interface
" "	1.27	Gold	4	100	31.5	Gold Deposit
Martin-Marietta[3]	.64	Gold	4	32	6	Gold-Mo Interface
" "	.64	Copper	4	37	2	Copper-Mo Interface
" "	1.27	Copper	5	37	14.5	Copper-Mo Interface
Magnetron Ion Plating	1.27	Copper	11	216	16.5	Copper Deposit

1 Bell Labs Process - degrease in perchlorethylene, fire in dry hydrogen (< 2 ppm H_2O) at 1000°C for 10 minutes, immerse in a solution containing four parts NH_4OH (28%) and one part H_2O_2 (30%) for 8 to 10 seconds at room temperature, rinse in distilled water, gold strike to deposit 0.15 to 0.62 mg/cm^2 (0.08 to 0.32 μm), rinse in distilled water, fire in dry hydrogen at 1000°C for 10 minutes, electrodeposit to desired thickness.

2 Alkaline Ferricyanide Process - degrease in perchlorethylene, immerse in a solution containing 300 g/l potassium ferricyanide, 100 g/l sodium cyanide for 10 seconds at room temperature, water rinse, electroplate to desired thickness, water rise and then dry, hydrogen fire at 800°C for 4 hours.

3 Martin-Marietta (Oak Ridge) Process - degrease in perchlorethylene, anodic etch in chromic acid (10 percent by wgt) and 71 percent nitric acid (10 percent by vol) at room temperature and 15.5 A/dm^2 (1 A/in^2) for 1 min, rinse, anodic treat in 50 percent aminotrimethylene phosphonic acid (ATMP)- 25 percent by vol, at room temperature and 15.5 A/dm^2 (1 A/in^2) for 4 min, rinse, remove oxide film in chromic acid (10 percent by wgt) and 71 percent nitric acid (10 percent by vol) at room temperature for 1 min, rinse, electroplate to desired thickness.

4 Magnetron Ion Plating Process (LLNL) - degrease in perchlorethylene, sputter etch in vacuum, magnetron ion plate with 6μm of copper, electroplate to final thickness with copper.

TABLE 2

SEMI-QUANTITATIVE SPECTROCHEMICAL ANALYSIS
OF
IMPURITIES IN CHROMIUM SPUTTER TARGETS

Element	Commercial (ppm)	Electroplated (ppm)
Iron	1300	130
Aluminum	1300	400
Copper	1300	--
Zinc	400	400
Silver	130	130
Calcium	130	40
Nickel	40	< 15
Manganese	40	< 5
Titanium	40	< 5
Magnesium	15	15
Antimony	< 400	< 400
Boron	< 40	< 40
Cobalt	< 40	< 40
Molybdenum	< 15	< 15
Beryllium	< 1	< 1
Strontium	< 1	< 1
Tin, Niobium, Cadmium Bismuth, Zirconium, Vanadium, Lead and Barium	< 130 (each)	< 130 (each)

FIGURE 1 - CATHODE CONFIGURATION OF SPUTTER TARGETS

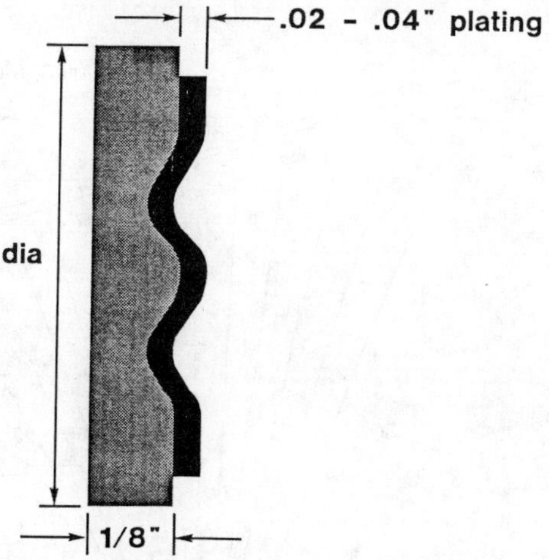

FIGURE 2 - ELECTROFORMED PINHOLE DESIGNS

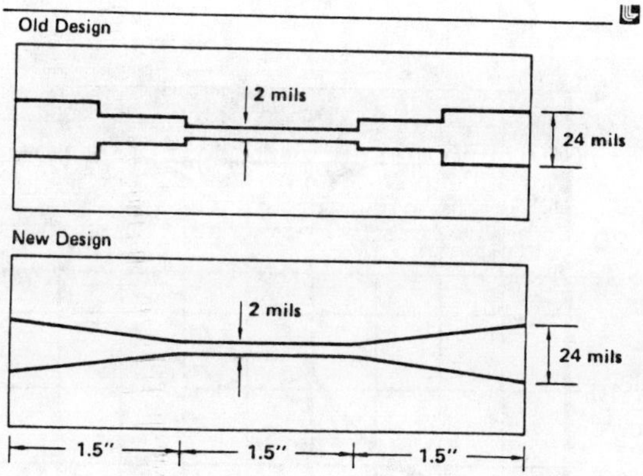

FIGURE 3 - COATING PROCESS SCHEMATIC FOR PINHOLE MANDRELS

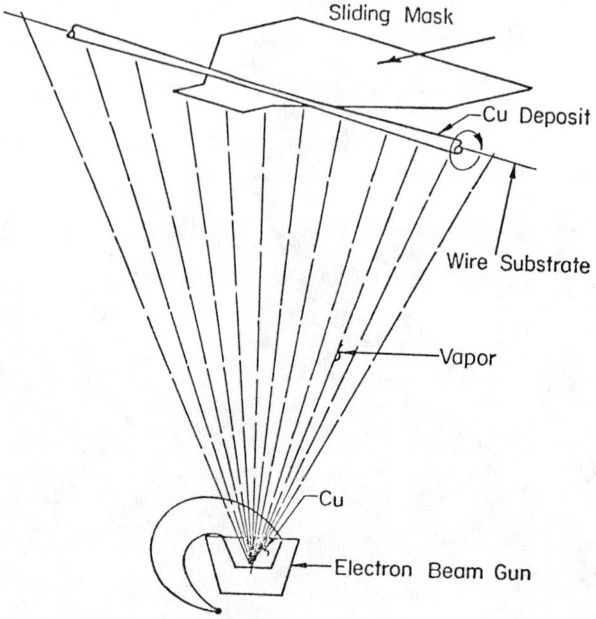

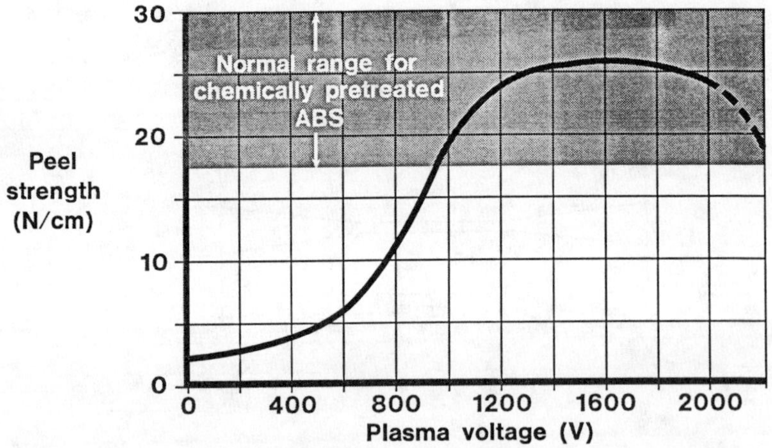

Fig. 4 - Peel strength vs. dc plasma voltage for ABS plastic (from reference 8).

1st International SAMPE Metals Conference
August 18-20, 1987

DETONATION GUN COATINGS - COATING CHARACTERISTICS AND APPLICATIONS

R. C. Tucker, Jr. and R. R. Chevalier
Union Carbide Corporation
Coatings Service Division
Indianapolis, Indiana

Abstract

Although a variety of new thermal spray coating processes and materials have evolved in recent years, detonation gun coatings, as a result of the inherent characteristics of the process and continual evolution of equipment and new materials, have remained the hallmark of the industry. This paper briefly reviews the process, both its advantages and limitations. More attention is paid to the characteristics of representative coatings including their microstructural characteristics and mechanical properties. Microstructural characteristics discussed include the chemical composition, crystallographic phases present, and distribution of phases and density. Mechanical properties that have been measured include hardness, moduli of elasticity and rupture, and bond strength.

A variety of current applications of detonation gun coatings selected from a wide range of industries are included. These illustrate not only the effective utilization of the coatings, but provide examples of the use of coating properties in the selection of coatings for various applications.

I. Introduction

Although a variety of new thermal spray coating processes and materials have evolved in recent years, detonation gun coatings, as a result of the inherent characteristics of the process and continual evolution of equipment and new materials, have remained the hallmark of the industry. Almost any metallic, ceramic or cermet material that melts without decomposing can be used to produce a coating. Usually coatings from about 0.002 (50 microns) to 0.020 inches (500 microns) thick are used. Although they are primarily used to resist wear, they have also been used to resist corrosion or for their electrical or thermal properties. Detonation gun coatings are used in almost all industrial environments, ranging literally from submarines to the space shuttle and including gas turbine engines, aircraft, textile mills, steel mills, medical instruments, oil and gas drilling equipment, chemicals and plastics industry equipment, etc. While even their direct cost is frequently less than competitive processes such as plasma deposition, the improved performance that they provide almost always results in the lowest cost when amortized over the longer life of the component, reduced down time, etc.

II. Deposition Process and Limitations

Detonation gun deposition has been described in detail elsewhere,[1] but a brief review may be appropriate. A detonation gun is shown schematically in Figure 1. It consists of a water-cooled barrel several feet long with about a 1" inside diameter. A mixture of oxygen and acetylene is fed into the gun along with a pulse of powder. The explosive gas mixture is ignited and the detonation wave traveling down the barrel accelerates and heats the powder. The powder exits the barrel at a velocity of about 2400 ft/sec (730 m/sec) or more and at a temperature near or above its melting point. The powder particles impact on the part to be coated forming a circle of coating about 1" (25 mm) in diameter and less than a thousandths of an inch (a few microns) thick. The barrel is then purged with nitrogen and the process repeated at a rate of about 4 to 10 times per second. The area to be coated is covered by traversing the part relative to the gun forming a precisely controlled pattern of overlapping circles. Since each circle is quite thin and the pattern of overlapping circles carefully controlled, a coating of uniform thickness over large areas can be achieved.

The powder used in the process is usually less than 44 microns in diameter. Since the powder particles are at or above their melting point and traveling at a very high velocity when they impact upon the surface, they flow out forming very thin lenticular particles (called splats) which are overlapping and intertwined. The splats are very well bonded to each other and to the substrate forming a very dense microstructure. *

An important attribute of the process is its ability to apply materials with very high melting points to metallic substrates without significantly heating the substrate; i.e., the temperature of the substrate can easily be held below 300°F (150°C). Temperature control is achieved through control of the pattern and rate of deposition and the occasional use of auxiliary cooling. As a result of this temperature control, it is possible to apply coatings to fully heat treated and/or work hardened components made of steel, aluminum, titanium, etc. without fear of changing their metallurgical condition.

The most important limitation of the detonation gun process is that it is a line-of-sight process. Coatings with the best properties (i.e., maximum bond strength, cohesive strength, and density) are obtained when the angle of deposition is close to 90°. Because of the very high velocity of the particles, the rate of change in coating properties as the angle of impact is diminished is less than for most other thermal spray processes. Little degradation in properties occurs down to about a 60° angle of impingement, and useful coatings can be made down to at least 45°. Another limitation is the size of the equipment. Because of this the detonation gun cannot be traversed inside cylinders. Nonetheless, it is frequently used to coat the inside surfaces of cylinders without traversing into the cylinder itself by utilizing angles of impact as low as 45°, i.e., to a depth about equal to the inside diameter.

III. Microstructures

Representative micrographs of some cross-sections of several typical detonation gun coatings are shown in Figure 2. As can be readily seen, detonation gun coatings consist of many layers of thin, lenticular particles (splats) typically about 5 microns thick and 50 microns in diameter. The very high density is due, in large part, to the very high velocity of the particles on impact. The cleanliness of the microstructure (i.e., lack of unintentional oxidation during deposition, etc.) is due to the very close control of the deposition parameters including the oxygen to carbon ratio in the gas mixtures.

The very small molten powder particles are quenched very rapidly on impact. It has been estimated that the quench rates for ceramic particles are in the range of 10^4 to 10^6 ^{o}C/second, while those of metallic particles are in the range of 10^6 to 10^8 ^{o}C/second. As a result, the phases present in the coatings after deposition may be thermodynamically metastable, as well as stable, and have an amorphous, microcrystalline, or very fine grained structure. For example, alumina coatings usually contain a large volume fraction of metastable gamma phase and may contain other high temperature phases as well as the thermodynamically stable alpha phase. Coatings made from (WC,W_2C)-Co powders usually contain $(W,Co)_6C$ and higher volume fractions of W_2C than the powder. These metastable phases remain unchanged indefinitely, however, unless exposed to very high temperatures. The performance of these coatings therefore does not change during service as a result of phase transformations unless the service temperature is very, very high. Differences in phase composition from that expected from thermodynamic equilibrium may, however, have a substantial impact on corrosion resistance; e.g., gamma alumina is not nearly as corrosion resistant as alpha alumina.

IV. Properties

A few selected properties of representative detonation gun coatings are shown in Table I. It should be kept in mind, however, that detonation gun coatings, like all thermal sprayed coatings, have very anisotropic as-deposited properties. This results, of course, from their lamillar microstructure. The micro-hardness values shown in Table I were measured on polished metallographic cross-sections. The other mechanical properties were measured on free-standing rings of coatings 1" (25.4mm) in diameter, 0.5" (12.7 mm) wide, and 0.010" (0.25 mm) thick. Since the specific microstructure and phase composition of a coating may be different on different parts because of the variation in angles of deposition and quench rates, properties such as those in Table I are best used only as guidelines. It is important to note that the strain-to-failure of virtually all as-deposited thermal sprayed coatings is quite low, usually less than 0.5%, even for pure metallic coatings.

The bond strength of a coating to a substrate is, of course, a significant parameter. The tensile bond strength of most detonation gun coatings exceeds the bond strength of the epoxy used in the modified ASTM test[2] usually used for this measurement; i.e., the bond strength exceeds 10,000 to 12,000 psi (69 to 83 MPa). A special test was developed several years ago to assess bond strength and/or cohesive strength of the coatings in a somewhat different manner. In this test hydraulic forces applied to the coating/substrate bond line. The force required to cause either rupture of the coating cohesively or separation at the coating/substrate interface is measured. Using this test, values exceeding 60,000 psi for as-deposited tungsten carbide based coatings have been recorded.

Although detonation gun coatings have very low porosity, usually less than 2% measured metallographically, some of that porosity is interconnected, as it is for all thermal sprayed coatings. The presence of interconnected porosity can provide a channel for corrosive media to attack not only the interior of the coating, but the substrate as well. This interconnected porosity and the presence of metastable phases, as mentioned earlier, complicates the corrosion characteristics of thermal sprayed coatings, including detonation gun coatings. For service temperatures less than about 300^oF, this interconnected porosity can often be effectively sealed with an epoxy. At very high temperatures in air or similar oxidizing environments, because the pores in the detonation gun coatings are quite small, a protective scale may form over the surface, protecting both the bulk of the coating and the substrate. At intermediate temperatures, however, not only must the coating and the substrate be resistant to the environment, but the combination must not form a destructive galvanic cell.

V. Applications

General

Before presenting a few examples of the use of detonation gun coatings, some general comments may be in order. Since thermally sprayed coatings are most frequently used for wear resistance, albeit often in a corrosive environment, hardness is frequently used as a primary criteria for selecting a specific coating. This approach may be useful, but usually only if comparing two materials of the same composition and type (thermal sprayed coating, electro-plating, or wrought or cast alloy) and if adhesive, low stress abrasion, or low angle erosion is being considered. For example, in Table II,[3] the wear resistance of the detonation gun tungsten carbide base coating is greater than that of its plasma analog, as might be expected from their hardnesses, but even the plasma coating is more wear resistant than the harder chromium electroplate (different type of material and different composition).

In general, when selecting a coating for a given application, not only must one be concerned with its wear characteristics, but one must also be concerned with the intrinsic characteristics of the coating, the effect of the coating on the mechanical properties of the substrate, and the effect of the substrate's properties on the performance of the coating. One of the earliest applications of detonation gun coatings illustrates the first two points. Detonation gun coatings have, for many years, been used on the midspan stiffeners (snubbers) of compressor blades to prevent impact and fretting wear between stiffeners. In selecting a coating for the initial application, hardness was used as a major criteria because maximum resistance to fretting wear was of primary concern. As a result, a tungsten carbide-cobalt coating with a nominal hardness of 1300 kg/mm^2 HV_{300} was selected. Preliminary testing, however, revealed that while the coating did not suffer from fretting wear, it did crack as a result of impact. Subsequently, a similar coating, but one containing more cobalt, was selected with a nominal hardness of 1075 kg/mm^2 HV_{300} which was found to have more than adequate wear resistance and sufficient toughness to withstand the impact; the first coating had a modulus of rupture of 72,000 psi, the second coating had a modulus of rupture of 120,000 psi.

The same application illustrates the possible effect of the coating on the performance of the substrate. The fillet area, shown in Figure 3, is susceptible to repeated high stresses due to cyclic loads applied to the midspan during service. If coating were present in this area and strained beyond its strain-to-failure as a result of deflection of the substrate it would crack. Such cracks might propagate into the substrate (because of the high bond strength of the coating) initiating a fatigue failure of the blade. Using careful masking techniques, deposition of coating or even overspray in this area is avoided. As a result, no fatigue failures of a blade have ever occurred in service.

The third material characteristic, the effect of the substrate's properties on the performance of the coating, is illustrated by the use of detonation tungsten carbide-cobalt coatings on guide rolls used in controlling the movement of sheet steel through a rolling mill. Initial field trials with a tungsten carbide-cobalt coating with a nominal hardness of 1150 kg/mm^2 HV_{300} and modulus of rupture of 110,000 psi were successful. However, after coating new rolls, some failures were noted in the form of spalling of the coating as a result of deformation of the substrate beneath the coating. This deformation, because of the high bond strength of the coating, forced the coating to deform beyond its strain-to-failure, resulting in cracking and spalling. It was found that earlier experience had been with guide rolls having a minimum hardness of 60 HRC while the failed rolls had a hardness of less than 50 HRC. Ensuring that all guide rolls had a hardness of not less than 55 HRC eliminated this problem.

Since detonation gun coatings are most frequently used for wear resistance, it might be well to illustrate their erosive, abrasive, and adhesive wear characteristics with some laboratory data using standard ASTM tests before providing a few applications to illustrate their use. Data from selected wear tests are shown in Tables IIIA, IVA, and VA. Test parameters for each of these tests are shown in Tables IIIB, IVB, and VB.

Gas Turbine Engines

The use of a detonation gun coating on midspan stiffeners of compressor blades has already been mentioned. There are many other components in gas turbine engines that are coated with detonation gun coatings to combat adhesive, abrasive, and erosive wear as well. An example of the use of these coatings for erosive wear is the application of a detonation gun tungsten-titanium carbide-nickel coating on the tips of all of the blades in a major commercial gas turbine engine to combat erosion caused by dust particles primarily picked up from runways during landing and take-off. In this case, the coating was selected not only for its excellent erosion resistance, but also for its relative smooth as-deposited surface which minimizes aerodynamic effects without expensive finishing. Similar applications exist on compressor blades in an auxilliary power unit engine, and tungsten carbide-cobalt coatings have been used on other engines for similar purposes, Figure 4.

Aircraft

Detonation gun coatings are used on a variety of components in helicopters and fixed wing aircraft. An example is shown in Figure 5 of a detonation gun tungsten carbide-cobalt coating that is used on the main rotor swash plate pivot sleeve and on the main rotor swash plate support. In the first application, wear is primarily due to sliding on both the O.D. of the sphere and in the slots, while in the second, fret-ting wear is the primary problem. In both cases, the use of the coatings allowed substantial extension of the time between overhauls.

Plastics Industry

In the plastics industry, extruder screws are subjected to extreme abrasive and corrosive wear from the wide variety of compounds being extruded, many of which are loaded with very abrasive materials such as oxides and glass fibers. Figure 6 shows an extruder screw coated with a detonation gun tungsten carbide-cobalt coating. Figure 7 illustrates the tremendous advantage of this coating over welded hardfacing.[7]

Textile Industry

The textile industry provides excellent examples of the use of detonation gun coatings where not only wear resistance is important, but also very specific and very tightly controlled friction characteristics. Fibers being conveyed through a textile mill are not only extraordinarily abrasive, primarily because of being filled with materials such as TiO_2, but also may be very corrosive because of the "finishes" used on the fibers. If the roll surfaces are not only wear resistant but have the right friction characteristics, tension in the fiber may be either too high or too low, resulting in either breakage of the fiber or huge tangles.

VI. Conclusions

In this brief paper, it has not been possible to illustrate all of the characteristics or potential uses of detonation gun coatings. It is hoped, however, that some sense of their microstructures and properties has been conveyed as well as a few guidelines to their use. More work is needed to fully understand the mechanisms of wear that apply to detonation gun coatings, but their success in a tremendous variety of environments and applications has been amply demonstrated.

Table I
Properties of Selected Detonation Gun Coatings

UCAR* Coating Designation	Principal Constituent(s)	Nominal Composition (Weight %)	Hardness (HV.3) (kg/mm^2)	(Rc)	Modulus of Rupture (10^3 psi)	Modulus of Elasticity (10^6 psi)
LW-1	Tungsten Carbide	89W-7Co-4C	1300	72	72	23
LW-1N30	Tungsten Carbide	85W-11Co-4C	1150	70	110	26
LW-1N40	Tungsten Carbide	85W-14Co-3C	1075	69	120	25
LC-1B	Chromium Carbide	65(92Cr-8C)+ 35(80Ni-20Cr)	700	60	100	21
LC-1C	Chromium Carbide	80(92Cr-8C)+ 20(80Ni-20Cr)	775	63	77	18
WT-1	Tungsten-Titanium Carbide	83(W,Ti)Carbide-17Ni	1200	71	55	20

*Trademark of Union Carbide Corporation

TABLE II

Adhesive Wear Rates of Several Coatings in a Block-on-Ring Test

Material	Hardness HV 0.3	Wear Rate Dry*	Wear Rate Lubricated**
D-Gun Tungsten Carbide-Cobalt	1150	35	10
Plasma Tungsten Carbide-Cobalt	750	80	23
Electroplated Chromium	900	3600	44

*Dry, disk-on-disk, 10^{-6} in./1000 ft. of sliding

**Lubricated, LFW-1, block-on-ring, 10^{-6} cm^3/5400 rev., 450 lb. load vs. steel

Table IIIA

Low Stress Abrasive Wear of Selected Coatings*

UCAR Coating Designation	Coating Type	Coating Composition**	Wear Rate (mm^3/1000 rev)	Coefficient of Variation %
WT-1	D-Gun	(W,Ti)C-17Ni	0.84	7
LW-1	D-Gun	W_xC_y-7Co	0.99	8
LW-1N30	D-Gun	W_xC_y-11Co	1.5	14
LW-1N40	D-Gun	W_xC_y-14Co	1.9	N/A
LW-11B	Plasma	W_xC_y-11Co	5.6	4
LC-1C	D-Gun	Cr_xC_y-20NiCr	4.0	11
LC-1B	D-Gun	Cr_xC_y-35NiCr	4.2	7

*See Table IIIB for test parameters.
**See text re actual phase composition.

Table IIIB
Dry Sand/Rubber Wheel Abrasion Test Parameters

Abrasive	50-70 Ottawa Silica Sand
Load	30 (pounds) (130 Newtons) ± 3%
Rubber Wheel	Chlorobutyl Rubber of Durometer Hardness A 58-60
Wheel Speed	200 rpm ± 2 rpm
Test Duration	3000 or 6000 wheel revolutions
Value Reported	Average Volume Loss (mm^3) per 1000 revolutions
Temperature	Ambient
Feed Rate	380 to 430 gpm in 6000 revolution test (Generally it is 410 ± 10 gms/min.)
Reference	ASTM

Table IVA
Erosion Wear of Selected Coatings*

UCAR Coating Designation	Coating Type	Coating Composition**	Erosion Rate (μm/g) Alumina 30°	Alumina 90°	Silica 30°	Silica 90°
WT-1	D-Gun	(W,Ti)C-17Ni	28.4	138	3.9	9.3
LW-1	D-Gun	W_xC_y-7Co	16.6	108	1.2	3.0
LW-1N30	D-Gun	W_xC_y-11Co	17.4	100	2.1	5.2
LW-1N40	D-Gun	W_xC_y-14Co	19.8	105	2.6	6.5
LW-11B	Plasma	W_xC_y-11Co	27.0	247	4.6	34.0
LC-1C	D-Gun	Cr_xC_y-20NiCr	30.9	117	5.1	11.0
LC-1B	D-Gun	Cr_xC_y-35NiCr	28.6	96.3	7.5	12.0

*See Table IIIB for test parameters.
**See text re actual phase composition.

Table IVB
Erosion Test Parameters

	Alumina Test*	Silica Dust Test**
Temperature	Ambient (∼25°C)	Ambient (∼25°C)
Erosive Medium**	Alumina	Silica
Size	Nominal 27 micron	Nominal 15 micron
Feed Rate	1.15 ± 0.15 gpm	5.5 ± 0.5 gpm
Concentration	0.233 mg/cm3 of carrier gas	
Velocity	91 ± 3 m/s	139 ± 7 m/s
Carrier Gas	Argon	Argon
Pressure	44 ± 1 psi	19 ± 1 psi
Flow Rate	10.5 ± 0.5 cfh	17-18 cfh
Nozzle	WC, 0.018 inches ID	WC, 0.0625 inches ID
Nozzle Stand-Off	0.400 + 0.002 inch - 0.000 inch	Same
Parameters Measured	Scar Depth	Scar Depth
Angle of Impingement	90°, 30°	90°, 30°
Values Reported	μm/gm	μm/gm
Reference	ASTM	

*Nominal Composition of Alumina: Al_2O_3: 97.2-96.2; TiO_2: 3.5-2.5; CaO: 0.5; SiO_2: 1-2; FeO_3: 0.3; MgO: 0.3

**Nominal Composition of Silica: SiO_2: 65-76; Al_2O_3: 11-17; CaO: 3-6; Se_2O_3: 2.5-5.0; Na_2O: 2-4; MgO: 0.5-1.5

Table VA
Adhesive Wear of Selected Coatings*

Block Material UCAR Designation	Coating Composition**	Block Scar Volume (mm³) $\bar{x}$	Block Scar Volume (mm³) v	Ring Scar Volume (mm³) $\bar{x}$	Ring Scar Volume (mm³) v	Final Dynamic Friction Low	Final Dynamic Friction High	Number Tested
LW-1N10	WC-Co	.023	18%	.389	18%	.119	.132	6
LW-1N40	WC-Co	.026	28%	.335	15%	.121	.133	6
LW-15	WC-Ni-Cr	.026	12%	.397	12%	.118	.136	6
LW-5	WC-Co-Cr	.026	11%	.324	17%	.120	.134	5
LC-1C	Cr_xC_y-NiCr	.019	26%	.425	7%	.121	.135	3
LA-2	Al_2O_3	.050	15%	.316	26%	.125	.140	6

*See Table IIIB for test parameters.
**See text re actual phase composition.

Table VB
Adhesive Test Parameters

Test Type:	Block-on-Ring (LFW-1)
Block:	0.25" wide AISI, 1020 steel with ≥ 0.010" thickness of designated coating with a ground surface finish
Ring:	1.389" diameter AISI 4620 steel with a surface hardness of 58-63 HRC and a ground surface finished to 5-15 microinch Ra
Load:	600 lbf
Velocity:	65 fpm
Test Duration:	5400 rev. (625 ft.)
Lubricant:	Mil H5606A Hydraulic Fluid

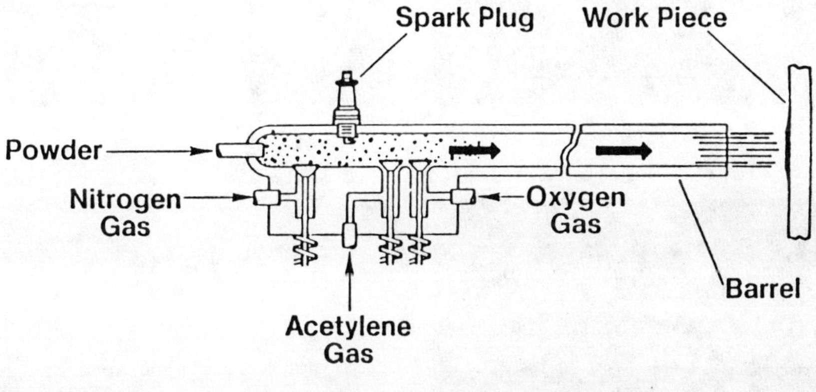

Figure 1: Schematic of a detonation gun

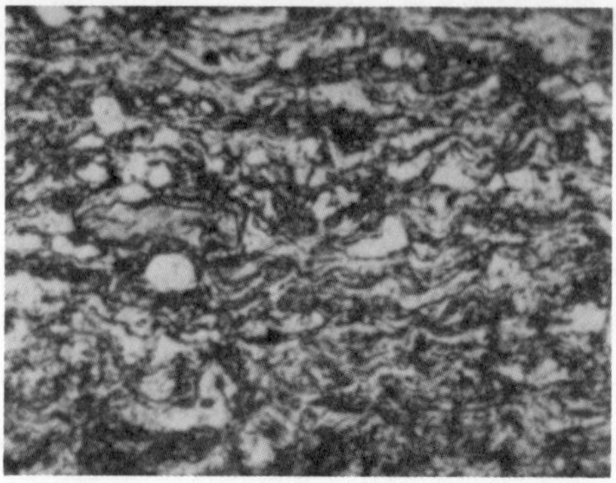

As-polished   Etched electrolytically in $KMnO_4$

Chromium Carbide-Nickel Chromium (UCAR LC-1C)

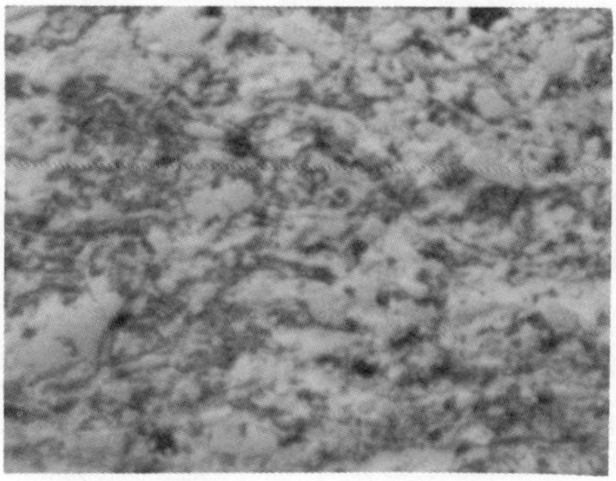

As-polished   Etched in modified Murakami's

Tungsten Carbide-Cobalt (UCAR LW-1N40)

Figure 2: Micrographs of several D-Gun coatings, 500X.

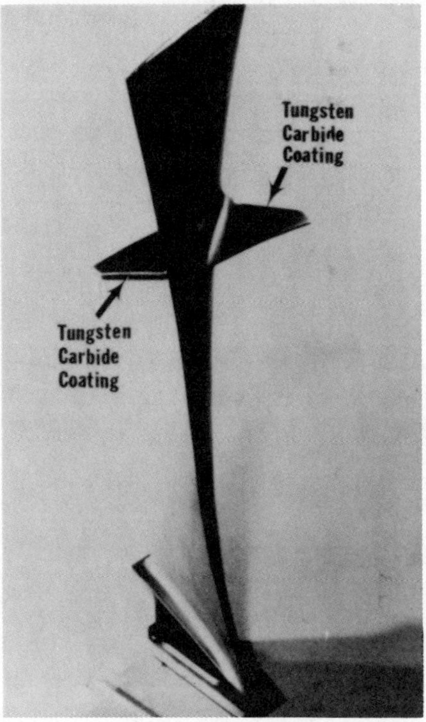

Figure 3: Typical compressor blade with coated midspan.

Figure 4: Compressor blade with coated test area illustrating the effectiveness of a detonation gun coating under service conditions.

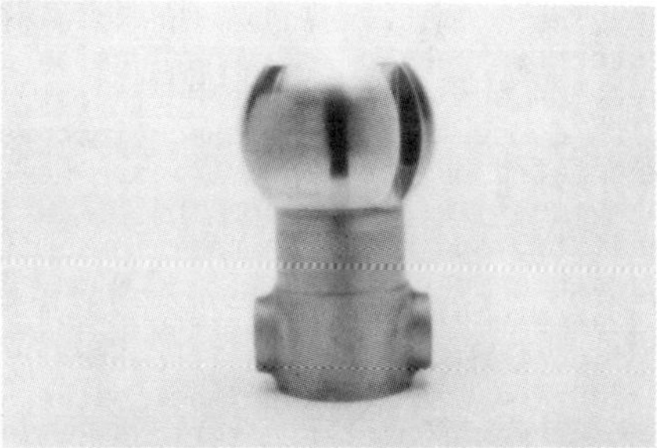

Figure 5: Helicopter main rotor swashplate pivot sleeve with a detonation gun tungsten carbide-cobalt coating.

Figure 6: Extruder screw with a detonation gun tungsten carbide-cobalt coating.

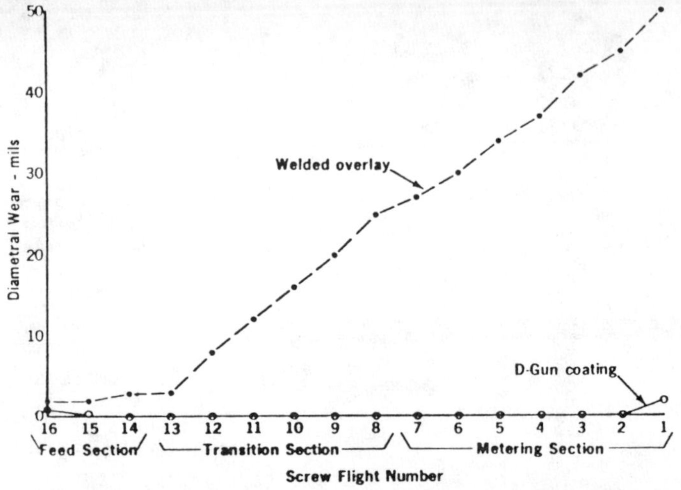

Figure 7: Wear profile of screw flights after extruding 400,000 lb. of PVC compound.

References

1. R. C. Tucker, Jr., "Plasma and Detonation Gun Deposition Techniques and Coating Properties," in Deposition Technologies for Films and Coatings, R. F. Bunshah, et. al, Noyes Publications, Park Ridge, New Jersey, USA, 1982, pp 454-489.

2. ASTM Bond Test

3. R. C. Tucker, Jr., "Detonation Gun Coatings and Their Applications," presented at the 1984 International Powder Metallurgy Conference, Toronto, Canada, 17-22 June 1984.

4. ASTM Sand Abrasion Test

5. ASTM Alpha Wear Test

6. ASTM Erosion Test

7. A. M. Kimball, "A Solution to Extruder Screw Wear in the Wire and Cable Industry," Conference Proceedings of the 51st Annual Convention of Wire Association International, Atlanta, Georgia, USA, 12-16 October 1981.

Richard R. Chevalier has a Bachelor of Science Degree in Engineering from the University of Bridgeport and an MBA from the University of New Haven. he is currently Sales Manager, General Engineering Markets, for Union Carbide Coatings Service. This includes the sale of detonation and plasma coatings as a service business in the Northeast with concentration on the helicopter and petrochemical industries. He previously held engineering, quality, and research positions with Warner Lambert, duPont, and Olin. He is a member of the American Helicopter Society and is on the Executive Committee of the Southern Connecticut Chapter, American Society for Metals.

Dr. R. C. Tucker, Jr. Associate Director of Technology for Union Carbide Corporation, Coatings Service Division, received his BS in chemistry/mathematics from North Dakota State University and his MS and PhD in metallurgy from Iowa State university. He is currently responsible for programs to develop materials resistant to high- and low-temperature wear, corrosion, high-temperature oxidation and sulfidation, thermal barriers and composite materials. This work currently utilizes plasma, detonation gun and chemical vapor deposition coatings. He is a member of ACS, AIME, ASM, ASTM, AVS, NACE, WRC and Sigma Xi. He has published over 50 papers and has 9 U. S. patents.

1st International SAMPE Metals Conference
August 18-20, 1987

ARALL® APPLICATIONS AND FUTURE

by
M. A. Gregory
and
L. N. Mueller

ARALL® laminates are a new family of high strength, hybrid composite laminates invented at Fokker Aircraft Company and Delft University and being commercialized for aerospace applications by Alcoa.

ARALL® laminates are composed of alternating layers of high-strength aluminum sheet and tough aramid/ epoxy prepreg. The primary benefits of ARALL® laminates are their improved fatigue and fatigue-crack properties relative to monolithic aluminum alloys. In fatigue tests, ARALL® laminates exhibit fatigue lives between 100 and 1000 times that of comparable monolithic aluminum alloys.

In addition to superior fatigue performance, ARALL® laminates have tensile strengths, 30 to 50% higher than the comparable high-strength alloy sheet products with equivalent stiffness and a density reduction of up to 20%.

Alcoa is currently producing first generation ARALL® laminate products. The standard products available and their compositions are outlined in Table 1. Typical mechanical properties for ARALL® 1 and ARALL® 2 laminates relative to alloys 7075-T6 and 2024-T3 are presented in Table 2.

In addition to the improved mechanical properties, ARALL® laminates offer improvements in acoustic damping and lightening-strike resistance relative to monolithic aluminum sheet.

These property improvements combined with the fact that, within limits, ARALL® laminate parts can be formed with conventional aluminum metal-working techniques, puts ARALL® laminates in a unique position within the structural materials' spectrum. ARALL® laminates offer some of the same property improvements relative to monolithic aluminum as graphite/epoxy composites, yet many of the unique forming characteristics of aluminum sheet are retained. With the inherent material property improvements of ARALL® laminates, combined with the fact that ARALL® laminates are an engineered material system offering the capability of tailored properties, ARALL® laminate applications seem limitless.

Due to the enhanced fatigue performance of ARALL® laminates, logical applications include lower wing skins, fuselage panels and tail structure. Weight savings of up to 33% have been documented by Fokker in their lower wing panel research program.

With the enhanced acoustic damping properties, ARALL® laminate applications on the rear fuselage panels in unducted fan-powered aircraft and parts exhibiting aeroacoustic fatigue in military aircraft are envisioned.

Finally, the improved lightening-strike resistance of ARALL® laminates relative to monolithic aluminum sheet offers significant weight savings in aircraft structure that is sized by lightening-strike requirements.

TABLE I

First Generation ARALL® Laminate Products

Product	Composition	Advantages
ARALL® 1 Laminate	0.3mm (0.012 in.) thick anodized and primed 7075-T6 sheet with unidirectional aramid fiber prepreg made from 121°C (250°F) cure epoxy resin. The laminate is stretched 0.4% after curing.	Excellent fatigue crack growth resistance, high strength
ARALL® 2 Laminate	0.3mm (0.012 in.) thick anodized and primed 2024-T3 sheet with the same prepreg as ARALL® 1 Laminate.	High damage tolerance
ARALL® 3 Laminate	The same as ARALL® 1 Laminate except it uses 7475-T76 sheet.	Improved damage tolerance with high strength
ARALL® 4 Laminate	0.3mm (0.012 in.) thick anodized and primed 2024-T8 with a 177°C (350°F) cure epoxy resin/ unidirectional aramid prepreg.	Higher temperature properties

TABLE 2

Typical Mechanical Properties for
ARALL® 1 and ARALL® 2 Laminates

Average Mechanical Properties		ARALL® 1 Laminate [(a)]	ARALL® 2 Laminate [(a)]	7075 -T6 [(b)]	2024 -T3 [(b)]
Tensile Ultimate Strength, mPa (ksi)	L	772 (112)	703 (102)	572 (83)	455 (66)
	LT	379 (55)	317 (46)	572 (83)	488 (65)
Tensile Yield Strength, mPa (ksi)	L	655 (95)	351 (51)	510 (74)	345 (50)
	LT	324 (47)	234 (34)	496 (72)	310 (45)
Compressive Yield Strength, mPa (ksi)	L	365 (53)	---	503 (73)	296 (43)
	LT	379 (55)	---	524 (76)	378 (49)
Elongation, Percent	L	0.6	1.3	11	18
	LT	6.9	12.4	11	18
Elastic Tensile Modulus, GPa (msi)	L	70 (10.1)	65 (9.5)	72 (10.4)	72 (10.5)
	LT	48 (7.0)	45 (6.5)	72 (10.4)	72 (10.5)
Ultimate Strain, μin/in	L	17,000	25,000	NA[(c)]	NA
	LT	78,000	136,000	NA	NA
Poissons Ratio	L	0.33	0.29	0.33	0.33
	LT	0.23	0.22	0.33	0.33
Density, g/cm^3 (lb/in^3)		2.29 (0.083)	2.29 (0.083)	2.79 (0.101)	2.79(0.101)

(a) 1.3mm (0.052 in.) thick laminate
(b) 1.5mm (0.060 in.) thick aluminum
(c) NA = not applicable

1st International SAMPE Metals Conference
August 18-20, 1987

THE PRODUCTION OF PREMIUM QUALITY ALUMINUM ALLOY SAND CASTINGS BY THE BOTTOM POUR PROCESS

Henry J. Proffitt
Technical Director
Haley Industries Limited

Abstract

Foundries supplying castings to the aerospace user are facing increasing challenges for the supply of castings which can effectively replace the components currently being produced by other more tedious production methods.

The key word in this statement is "effectively" by which is meant:

1) effective from the performance point of view offering comparable or better mechanical properties, fatigues and creep resistance etc.
2) components which show a good consistency in terms of casting quality and properties.
3) with optimum resistance to corrosion
4) where the capability exists to produce highly complex casting configurations and where weight saving can be more easily produced than by other methods.

and

5) which satisfy all of these requirements in the most cost effective manner.

Fundamental to the above is the provision of molten aluminum alloy to the casting process which is of the highest quality attainable in a completely consistent manner.

In order to achieve this, the author's company decided to re-evaluate the whole aspect of metal preparation and pouring methods used. This included a detailed study of the effects of turbulence in the running system of castings being produced by the "gravity" pour system.

A marked improvement was immediately seen when a changeover was made to a pouring system where the metal was introduced into the mould from below in a controlled non-turbulent manner.

The present paper describes the many advantages which have accrued from this method of pouring.

There are many instances of cast parts which have been made where production by the conventional gravity method would be very difficult, if not impossible.

1. INTRODUCTION

The aerospace industry has always been the prime target for that section of the light alloy foundry industry willing and able to supply true premium grade castings to meet their current needs.

All too often, the user-designer finds himself in a situation where the use of a monolithic cast component, substituting for a very expensive fabrication of extrusions, forgings, sheet and weldments could save considerably on total cost.

The question arises therefore "why is this avenue not followed more often?"

A lot of attention is currently being paid to this particular question. One problem has stood in the way of this in the past. To date, casting users have applied a "casting factor" when using castings. This factor weighs heavily against the casting compared with the fabrication where much lower factors are normal.

The whole question therefore boils down to a basic question of how far a casting can be trusted so that this factor can be reduced.

The onus is therefore clearly on the foundry to demonstrate that castings can in fact be trusted and that "casting factors" substantially lower than at present can in fact be applied.

This brings us to the "premium quality" casting grade which is a step forward in this path.(ref. #1) Premium quality is not just a name for a casting which is somewhat better than normal,"run of the mill" castings,in a given alloy -far from it!

In a premium quality casting, the general level of tensile properties must be significantly higher than with normal grade.

Secondly, this higher level of properties must be obtained consistent throughout the casting. The normal test carried out to demonstrate that this is so, is the tensile test employing test pieces cut from thick and thin sections of the casting. The casting must be acceptable by radiography, fluorescent penetrant and ultrasonic non-destructive testing.

For the required "designated" level of premium quality, the casting must be Grade B to ASTM E155. There should be no defects present in the designated areas of the casting in excess of the "B" level. Even in

Grade B areas there is no guarantee that tensiles will pass since there are other factors to be considered. Tensile tests on actual castings are necessary.

To satisfy all of the stringent quality requirements, several factors must be satisfied. These are:

1) The casting must satisfy all dimensional tolerance requirements with consistent thin wall sections, good surface finish and no distortion.
2) The casting process must be supplied with molten alloy of optimum composition at the correct temperature and grain refined, free from oxides or other detrimental non-metallic inclusions, gas-free and where applicable correctly modified.

   Later the casting will also have to be given an optimum heat treatment to satisfy the specified requirements.
3) A good pouring system without turbulent metal flow and where precise control can be exercised over the rate of flow of the metal during the casting process must be used.

   The low pressure bottom-pour system developed by the writer's company has proven ideal for this purpose and is the subject of this talk.
4) Direct consultation between the part's designer and the foundry methods engineer is necessary at an early stage.

Computer simulation of the solidification pattern is coming into greater use to indicate to the foundry methods engineer areas of probable defect. This will serve as a reliable guide in the positioning of gates, risers and chills. The use of such a system, replacing the older, laborious trial-and-error system on the foundry floor, will do much to improve economics and shorten development time.

An important factor to be borne in mind, is the significant influence of dendritic arm spacing in alloys of the 300 series and of grain size in the 200 series of alloys on the tensile properties of the final casting.

Appropriate measures are needed to satisfy these two requirements to give the high level of properties where required.

2.0 BOTTOM POURING METHOD

The "conventional" system of gravity pouring where the metal from the pouring cup passes down the vertical sprue is not ideal.

Any turbulence induced during the pour can result in small oxide films or inclusions which can be very detrimental to tensile properties.

Surface oxides and skins on the molten alloy in the pouring receptacle can find their way into the running system.

Steps are normally taken by "inmould" filtering, to remove these and prevent their entry into the mould cavity.

The control over the flow of metal as it fills the mould is also not very precise. The only effective control is the diameter of the down sprue.

It was for these reasons that the writer's company initially undertook the development of the bottom pouring system of filling the mould to answer these deficiencies. The general layout of the bottom-pour system is shown on the next slide.

The bottom pour system consists of the following (see Figure 1 - Slide C33):

i/ The molten alloy from electric resistance crucible bulk melters is transferred into specially designed electric resistance heated crucible holders, mobile and very suitable for this purpose. (see Figure 2-Slide B71)

ii/ The molten alloy is processed by the optimum metal treatment techniques and brought under precise temperature control. From this time, until any excess molten alloy is ingotted, there will be no decanting of the ladle or undue disturbance of the melt producing turbulence.

This particular system was adopted because it provided the best compromise between maximum flexibility for alloy change with economic, clean melting conditions and the ability to adequately control metal temperature.

After metal treatment, the complete holding unit is lowered into the below-ground chamber which can be very accurately pressurized. (see Figure 3 - Slide A47).

By this means the molten alloy can be displaced by low pressure air up a tube and into the mould. The molten metal for each pour is taken from well below the surface of the metal in the crucible; an additional precaution against the possible transfer of any oxide on the melt surface.

The pressurization of the casting furnace is very closely controlled during the casting process by a computer which follows the developed parameter very closely.(See Fig. 4 -Slide A35) The whole procedure of transfer of moulds to the pour position, effecting of a seal between the top of the riser tube and the entry into the running system, verification of casting temperature as well as the filling of the mould to predetermined parameters of pressure and time are all precisely controlled by the computer dedicated to this task.

The moulds made from conventional mould materials are mounted on a steel or thick aluminum alloy plate which carries a nozzle through which the molten metal passes from the riser tube into the running system.

The whole process sequence is illustrated graphically on a CRT as it is occurring. (see Figure 5 -Slide A36).

The optimum parameters regarding temperature times and pressures are determined experimentally. These data are then entered into the machine's memory and can be referenced each time a particular type of mould is presented to the machine for pouring with complete consistency.

A particular feature of this pouring system is the ability of the machine to control the metal flow rate.

The fill mechanism differs from conventional gravity pouring and enables generally thinner wall sections to be poured.

Having filled the complete mould, including the gravity-type feeders, the mould is maintained full of metal until the bottom runner gates freeze and allow the normal process of solidification to take place.

The casting solidifies progressively from the bottom of the casting towards the top as with gravity poured castings.

One feature of castings produced by this bottom pour method which was immediately apparent was a big reduction in surface imperfections in the castings or Zyglo.

The general improvement in casting quality is reflected in the tensile properties of test pieces taken from cut up castings.

Some examples of different castings produced by this bottom pour system which illustrate some of the present potential of the method are illustrated in the following slide. These are annotated to illustrate their pertinent features.
(see Figure 6 - Slide A38 -
Thinner Wall Sections
Figure 7 - Slide A37 -
Improvements in Scrap Rate

Figure 8 - Slide B38 - Improve Mould Yield
Figure 9 - Slide A39 - Decrease in Mould Size
Figure 10 - Slide B39 - Less Sand, Binder and Chemicals

SAND ECONOMY

The running and risering systems which have been developed for this process result in moulds which can be appreciably smaller than those of the conventional gravity system. This is because the down sprue in the older system and the customary runner channel around the periphery of the casting is eliminated.

There is therefore an appreciable saving in sand, binder materials etc. and the moulds are themselves much easier to transport to the casting machine.

CASTING YIELD

The overall casting yield, ie. the ratio between the casting weight and the total metal which must be poured to produce that casting, can also be improved.

This can have quite a significant effect on the volume of recycled process scrap and therefore an effect on the melting economics.

THINNER WALL SECTIONS

The ability to pour thinner wall sections has already been mentioned but obviously this is only possible if the dimensional quality of the moulds is also very good.

Fortunately, the parallel developments which have taken place in mould and core making materials and processes and in means of setting cores with greater accuracy has enabled this to be so.

LIMITATIONS

At the present time we have restricted the use of this new casting system to the production of aluminum alloy castings.

However, the possibility of employing this general technique of bottom pouring to the magnesium base alloy castings which we produce is an interesting possibility. Regarding the present process we have not recognized any real size limitations other than possibly the height of the mould.

Any size limitations, if they do arise, are very much a question of physically handling the moulds and the castings themselves within the plant during subsequent processing operations. The equipment itself is flexible, allows for easy alloy change and is highly controllable with ease in all operational aspects.

In conclusion, Mr. Chairman, gentlemen, I would like to thank you for your kind attention to my presentation which I hope you have found of interest.

FIGURE 1

FIGURE 2

FIGURE 3

FIGURE 4

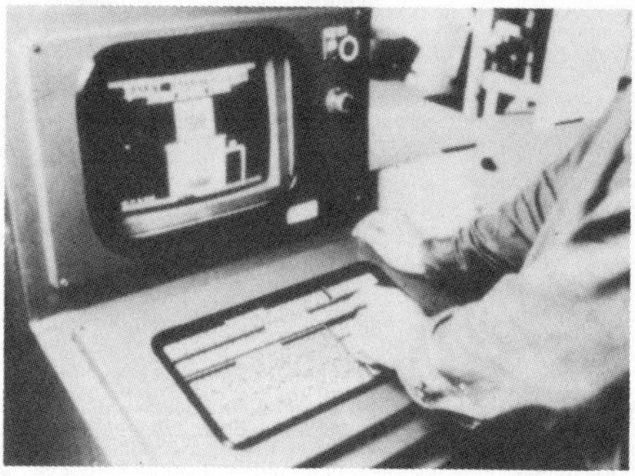

FIGURE 5

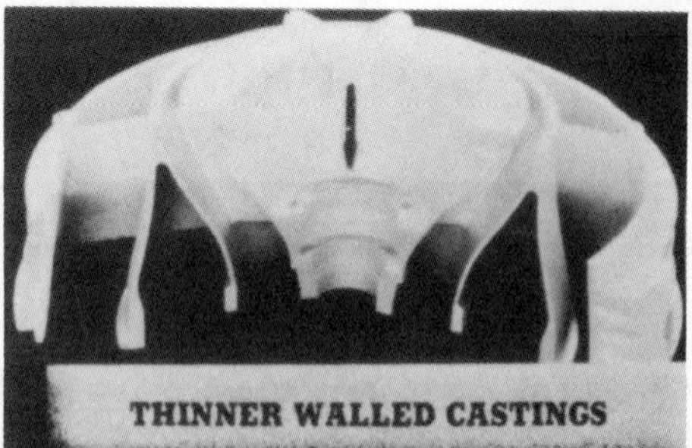

FIGURE 6

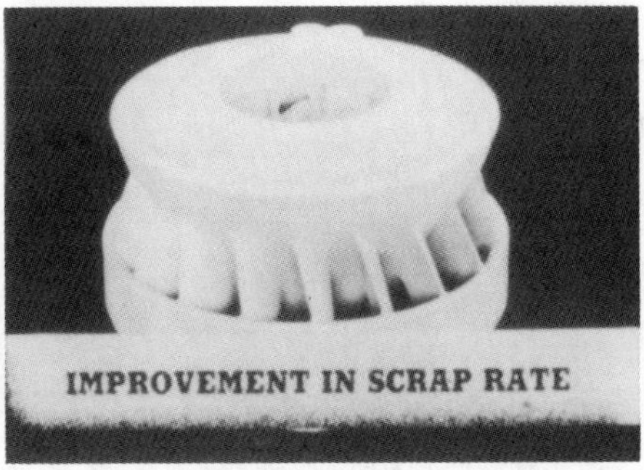

FIGURE 7

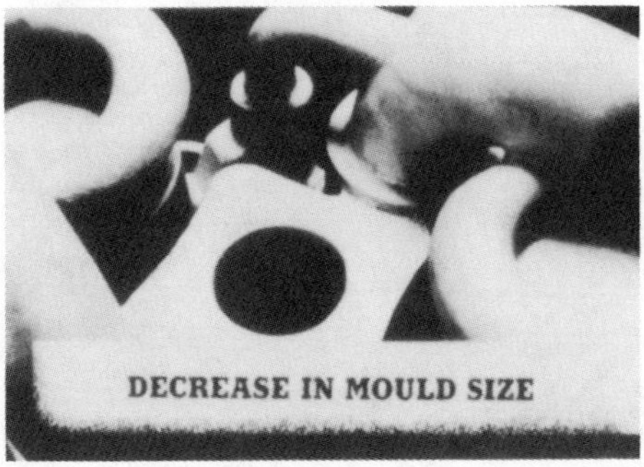

FIGURE 9

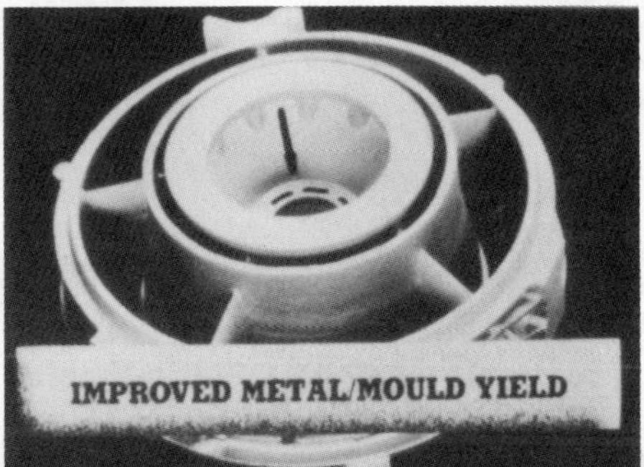

FIGURE 8

FIGURE 10

1st International SAMPE Metals Conference
August 18-20, 1987

RECENT DEVELOPMENTS IN SUPERPLASTIC FORMING

Prepared By: Michael J. Reynolds
Presented By: Andrew Poole
Superform USA Inc.
P.O. Box 5375
Riverside, California 92517

Abstract

During the first half of this presentation, the forming characteristics of superplastic alloys will be reported with suitable photographic examples of their available superplastic levels. In addition, we will be reporting on the opportunities for forming these alloys resulting in excellent control of component thickness and geometry.

1.0 BACKGROUND

Superform was established in England in the early seventies, and America in the early eighties, to commercially exploit the potentials of superplastic sheet aluminum forming, a rapid progression of material and technology developments has followed.

Qualified 2000 series SPF alloys are now widely accepted in many diverse market sectors. During the last three years, alloys with proved mechanical and physical properties have been developed within the British Alcan Corporation.

Superform has been working closely with its parent company, British Alcan Aluminium, and are currently addressing the practical aspects of forming, heat treating and postforming fabrication of SPF aluminum lithium alloys.

Additionally, we can report technological advancements in hot forming "off-the-shelf" aerospace sheet alloys.

2.0 SPF APPLICATIONS

Figure 1 shows a range of airborne components manufactured from Superform's 2000 series alloys. Since 1974, many hundreds of thousands of superplastic parts have been produced in a wide range of applications, including fairly extensive penetration into the aerospace and defense sectors.

Figure 2 shows typically complex configurations readily manufactured from SPF aluminum alloys utilizing machines and techniques developed and patented by Superform.

Figure 3 is a complex assembly manufactured by Westland Helicopters dramatically reducing parts count and improving accuracy and repeatability.

3.0 SPF ALUMINUM LITHIUM

During the last three years, Superform has been developing with its parent company, British Alcan Aluminium, a superplastic grade of Alcan's 8090 aluminum lithium alloy. This alloy has now been successfully scaled up from laboratory work to preproduction quantities, and Superform has been forming many components for internal evaluation, and have currently completed several R&D programs with aircraft mainframe manufacturers. This paper will report on two of those projects:

Project 1 - General Dynamics, Ft. Worth

Project 2 - Fairchild Republic, Long Island

For General Dynamics, we produced a forming tool simulating a bulkhead fairing on the F-16. A half section of this 24" square component is shown on Figure 4.

The forming profile for this fuel cell component is described in Figure 5.

One can note a preheat of five minutes, followed by tool movements into the clamped sheet. As the tool approaches full height, the press chamber is pressurized to a maximum of 102 psi, during an eleven minute time ramp. A positive pressure was applied to one side of the sheet only. The back pressure environment was not created for this program of work.

Figure 6 tabulates the properties achieved in the superplastically formed G.D. component given a variety of aging conditions. These results are compared with the less superplastic shelf aluminum-lithium 8090 alloy. One can observe that essentially the properties compare favorably. The areas of improved values can be attributable to both different process lots and strain hardening.

The most marked effect is that of elongation in the transverse direction. This was not entirely surprising given the history of SPF effects on other proven alloys.

Figure 7, information generated by British Alcan on the effects of processing aluminum-lithium at 500°F to 530°F. At Superform, we process sheet at this temperature for up to 30 minutes, and would, therefore, suggest that this has a minimal effect on lithium loss.

For Fairchild Republic we manufactured a wing tip forming tool, over which we superplastically formed British Alcan 8090 sheet. Figure 8 shows the resultant pressing. Good uniformity of thickness was achieved. Fairchild subsequently tested these components, and compared the results with pseudoformed sheet.

Tensile specimens could only be taken from the sides of the pressing due to the curvature in the opposite direction.

Figure 9 tabulates these results.

Supplementary to British Alcan's development program on superplastic aluminum-lithium, Superform has been developing innovative forming techniques for standard aluminum alloys.

Though the process is still believed to have few applications outside the aerospace industry, it could reduce costs and speed up tooling for some aerospace parts requiring standard alloys.

Suporform has devised similar SPF related processes, which they respectively call diaphragm and membrane processes, and which make use of "throwaway" standard SPF sheet as well as non-SPF sheet. In each process, while the SPF sheet is being molded through standard SPF processes, the non-SPF alloys are drawn into the shape of the SPF mold. Because the material is not fixed to the edge of the SPF molds, the non-SPF material is not stretched into the desired shape, but is actually drawin into shape. Figure 10.

Opinions vary on the potential of the new technique, but there seems to be a growing consensus of officials that the development will find a wide range of applications in the aerospace industry. While it seems unlikely these applications will be for critical parts, they are expected to include applications in areas where standard aerospace alloys are required, such as ducts, wing tips and fairings.

The development of the SPF-related technologies, which are already beginning to have some commercial success, is an outgrowth of the work done in traditional SPF areas. SPF technology has been available for at least 20 years, but now about 100,000 parts are being made annually by Superform for a variety of applications.

Figure 11 shows a variety of components formed by the diaphragm techniques. The intimate contact of the SPF slave sheet eliminates the propensity for wrinkles to form inside the nett trim line. The resultant pressings have very consistent thickness distributions with little to no reduction in stock sheet thickness.

Subsequent testing reveals no degredation in mill handbook properties. This opens up a whole new field of applications where 2004 or 7475-02 SPF alloys cannot be specified. Now Superform can process almost any standard alloy, including available aluminum-lithium alloys over single surface forming tools using a thin diaphragm of SPF 2004 alloy. These new techniques could readily allow designers to take full advantage of aluminum-lithium alloys without the current concerns of cavitation and complex forming cycles.

5.0 CONCLUSIONS

1. A truly superplastic version of 8090/8091 is deemed to be important if for no other reason than to achieve truly isotropic properties.

2. Even with an SPF grade, we conclude that for geometry of medium to high equivalent thickness strains, a back pressure environment of 300 to 500 psi will be required to restrict the formation of microcavities.

3. In the absence of this back pressure, techniques of hipping may be employed.

4. Further development is required to evaluate the need for clapping on these SPF alloys. Superform has processed clad aluminium-lithium sheet and have found the cladding to be an adverse influence.

5. Diaphragm forming may develop into a very useful production route rather than an intermediary step.

6. Our work on post forming processes will continue. Addressing areas such as welding, heat treating and surface treatment.

7. Our ideal conclusion to the next few years of development would be to offer a superplastic aluminum-lithium alloy or alloys which are capable of being formed in minutes rather than hours and that have a forming temperature compatible with the solution treatment temperature such that the formed sheet can be quickly transferred from the press to an air blast or liquid quench. Meanwhile, the many advances made to date are positive moves towards implementation of aluminum-lithium on real applications.

1st International SAMPE Metals Conference
August 18-20, 1987

AGE CREEP FORMING ALUMINUM AIRCRAFT SKINS

Denise M. Hambrick
Avco Aerostructures Textron
Nashville, TN 37202

Abstract

Avco Aerostructures Textron has advanced the technology of age creep forming by developing the ability to form aircraft structures which, previously, were difficult or impossible to form. The Autoclave Forming Aluminum process, developed by Avco, has formed smooth airfoil contours in 15 m (50 ft.) long integrally stiffened wing skins of alloys 2124 and 2419. The process produces smooth contours even across abrupt changes in thickness. The age creep forming process has also proven its ability to form compound curvatures. A data base has been developed by forming single curvature parts of seven aluminum alloys. This paper reviews development of a data base, forming compound curvature parts of alloy 7075, and applications of age creep forming aluminum.

1. INTRODUCTION

Forming large aircraft skins usually involves forming airfoil contours. Often those contours have a compound curvature shape which is one of the most difficult shapes to form. Conventional methods of forming aircraft contours, such as shot peen forming, brake forming, bump forming (room temperature or heated), and stretch forming, tend to result in a buckle or a bubble near areas of transition. Efforts to re-form to correct the problem usually move the bubble but fail to eliminate it.

The process of age creep forming offers inherent benefits compared to other methods of forming age hardenable alloys. The method generates contour by restraining a part to a developed shape during precipitation heat treatment. Time at the aging temperature allows the part to relieve stresses while creeping to contour. The formed part is dimensionally stable with low residual stress. Therefore, warpage during post machining, assembly, or service is reduced to a minimum.

Age creep forming results in a smooth, repeatable contour. The developed tool controls the resulting panel contour, thereby providing the mechanism to correct any discrepancies which might appear. For that reason, age forming is operator independent.

The aircraft industry has used age forming to form fuselage sections which have a thin uniform wall thickness with stiffening members perpendicular to the forming direction. Typically, the curvature has been a simple radius with a minimum amount of forming required. For example, the Boeing Company, Wichita, Kansas age formed integrally stiffened plates of 2219 aluminum used to fabricate fuel tanks for the Saturn V space vehicle.(1)

Previous experience did not include forming of thick sections nor forming of sections with varying thickness. However, the potential for the process to form such complex parts has been demonstrated.

Under a joint development program with Rockwell International - North American Aircraft Operations and the U.S. Air Force Systems Command, Avco Aerostructures Textron developed the technology to age creep form two integrally stiffened skin structures for the B-1B long range combat aircraft.(2)(3)(4) The skins of aluminum alloys 2124 and 2419 were formed to specific airfoil contours. Skin thickness varied from 2.5 to 63.5 mm (0.1 to 2.5 in.) over a total length of 15 m (50 ft.). The skins were accurately formed to smooth airfoil contours showing no effects of the abrupt changes in thickness. Tests confirmed that the formed skins met specified material property requirements including fracture toughness. Also, the residual stresses were substantially lower than those resulting from the previous forming method, thereby yielding a more dimensionally stable structure.

Process implementation for this specific application enabled Avco to achieve and maintain a schedule of 4 ship sets (16 formed panels) per month and reduce final assembly time.

The success of that forming program led to a research project to develop an empirical data base of age creep forming age hardenable aluminum alloys. The ultimate goal is to create a design guide for age creep forming and thereby eliminate the need for custom forming development.

This paper discusses development of a data base by forming single curvature parts, forming compound curvature parts of alloy 7075, and applications of age creep forming.

2. DATA BASE DEVELOPMENT

Specimens of eight alloys have been age formed using the aging cycles specified in military specification, MIL-H-6088. The alloys tested and their age cycles are listed in Table I.

Forming tests on specimens t x 76 x 762 mm (t x 3.0 x 30.0 in.) were used to determine empirical springback relationships. Specimen thickness ranged from 1.6 to 25.4 mm (0.063 to 1.00 in.). Three different tool radii were selected to generate panel contours which are representative of those typically found in aircraft skins. Mechanical clamping fixtures were built to individual matched die radii of 1.27, 3.81, and 7.62 m (50, 150, and 300 in.).

Parts were clamped into the tool shape using a 9072 kg (10 ton) Enerpac press and frame. Pressure was maintained on the tool while bolts were secured on both ends. The fixture was then placed on a dolly with the other two fixtures and loaded into a preheated furnace. After being held at the aging temperature for the aging time, the parts were air cooled to room temperature before being released from the fixtures.

Temperature was continuously recorded throughout the aging cycle. In addition, hardness, electrical conductivity, and gap between the loaded part and tool were measured before and after age forming.

Precise contour measurements were achieved using a coordinate measuring machine (CMM). Controlled by a computer program, the CMM measured each part at 25.4 mm (1 in.) intervals along two rows of the outer surface, printed the contour coordinates, and plotted graphs representing the contour. The data was later analyzed using a personal computer and spreadsheet software.

Springback as a function of thickness is illustrated in Figure 1 for transverse specimens of alloy 7075 formed in the mechanical fixtures. Springback for each tool radius is shown as a computer generated best fit curve. Forming data for the other alloys tested resulted in similar springback curves.

In all cases, a peak in percent springback exists at approximately 6.4 mm (0.25 in.) thickness. According to bending stress theory, as thickness is reduced, stress in the outer fibers will decrease leading to more springback, i.e., less forming.[5] The cause of the peak is not completely understood. However, it appears that as thickness is lowered, the predominant mechanism of forming changes from mechanical bending to metallurgical creep.

The results demonstrate the feasibility of age creep forming thin parts, 1.6 to 6.4 mm (0.063 to 0.25 in). Contrary to expectations, the springback did not continue to increase with reductions in part thickness. Conventional forming of thin material to airfoil contours results in springback values up to 100%, i.e., no forming. Age creep forming makes it possible to form contours typical in aircraft, which would otherwise be difficult to achieve.

Tool contour development for a part with varying thickness and changing contour requires a relationship which combines the major variables. That relationship must incorporate the interactions between part thickness, required radii and tool radii. The forming data demonstrated a logarithmic relationship of tool radius divided by thickness R_T/t) as a function of part radius divided by thickness (R_P/t). Figure 2 illustrates the contour relationship for alloy 7075, T6 to T73 temper. A similar relationship exists for each of the other alloys tested.

3. COMPOUND CURVATURE FORMING

Capability of the age creep forming process to form a compound curvature was tested by forming a section of a typical wing skin as shown in Figure 3. The curvature has a transition between two separate airfoil contours creating a dihedral break referred to as the transition vertex. The result is a "saddleback" shape. A section of the panel representing 6 feet on either side of the transition vertex was selected for testing. Then the scale was reduced to 1/3 of full size for convenience and cost savings.

Test panel thickness was 2.0 mm (0.078 in.). Part length was 1.27 m (50 in.) and width was 1.04 m (41 in.) on the inboard end and 0.79 m (31 in.) on the outboard end.

3.1 Tool Design and Contour Development

The compound curvature tool design consisted of a pressure box with contour boards. The box structure was welded from 6061-T6. A 6.4 mm (0.25 in.) thick caul plate was used to adequately create the tool surface and to prevent part deflection between boards during forming. The caul plate was roll formed to the developed contour in two sections and welded at the transition vertex. The completed tool is shown in Figure 4. With the aid of the straightedge, one can see the bend at the transition vertex.

Contour development for the tool followed analysis of the single curve data. The desired part contour was scaled down using a combination of CATIA and CADAM computer software packages.

In the selected panel, the two modes of curvature conflict. Stretching of the material is required to achieve the desired part contour. Also, after achieving the required stretch, springback of the chordwise contour will tend to cause overforming of the spanwise contour. Therefore, the expected springback for single curvature forming was significantly reduced for the compound curvature tool. The above considerations lead to the selected tool contour which would allow for the anticipated springback to the desired shape.

An eggcrate style check fixture was fabricated to measure formed part contour. The check fixture contour matched that of the desired shape. Contour board position was the same as on the tool.

3.2 Forming Procedure

After loading the panel in the tool, reinforced silicone rubber was used to cover the edges of the part to protect the bag. A flat silicone rubber bag was clamped to the tool using an aluminum "picture frame" before pulling a vacuum. Then the tool was placed in the lab autoclave.

Vacuum lines were used to connect the tool to the vacuum pump and to a gauge line. Six thermocouples were used to measure temperature of the tool and atmosphere. Since the part was much thinner than the tool, part temperature was considered equal to tool temperature.

The autoclave procedure was as follows:

1. Load part and pull vacuum of 34 kPa (5 psi).
2. Place tool in autoclave and apply 34 kPa (5 psi) pressure.
3. Vent vacuum to atmosphere.
4. Heat part to 163 ± 5°C (325 ± 10°F).
5. When part reaches 157°C (315°F), time = 0.
6. Increase pressure to 241 kPa (35 psi) and pull full vacuum for total pressure differential of approximately 345 kPa (50 psi).
7. Maintain pressure and temperature until time = 24 hours.
8. At time = 24 hours, cool the part to 66°C (150°F).
9. Release pressure and vacuum.

The formed part was measured with feeler gauges for deviation from the check fixture.

3.3 Results

The formed panel is illustrated in Figure 5 with a view of the outer mold line. The straightedge in front of the panel helps to demonstrate the airfoil contour. The line on the table surface next to the panel's forward edge displays the break.

Two panels were formed. The first panel had a maximum deviation of 4.04 mm (0.159 in.) at the inboard edge. However, average deviation of all 45 points measured was only 0.74 mm (0.029 in.). The second panel exhibited a maximum deviation of 3.30 mm (0.130 in.) and an average deviation of 0.66 mm (0.026 in.). Contour of the second panel repeated that of the first within 0.76 mm (0.030 in.).

Mechanical properties of the first panel were tested to confirm attainment of T73 temper. Six longitudinal and 6 long transverse specimens were cut from the formed panel. Also, control specimens from the same T6 material were aged to T73 and tested to compare the properties of material aged without forming. Rockwell B hardness and conductivity were measured on each specimen.

All property values met specification requirements. Comparison of panel specimen results to those of the control specimens revealed no measureable difference. Therefore, age creep forming has demonstrated the ability to achieve desired material properties.

Forming the panels was a complete success. Uniform pressure eliminated buckling, resulting in smooth compound contours. Also, the panels demonstrated trends of forming conflicting contours.

4. APPLICATIONS OF AGE FORMING

Current standard forming processes necessitate assembly of many small parts of simple design in order to produce large complex structures. The resulting structures contain multiple joints which may allow fuel leaks as well as increase the weight of the structure due to lap joints and fasteners. The age creep forming process offers the ability to form complex structual aircraft parts, such as integrally stiffened wing skins, compound curvature skins, spars and other critical structural members, to close tolerances. The result is more efficient fabrication and assembly as well as improved longevity which is applicable to many aircraft.

Forming of the B-1B skins demonstrated the ability to form large integrally stiffened parts as a single unit rather than as an assembly. Similar applications at the design stage will reduce the number of parts and joints, help avoid fuel leaks, reduce weight, and reduce acquisition and maintenance costs.

The ability to form compound curvature parts to a smooth airfoil without buckling offers additional benefit to modern aircraft. Complex contours will not have to be designed as multipiece structures including extra doublers for stiffness at joints.

Another application of age forming is forming of waffle-like skin panels with either internal or external stiffeners. Parts of alloys 7075 and 2419 with machined ribs arranged in a diamond pattern as well as ribs parallel to the forming direction have been successfully formed using age creep forming.

Also, preliminary tests on 12.7 mm (0.5 in.) aluminum-lithium alloy 2090 have been successful. The tests indicate that age creep forming will be an excellent method of forming aluminum-lithium alloys.

5. SUMMARY

An age creep forming data base of seven aluminum alloys has been developed through extensive forming tests. Two compound curvature skin panels of alloy 7075 were formed to a typical aircraft compound curvature.

The age creep forming process has proven the ability to eliminate buckling and erratic contours which often occur with other methods of forming. Age creep forming can successfully form compound curvatures.

6. REFERENCES

(1) Layton, R.E., Am. Mach/Metalworking Manuf., May 27, 1963, pp. 67-70.

(2) Hambrick, D.M., Age Creep Forming in an Autoclave, 30th National SAMPE Symposium and Exhibition, Vol. 30, Advancing Technology in Materials and Processes, 1985.

(3) Hambrick, D.M., Age Forming Technology Expanded in an Autoclave, SAE 1985 Transactions, 1986.

(4) Hambrick, D.M., Age Forming Aluminum in an Autoclave, Aluminum Alloys - Their Physical and Mechanical Properties, Vol. II, 1986.

(5) Juvinall, R.C., Engineering Considerations of Stress, Strain, and Strength, McGraw-Hill Book Co., 1967.

7. BIOGRAPHY

Denise Hambrick is a Senior Research and Development Engineer at Avco Aerostructures Textron in Nashville, Tennessee. From 1982 to 1985 she served as principal investigator on the Autoclave Forming Aluminum program for forming B-1B wing skins. Since then, she has developed a data base for age creep forming aluminum. Also she has developed manufacturing methods for aluminum-lithium alloys.

Prior to joining Avco, Denise was employed by General Dynamics' Convair Division in San Diego, California. As a Manufacturing Development Engineer, she was assigned to develop and improve forming of aircraft alloys.

Ms. Hambrick received her B.S. degree in Metallurgical Engineering from the Ohio State University in 1980.

TABLE I
AGE CYCLES FOR ALLOYS TESTED

ALLOY	INITIAL TEMPER	AGE CYCLE			FINAL TEMPER
		TEMPERATURE °C	(°F)	TIME HOUR	
2024	T3	190	(375)	12	T8
2124	T3	190	(375)	12	T8
2219	T3	177	(350)	18	T8
2419	T3	177	(350)	18	T8
6013	T4	190	(375)	4	T6
7075	T6	163	(325)	15	T76
7075	T6	163	(325)	24	T73
7475	T6	163	(325)	10	T76

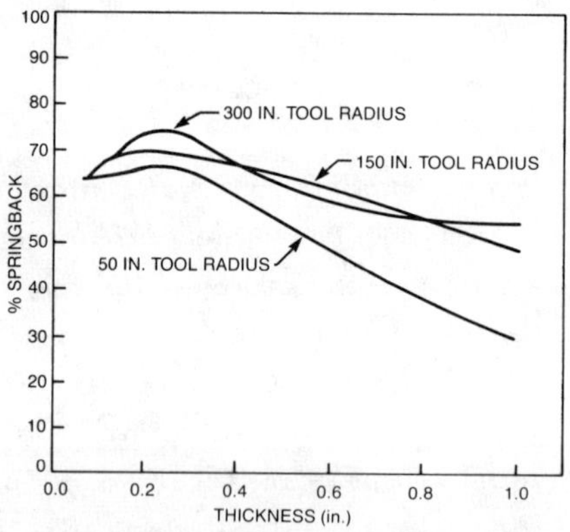

Figure 1 SPRINGBACK AS A FUNCTION OF THICKNESS FOR 7075-T6 AGE FORMED TO T73

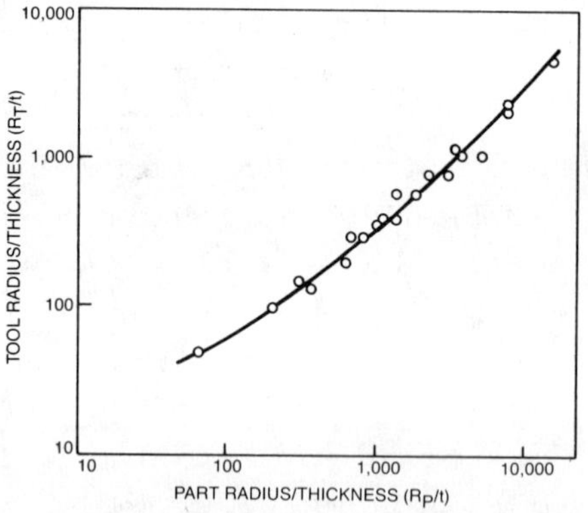

Figure 2 CONTOUR RELATIONSHIP FOR ALLOY 7075 AGING FROM T6 TO T73 FOR THICKNESSES FROM 0.063 TO 1.0 IN.

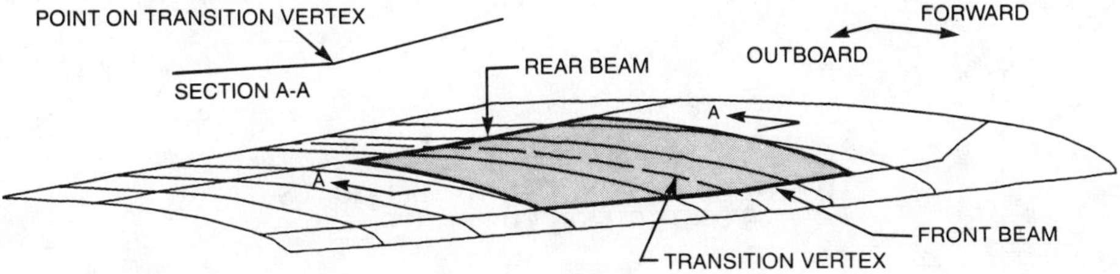

Figure 3 TYPICAL WING SKIN OF COMPOUND CURVATURE SHOWING PANEL SELECTED FOR TESTING

Figure 4 COMPLETED TOOL

Figure 5 OUTER MOLD LINE OF FORMED PANEL

1st International SAMPE Metals Conference
August 18-20, 1987

ADVANCED CONCEPTS AND MATERIALS FOR HIGH-SPEED FLIGHT

by

J. E. Fischler
Douglas Aircraft Company
McDonnell Douglas Corporation
Long Beach, California

ABSTRACT

The combination of "New" metal matrix composites, reinforcing "New" SPF/DB titanium alloys of base sandwich concepts, has made it possible to advance into the high Mach environment with structural mass ratios that are competitive with low Mach civil transport values. These studies are still in the "research" and "development" stages but show large benefits. These benefits have been accomplished from ideas initiated from the results of NASA contracts mentioned in the following paragraphs.

In NASA CR 178006 metal matrix reinforcements, Task VI, and crack growth characteristics of SPF/DB titanium with reinforced panels, Task VII, were obtained. Also, Contract NAS1-16147, studies of benefits of metal matrix composites for supersonic aircraft, now being done, will be completed before the symposium. Therefore, this symposium would be a good forum to acquaint people in the materials community with our progress in this area. Some of the subjects discussed will be:

1. Significant weight savings are accomplished for MMCs when joined to an SPF/DB titanium sandwich.
2. The initial costs of the concepts have been compared to the value of airframe weight. Significant direct operating cost savings are gained with the new materials and concepts.
3. Using MMCs to reinforce panels of SPF/DB titanium substantially increased the crack propagation life.
4. Using rapid solidification rate titanium increased the ability to provide a base SPF/DB titanium sandwich for high-temperature applications.
5. Evaluation procedures for advanced materials required many material characterization tests, process optimization, and allowables tests which are described.
6. Concluding remarks.

Significant savings in weight, increases in life, damage tolerance, and stiffness has been achieved by improving the metal matrix composites and joining them to new titanium alloys in the form of an SPF/DB sandwich. The higher cost per pound was found to be less important than the weight saving for high-temperature civil transports.

INTRODUCTION

The results of three NASA-sponsored programs, References 1, 2 and 3, plus some company-sponsored research and development are reported herein.

After Douglas developed the four-sheet sandwich spot-welded process called the SPF/DB 6AL-4V annealed titanium sandwich (Figure 1) it became apparent that much of the supersonic transport's forward wing and fuselage were minimum gage. Soon after, a further development by Douglas using a three-sheet spot-welded SPF/DB titanium sandwich, which saved additional weight for the minimum gage over much of the aircraft, was also explored (see Figures 2 through 4). A successful small panel was fabricated, and NASA then sponsored (Reference 1) the fabrication of a large 3- by 3-foot forward wing panel (Figure 5).

Further analyses indicated that using an SPF/DB titanium sandwich provided high-compression allowables for much of the supersonic transport with more weight savings, especially for the fuselage, where the compression and damage tolerance of an SPF/DB sandwich is very efficient. Also, by laying up the doublers as necessary, at the edge of the panels, and making large panels, the cost of wing and fuselage structure was reduced because the number of parts, and the weight, were simultaneously reduced (see Figures 6 to 9).

Notice that the SPF/DB process allows the designer to take advantage of joining, by diffusion bonding, the spar and rib caps to the tailored SPF/DB sandwich, thus reducing the number of parts and attachments. Tubular truss core SPF/DB struts and end plates can easily be fabricated and attached to the tees that form the rib and/or spar caps (Figures 10, 11 and 12).

The development of metal matrix composites (MMCs) for lightweight reinforcement further enhanced the advantages of using the SPF/DB titanium sandwich. After studying the various ways to reinforce the SPF/DB titanium sandwich in Task VI of Reference 2, and after investigating the crack propagation characteristics of MMC reinforcements in Task VII (also Reference 2) a large effort was made to develop an ultra-lightweight SPF/DB sandwich out of thin-gage titanium. Many difficulties were overcome (Figure 13). These thin-gaged SPF/DB titanium sandwiches were used as the ultra-lightweight stabilized core sandwich, which were then brazed or diffusion bonded to MMC face sheets. These face sheets had an aluminum or titanium matrix with continuous fibers of B_4C, SCS-2, SCS-6, SCS-8, or powdered SiC particulates, or whiskers pressed or cast at high temperatures. These concepts were first developed in Reference 2 (Figures 14 to 21) and then explored further in Reference 3. Figures 22 and 23 show the various concepts and materials that were developed, almost all of which are Douglas proprietary information.

In Reference 3, the analyses of four candidate concepts and materials were compared to the basic 6AL-4V annealed SPF/DB titanium sandwich up to a Mach 4 supersonic transport. Figure 23 shows the relative idealized structural finite element structural weights for membranes, bars and shear panels that were optimized and resized using the structural models shown in Figures 24 and 25. Approximately 30 cases were run to obtain the idealized weight versus the Mach number curve in Figure 23. For each case, the allowables were calculated considering the reductions for thermal degradation, thermal gradients, fatigue, damage tolerance, coefficient of variance of the materials, and some reductions for aerothermoelastic deflections and creep limitations. Each case had a minimum size for the membranes, bars, and shear panels. The combined Young's elastic modulus and shear modulus were corrected for temperature and reflected the minimum sizes. A 530-page computer input for each case indicated how well the elements were optimized, what portion of the elements were minimum size, and what the weights were for each element. Each airplane component weight was summed and Figure 23 shows the totals for the full structural model.

The McDonnell Douglas Research Laboratory (MDRL) has investigated rapid solidification rate titaniums and the allowables for this material is considerably better than 6AL-4V material (Figure 26). Therefore, two of the five candidate concepts and materials utilized the Rapid Solidification Rate (RSR) process with the alloys developed at MDRL, concepts No. 3 and 4.

These four candidates show a very significant weight saving over the current SPF/DB 6AL-4V annealed titanium sandwich idealized structure, especially at Mach numbers above 3. A preliminary cost benefits analysis was done for the four best candidates compared to the base 6AL-4V annealed SPF/DB titanium concept. Tables 1 to 5 show these results. The value of a pound of weight saving for the high Mach number transport aircraft are estimated to be more than $1,240 per pound. Therefore, with the large weight savings for the five concepts, the initial costs are less, in spite of the high initial metal matrix composite cost. Also, there are significantly lower operating costs for the economic lifetime per aircraft (see Tables 1 through 5).

Tests of the MMCs reinforcing the SPF/DB titanium sandwich has increased the lifetime after a crack appreciably. The 1-g stress level, and the spacing of the MMC reinforcements, are two significant factors. If the 1-g stress level is low, the crack size will be relatively small after one full lifetime, and the increased life from the reinforcement will be significant. Using a 1-g stress for titanium of 25,000 psi, which is 1.85 times the DC-10 wing 1-g stress of 13,500 psi for aluminum, and correcting for density, yields 1.85/1.6, which is 1.16 more for the relative specific strength of titanium compared to aluminum. Using the 25,000 psi 1-g stress and an SPF/DB titanium sandwich with an initial crack on the outer face sheet results in a crack growth that slows down significantly when the crack reaches the MMC crack stopper which is brazed to the SPF/DB titanium panel (Figure 27). At higher 1-g stresses than 25,000 psi the crack growth rate is moving too rapidly for the crack stopper of MMC reinforcement to be very effective when placed 6 inches apart. Therefore, at higher 1-g stresses, the crack stoppers should be placed closer, and/or the crack stopper size increased.

The evaluation procedures for determining if a material is acceptable are described. Some of the background for evaluating a new concept and material comes from Reference 2 where the material characterization, design allowables type of data, process optimization, and crack growth rate and lifetime after a crack, were obtained on the baseline 6AL-4V annealed titanium SPF/DB sandwich. The "New" concepts and materials can be compared to the Reference 2 data to determine the relative benefits of the new concepts and materials compared to the SPF/DB titanium 6AL-4V base sandwich.

SIGNIFICANT WEIGHT SAVINGS FOR MMCs JOINED TO SPF/DB SANDWICH

Weight savings can be illusive. It is not sufficient to compare only the strength allowables of the new concepts and materials. By dividing by the average density of the "New" concepts and base concepts, the specific strengths can be compared. This helps to initially select the systems with the largest benefits. Figures 19, 20 and 21 show some of these comparisons.

However, specific strength is only one parameter. Other modes of failure must be considered, especially for a supersonic transport that must have two lifetimes of adequate fatigue strength (100,000 hours) and must sustain the thermal gradient stresses, thermal degradation of properties and have adequate damage tolerance (after failure of a major structural element, must sustain limit load). These requirements, plus the creep and aerothermoelastic considerations at the higher Mach numbers must all be accounted for.

In Reference 3, the allowables were computed for the basic 6AL-4V titanium SPF/DB sandwich structure and four other leading concepts and materials throughout the Mach 2 to Mach 4 range. These allowables accounted for the modes of failure mentioned previously and were input into the finite elements shown in the structural model of Figures 24 and 25. Minimum gage structure for each concept and the material for each Mach number was specified for the membrane (sandwich), bar and shear panels. Using the Von Mises criteria for the membrane allowables, the resultant sums of the idealized structural weight were obtained after a number of optimization cycles going to the fully stressed design. Tables 1 to 5 show the idealized structural weights of the finite element model of the wing, fuselage and tail structure corrected for nonoptimum factors and adding engine, nacelle, pylon, systems and landing gear weights.

INITIAL COST SAVINGS

Structural concepts and materials Number 2, 3, 4 and 5 are described in Figure 23 and are illustrated in Figures 16, 17 and 22. Those concepts that use metal matrix composites for the face sheets that are joined to an SPF/DB titanium sandwich, namely concepts Number 2 and 4 have the highest weight and cost benefits. Concept 2, which has been successfully fabricated in Reference 2, has an aluminum matrix, which when reinforced by the SCS-2 continuous fibers, can sustain high allowables up to the Mach 3.5 to 4.0 range. The aluminum face sheet matrices of the sandwich have a density of only 0.10 pound per cubic inch, and when reinforced with strong SCS-2 fibers with only 0.103 pound per cubic inch density, results in a specific strength that is very high due to the fiber strength of over 200,000 psi up to temperatures of 2,000°F (see AVCO data in Figures 28 and 29). This strong combination in the face sheets, stabilized with the SPF/DB ultra-lightweight titanium sandwich, yields high specific strength until the Mach number occurs which causes the base 6AL-4V SPF/DB titanium sandwich to begin to soften. These internal core temperatures begin to be critical at Mach numbers of 3.5 to 4.0. If it is possible to make an SPF/DB titanium sandwich out of rapidly solidified titanium, now being developed by MDRL, the allowables will not deteriorate as much as SPF/DB 6AL-4V annealed titanium (Figure 26). Therefore, it would be possible to achieve higher Mach numbers than the Mach 3.5 to 4.0 especially with Concept 2, if the Rapid Solidification Rate (RSR) titanium is substituted for the 6AL-4V annealed base SPF/DB sandwich. This would require further development, but analytically it seems possible.

Concept 4 also uses RSR titanium for the base three-sheet SPF/DB sandwich, and is expanded and diffusion bonded to MMC continuous fibers in a RSR titanium matrix for the face sheets. This combination holds up very well at Mach 4 and above. Again, the continuous fibers in a RSR titanium matrix for the face sheets, diffusion bonded to a RSR titanium base four- or three-sheet SPF/DB sandwich seems to be the best from the weight and cost standpoint. This concept has such high allowables because the continuous fibers, which have very high strength, even at high temperatures, are pressed into a matrix of RSR titanium for the face sheets of the sandwich. The RSR titanium maintains stable properties that do not deteriorate substantially till about 1,400°F. These face sheets are diffusion bonded to a core SPF/DB sandwich of RSR titanium which is also thermally stable and therefore has high crippling strength to support the very high allowables of the face sheets so that large compression panels of this concept can be fabricated. Large panels mean low fabrication labor costs and also weight savings by avoiding the edge doublers and attachments of smaller panels.

Poe, from NASA, has determined that with a relative stiffness of $\mu = 0.41$ for the reinforcing stiffener compared to the sheet, the rate of crack growth was substantially reduced, see Figure 30. The high modulus of the Metal Matrix Composite, (B_4C in 6061 Aluminum) helped to obtain a relative stiffness of $\mu = 0.3083$ when the 2-inch-wide MMCs were placed 3 inches from the centerline of the SPF/DB titanium 6AL-4V sandwich (Figure 31). When the crack reached the MMC doublers (Figure 27, Reference 2), the cracks on the left and right sides slowed considerably. A crack on the back of the sandwich occurred at 1.77 lifetimes. This compares to a single-sheet failure at 0.66 lifetimes, and a sandwich without MMC reinforcement at 1.47 lifetimes. Therefore, a 20-percent longer life was obtained with metal matrix composite reinforcements. A higher stiffness factor (μ), and/or a closer spacing than the 6 inches can both substantially improve the fatigue life. Alternatively, the 1-g stress can be increased to save weight since only one lifetime after a crack is required. All of the above was investigated with a 1-g stress of 25,000 psi.

USING MMCs TO REINFORCE PANELS OF SPF/DB TITANIUM

However, when the 1-g stress was increased to 30,000 psi, the crack cut through the doubler of MMCs on the left side, with hardly any rate of crack growth change from the addition of the MMC doubler (Figure 32, Reference 2). Therefore, at the higher 1-g stress, the crack stoppers should be spaced closer together. Notice that the right-side crack did slow down the crack considerably. Notice also for comparison, in Figure 33, Reference 2, that a 1-g stress of 30,000 psi and a full spectra on a single sheet would fail from critical crack growth after only approximately 6,000 flights, after initiating a quarter-inch crack. The SPF/DB titanium sandwich with MMC doublers will take 21,815 flights for failure (Figure 32, Reference 2). This is 1.034 lifetimes which is acceptable (but marginal) since only one lifetime after a crack is required. However, if the doubler spacing was reduced from the 6 inches between centerlines of reinforcement to 5 inches, the energy might have been absorbed by the left-side MMC doubler and the total life could have been increased substantially. The 1-g stress increase to 30,000 psi from 25,000 psi could save 16.7-percent weight. This theoretically could be accomplished by a 3-percent weight increase of the total panel by reducing the MMC doubler spacing from 6 to 5 inches, a net 13.7-percent weight gain, (16.7 percent minus 3.0 percent). This theoretically may be obtained with more than a 1.034 lifetime. Further tradeoff of weight savings, 1-g stress, crack growth lifetimes to critical crack length after initiating a crack, and the characteristics of the MMC crack stopper doublers should be analyzed and tested for verification. Having the unidirectional high modulus of the MMCs for crack stoppers is one of the major advantages, especially for high temperature structures (where the MMCs properties are stable).

USING RAPID SOLIDIFICTION RATE (RSR) TITANIUM

The McDonnell Douglas Research Laboratory (MDRL) discovered that certain dispersoid-forming elements, in rapidly solidified titanium, produce large increases in modulus and strength at room and elevated temperatures. The material characteristics are shown in Figures 26, 34 and 35. Because of the small grain sizes of Rapid Solidification Rate (RSR) titanium, it may be possible to superplastically form and co-diffusion bond this material. If this is possible, lightweight sandwich structure of RSR titanium can provide a "New" base sandwich structure that will be stable up to approximately 1,400°F before the material properties start falling rapidly. This is significantly higher than the 6AL-4V annealed titanium sandwich that has been evaluated thoroughly in Reference 2.

The RSR titanium should also be machine rolled down to 6 to 10 mils. These laminations shall form the base for the addition of continuous fibers of SCS-6 or SCS-8. This is expected to result in the strongest, most rigid face sheets. Using these face sheets, diffusion bonded to SPF/DB RSR titanium SPF/DB sandwich, Concept 4 in Figure 23, is expected to result in the highest weight and cost savings (see Table 4).

EVALUATION PROCEDURES FOR ADVANCED MATERIALS

The evaluation procedures for any "New" or advanced materials followed the pattern used for evaluating the 6AL-4V titanium SPF/DB sandwich in Reference 2. Because the temperature range desired is much higher for the materials now being evaluated, the range of temperatures considered in Reference 2 have been increased and the additional creep-rupture, creep-fatigue, aerothermoelastic-plastic modes of failure, probability of structural failure, initial cost and direct operating cost, and value of a pound saving considerations have been added to the evaluation. With these additions to Reference 2, the evaluation tasks are updated to be the following:

Task I	–	Selection of Structural Concepts and Materials to be Evaluated
Task II	–	Process Optimization and Specimen Fabrication
Task III	–	Material Characterization
Task IV	–	Design Allowables Type of Data
Task V	–	Structural Panel Performance
Task VI	–	Metal Matrix Reinforcement
Task VII	–	Crack Growth Characteristics
Task VIII	–	Creep-Rupture and Creep-Fatigue Analyses and Tests
Task IX	–	Aerothermoelastic-Plastic Modes of Failure
Task X	–	Probability of Structural Failure

Task XI – Initial Cost, Direct Operating Cost, and Value of a Pound-Saving Considerations

We have come a long way from selecting duraluminum for the DC-3 and only "stressing it" to a 1-g stress of 5,000 psi (References 4, 5 and 6). Spars or longerons could be shot away – or big holes in the skin could occur – and still the aircraft could return safely since the spar caps or longerons and partial skin could carry all the 1-g load separately and not fail.

Fortunately, the (infinite life) endurance stress was above 5,000 psi so we had no primary component fatigue failures. We can take credit for a great design, but perhaps we were also very lucky. The DC-3 payload and range was very low by today's standards. Today, the economics are pushing the payload-range and speed so that materials must be much more efficient. Therefore, all modes of failure need to be considered *before* the design is committed to final design and production. The 11 tasks outlined above are reminders of what has to be considered. Many of these tasks are discussed in detail in Reference 2. However, Tasks VIII to XI are added tasks for the high-temperature concepts and materials application. Therefore, in this paper, only some concerns will be addressed for these additional tasks.

The creep-rupture curve of Figure 36, of Reference 7, shows that Inconel 718 alloy will rupture at a little over 1,600°F with an 80,000-psi stress and exceeds the proportional limit (just begins to yield) at about 1,300°F. At 1,400°F, and at about 65,000 psi, after about 10 minutes, the percent strain increases very quickly to the critical strain growth rate that can be considered the failure rate (Figure 37 of Reference 7). If low percent creep is desired, e.g., 0.2 percent, then the short time tension creep shown in Figure 38 of Reference 7 should be considered. For Inconel 718, at 1,400°F, and 80,000 psi, maximum stress might be too high for a high Mach transport that would have a cumulative time at elevated temperature of more than 600 seconds at 1,400°F. If the ultimate strength is 120,000 psi, then the 1-g stress is 32,000 psi (120,000 psi divided by 3.75 g). If the density of Inconel 718 is 0.277 lb/in.^{3}, then the specific strength at 1-g, is 32,000 divided by 0.277 which is 115,523 psi. This compares to a DC-10 wing with a 1-g of 13,500 psi, at room temperature, and an aluminum density of 0.10 lb/in.^{3}, which results in a specific strength of 135,000 in. (at 1-g), which is 17 percent better than Inconel 718 at the elevated temperature. If the maximum aircraft dive upset is at 2.5 g and the maximum stress is at 80,000 psi, then these dives throughout the vehicle's lifetime should not accumulate more than 600 seconds (10 minutes) to be within the creep desired values. Note that after 200 seconds the creep is 0.45 percent elongation which is probably all that can be tolerated before the elongation becomes nonlinear (Figure 38 of Reference 7).

The above examples were discussed (using steel) to show how critical the creep limitations can be in selecting the specific strength and therefore the material selection at elevated temperature. The creep data for metal matrix composites using RSR titanium for the basic SPF/DB sandwich and bonded to face sheets with SCS-2, SCS-6, and/or SCS-8 continuous fiber in aluminum or titanium are not yet available in enough detail to compare to the more complete 6AL-4V data. However, separately, the RSR titanium and MMC continuous fibers in the aluminum and titanium matrices show high stability (less creep at elevated temperatures for the desired stress levels that yield efficient structure). This data, however, is very sparse. Reference 8 contains further discussions of the advances in technology of many other disciplines.

RECOMMENDATIONS AND CONCLUDING REMARKS

A research development test program should be planned to test the concepts and materials suggested above throughout the stress temperature-time spectra of the high temperature, high Mach environments anticipated. The testing should follow the procedures of Reference 2 with adequate temperature and time at heat soak with sufficient creep-rupture and creep-fatigue spectra testing. It is anticipated that the results will confirm the large economic benefits of the preliminary results shown in Tables 1 to 5.

In conclusion, the studies of the benefits of MMCs to the supersonic transport aircraft were investigated in Reference 3 and some of the highlights were presented in this paper. Large weight and cost benefits were obtained using advanced concepts and materials. The conclusions were based on analyses and tests for a limited temperature range. Potential benefits are shown for application up to Mach 4. Some of the concepts and materials investigated also show large potential for application to vehicles above Mach 4 for sustained supersonic and hypersonic flight. Further analyses and development testing are recommended with the concepts and materials suggested in order to reduce the risks and verify the potential benefits shown.

REFERENCES

1. W. T. Rowe and J. E. Fischler – Technology Study for Advanced Supersonic Cruise Vehicles Structures Technology, NASA CR172312. Prepared by the Douglas Aircraft Company, McDonnell Douglas Corporation, under Contract NAS1-16147, for Langley Research Center, January 1984.

2. J. E. Fischler, – Evaluation of Superplastic Forming and Co-Diffusion Bonding of Ti-GAl-4V Titanium Alloy for Supersonic Cruise Aircraft Research, NASA CR178006. Prepared by the Douglas Aircraft Company, McDonnell Douglas Corporation, under Contract NAS1-15527, for Langley Research Center, December 1985.

3. J. E. Fischler – Technology Study for Advanced Supersonic Cruise Vehicles Structures Technology – Studies of Metal Matrix Composites for Supersonic Aircraft. Final draft prepared by the Douglas Aircraft Company, McDonnell Douglas Corporation, under Contract NAS1-16147, for Langley Research Center. Final draft was submitted.

4. J. E. Fischler – Opportunities for Structural Improvements for Advanced Supersonic Transport Vehicle. Presented to NASA-sponsored Conference on Supersonic Cruise Research, Langley Research Center, Hampton, Virginia, on 13-16 November 1979, Douglas Paper 6868.

5. J. E. Fischler – Advanced Design Structural Considerations When Introducing New Materials and Construction Methods. Presented to First Canadian Symposium on Aerospace Structures and Materials, Canadian Aeronautics and Space Institute, Sutton Place Hotel, Toronto, Ontario, Canada, 14-15 June 1982, Douglas Paper 7183.

6. J. E. Fischler – Advances in Joining Techniques Used in Development of SPF/DB Titanium Sandwich Reinforced With Metal Matrices. NASA Conference Publication 2387, Welding, Bonding, and Fastening 1984. Proceedings of a symposium held at Langley Research Center, Hampton, Virginia, 23-25 October 1984.

7. J. E. Fischler, et. al. – Modes of Failure of a Hypersonic Re-Entry Glide Vehicle, Aerospace Engineering, Institute of the Aerospace Sciences, Volume 20, Number 9, Part 1, September 1961.

8. G. G. Kayten, C. Driver, D. J. Maglieri – The Revolutionary Impact of Evolving Aeronautical Technologies. NASA Headquartrs, Washington, D.C., and NASA Langley Research Center, Hampton, Virginia, presented at AIAA/AHS/ASEE Aircraft Design Systems and Operations Meeting, October 31-November 2, 1984, San Diego, California, AIAA-84-2455.

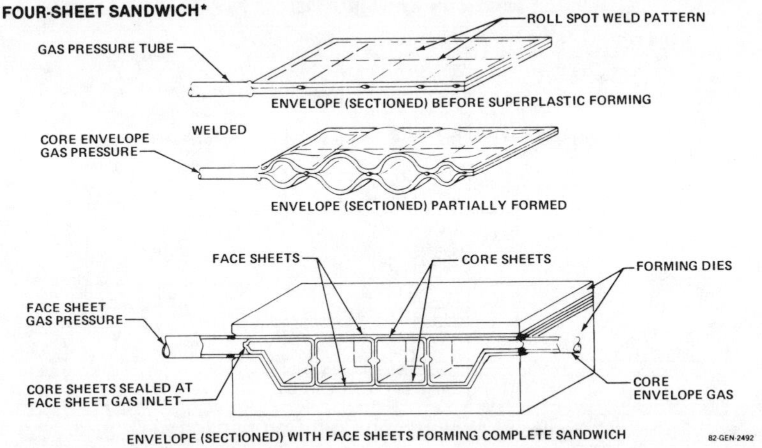

*MCDONNELL DOUGLAS PATENTED PROCESS

Figure 1. SPF/DB Phases of Fabrication

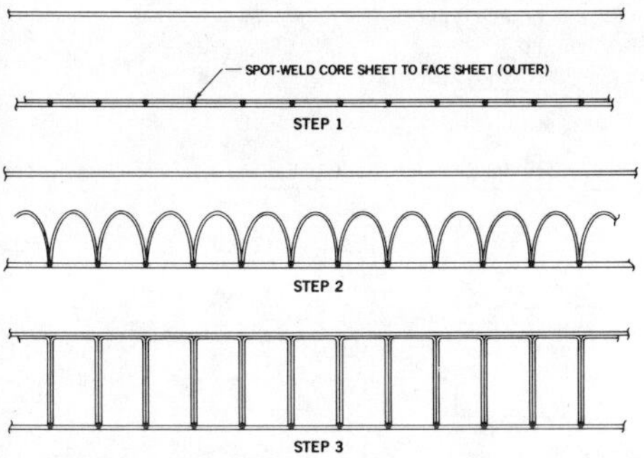

*MCDONNELL DOUGLAS PATENTED PROCESS

Figure 2. Three-Sheet Sandwich*

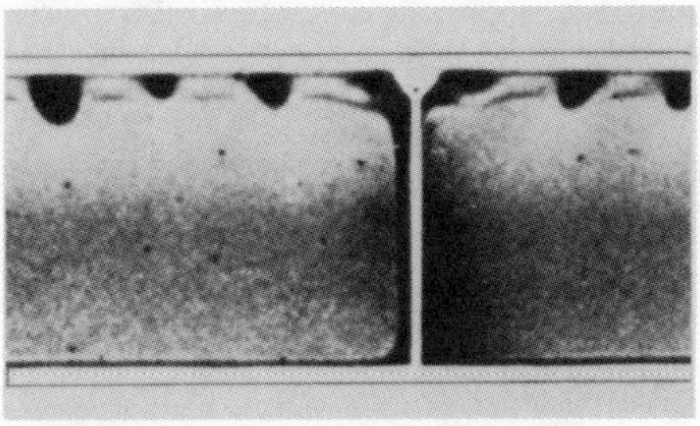

*MCDONNELL DOUGLAS PATENTED PROCESS

Figure 4. Three-Element Panel Section*

OPTIMUM GAUGES FOR MINIMUM WEIGHT

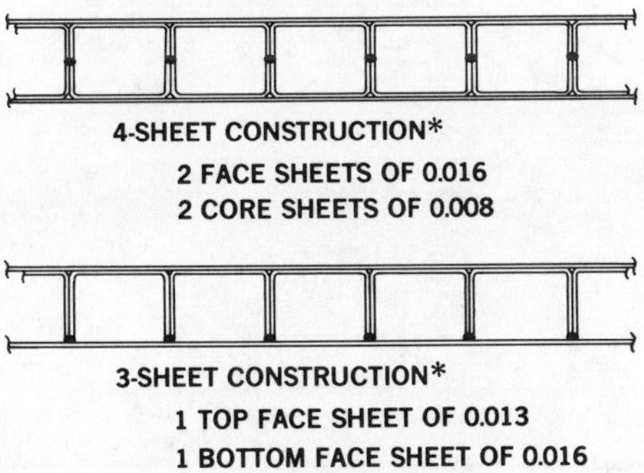

4-SHEET CONSTRUCTION*

2 FACE SHEETS OF 0.016
2 CORE SHEETS OF 0.008

3-SHEET CONSTRUCTION*

1 TOP FACE SHEET OF 0.013
1 BOTTOM FACE SHEET OF 0.016
1 CORE SHEET OF 0.008

*MCDONNELL DOUGLAS PATENTED PROCESS

Figure 3. Wing Panel Detail

Figure 5. Inner Skin Side, Forward Wing

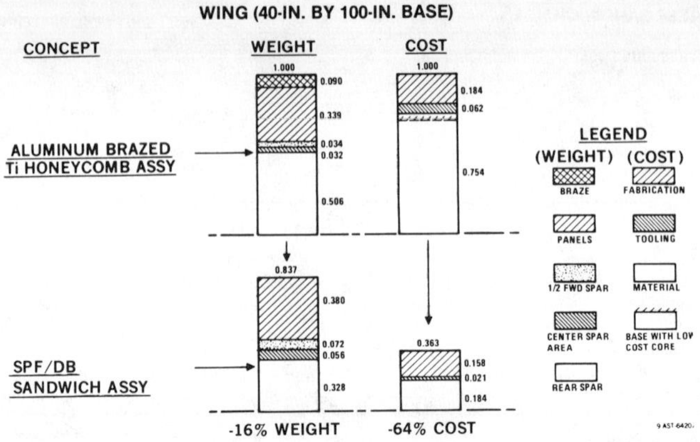

Figure 6. Weight and Cost Comparisons

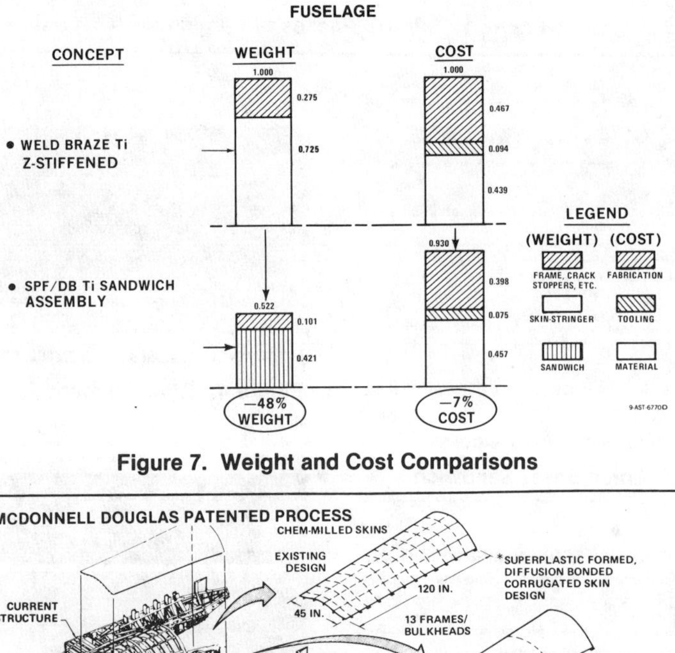

Figure 7. Weight and Cost Comparisons

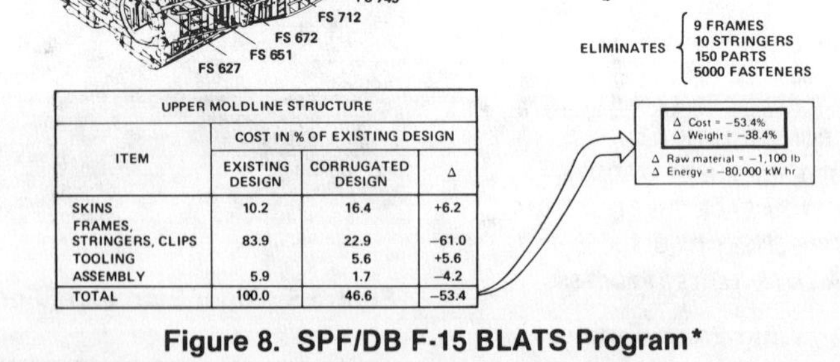

UPPER MOLDLINE STRUCTURE			
ITEM	COST IN % OF EXISTING DESIGN		
	EXISTING DESIGN	CORRUGATED DESIGN	Δ
SKINS	10.2	16.4	+6.2
FRAMES, STRINGERS, CLIPS	83.9	22.9	−61.0
TOOLING		5.6	+5.6
ASSEMBLY	5.9	1.7	−4.2
TOTAL	100.0	46.6	−53.4

Figure 8. SPF/DB F-15 BLATS Program*

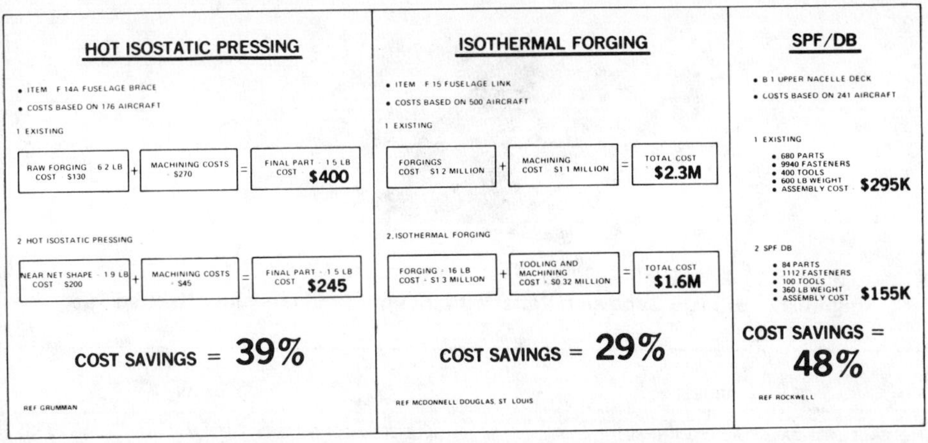

Figure 9. Cost Savings Using New Technology

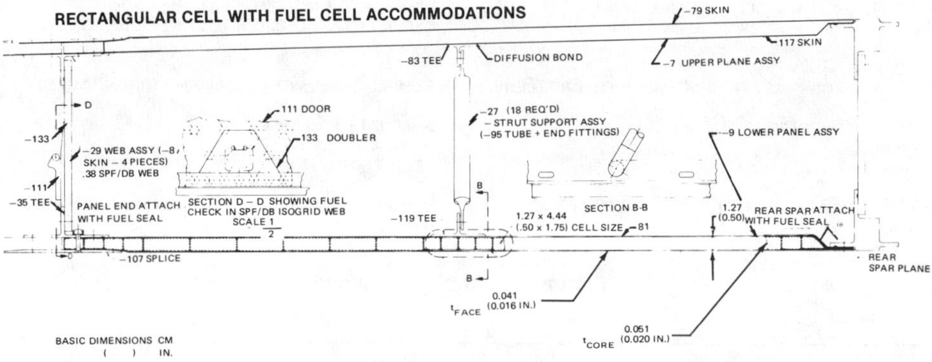

Figure 10. -501 Box Assembly — SPF/DB Sandwich Ti Concepts

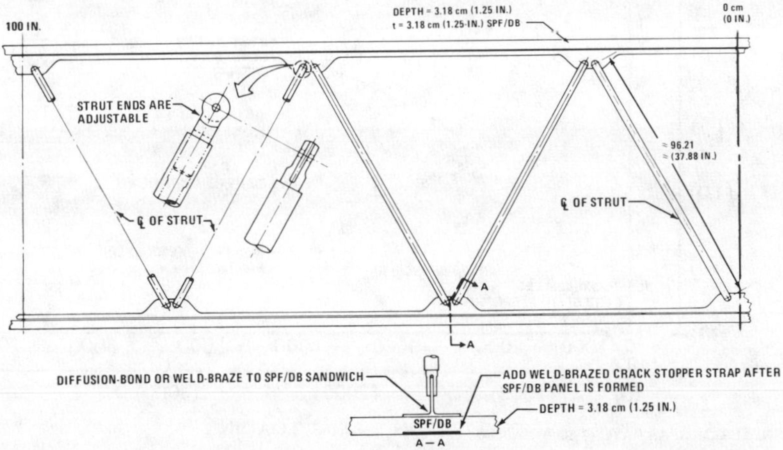

Figure 11. Wing Spar — Tube Struts Attached to Tee Spar Cap and SPF/DB Cover Panels

*MCDONNELL DOUGLAS PATENTED PROCESS

Figure 12. SPF/DB Sandwich Panel with Integral Doublers and Attached Tee*

PROBLEM	SOLUTION
NONUNIFORM MATERIAL THICKNESS	SELECTED CORE STOCK WITH MATCHING GAGE AND UNIFORM THICKNESS
TEMPERATURE NONUNIFORMITY	REVISED HEATER DESIGN TO MINIMIZE EDGE EFFECT OF TEMPERATURE
CORE DEPTH AND CELL CONFIGURATION	DEVELOPED TIME, TEMPERATURE, AND PRESSURE CYCLES TO ACCOMMODATE CONFIGURATION
LEAKS • EXCESSIVE STRAIN/STRAIN RATE BRITTLENESS	EXPERIMENTED WITH CYCLE AND MINIMUM GAGE
POOR FLOW THROUGH CLEAVAGE	ELIMINATED LEAKS
SURFACE CONTAMINATION	SELECTED CONTAMINATION-FREE STOCK
POOR AVAILABILITY OF PROPER MATERIAL SIZE — WIDTH AND GAUGE	MILL PROBLEM — WILL IMPROVE WITH INCREASED USAGE

Figure 13. Ultralight Sandwich

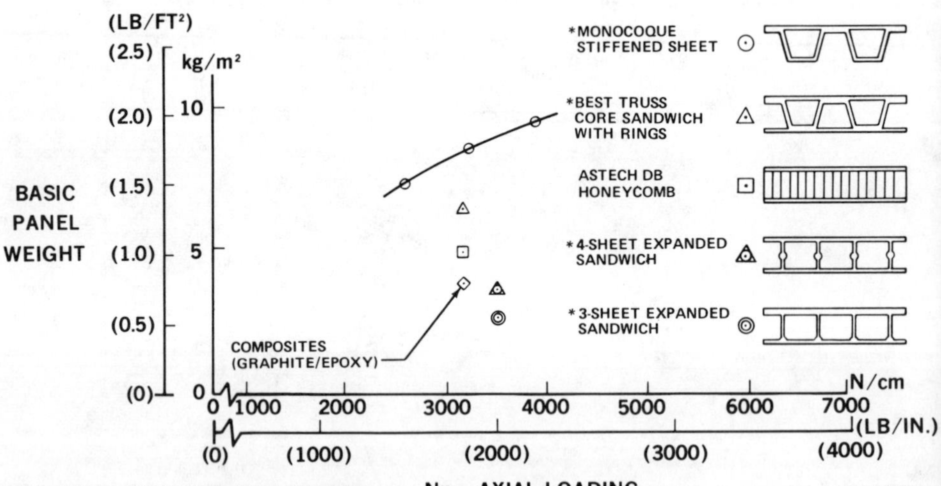

*MCDONNELL DOUGLAS PATENTED PROCESS

Figure 14. Structural Concepts

*MCDONNELL DOUGLAS PATENTED PROCESS

Figure 15. 3- by 5-Foot Wing Panel Prior to Testing*

*MCDONNELL DOUGLAS PATENTED PROCESS

Figure 16. SPF/DB Expanded Sandwich with Brazed Metal Matrix Composite*

*MCDONNELL DOUGLAS PATENTED PROCESS

Figure 17. Edgewise Compression Test Specimens*

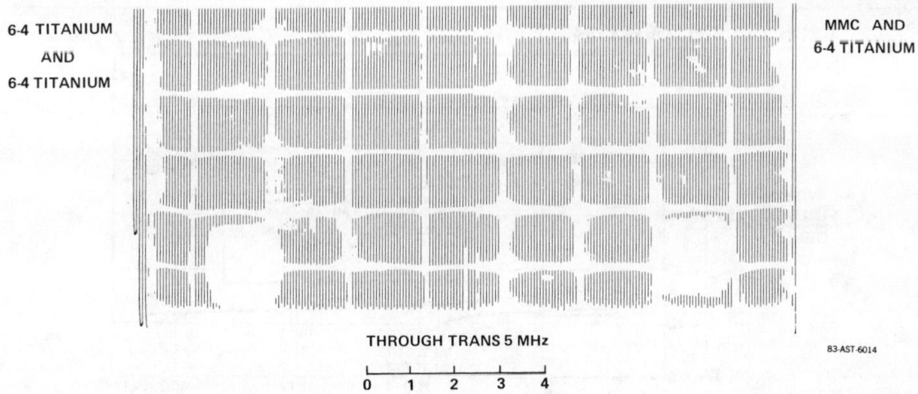

*MCDONNELL DOUGLAS PATENTED PROCESS

Figure 18. Ultrasonic Check of Brazing of MMC to SPF/DB Titanium Sandwich Panel*

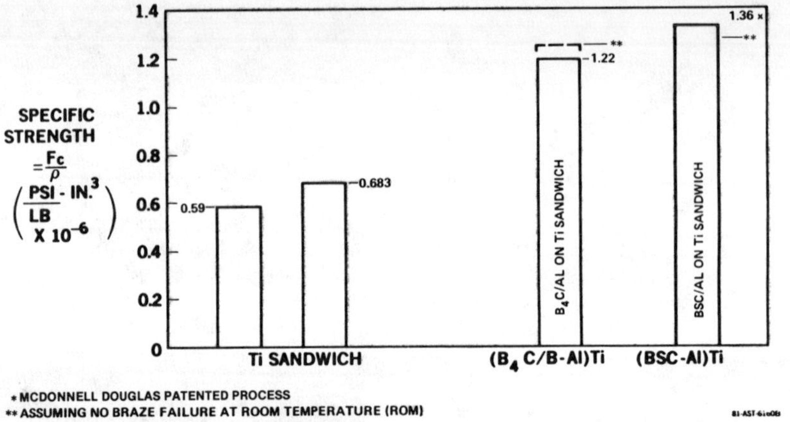

Figure 19. SPF/DB Titanium Sandwich and MMC* Test Results

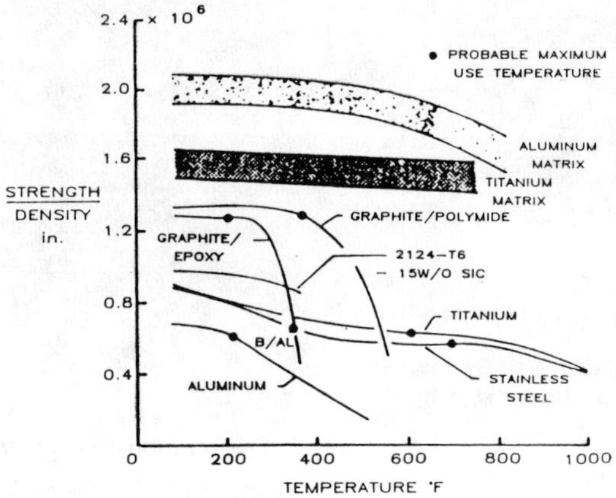

Figure 20. (From Reference 7) Estimated Specific Strengths of Various Structural Materials

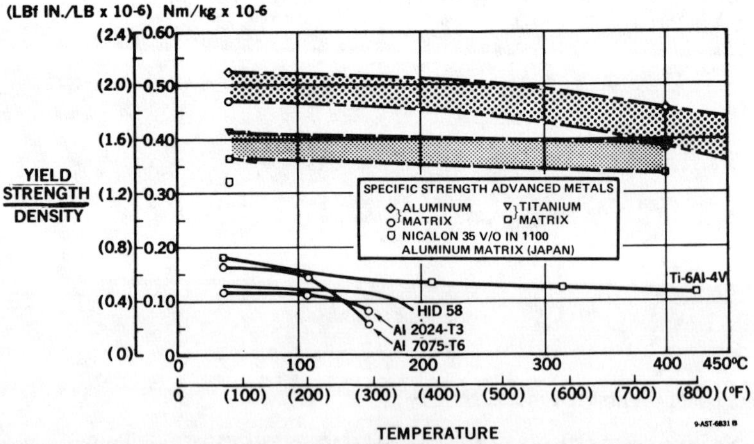

Figure 21. Specific Strength

CONCEPT XXV*
COVER SHEETS OF SCS-2/Al BRAZED TO A FOUR-SHEET SPF/DB RSP TITANIUM SANDWICH (RSP = RAPIDLY SOLIDIFIED TITANIUM [FROM POWDER])

CONCEPT XXVI*
COVER SHEETS OF SCS-2/Al IN A TITANIUM SHEATH OF 15-3-3-3 BRAZED TO A TWO-SHEET RSP TITANIUM SPF CORE

CONCEPT XXVII*
COVER SHEETS OF SCS-2/Al BRAZED TO A THREE-SHEET RSP TITANIUM SANDWICH

CONCEPT XXVIII*
COVER SHEETS OF SCS-6 FIBERS IN RSP TITANIUM WITH A TWO-SHEET RSP TITANIUM CORE DIFFUSION BONDED AFTER SPF

CONCEPT XXIX*
COVER SHEETS OF SCS-6 FIBERS IN RSP TITANIUM WITH A ONE-SHEET RSP TITANIUM CORE DIFFUSION BONDED AFTER SPF

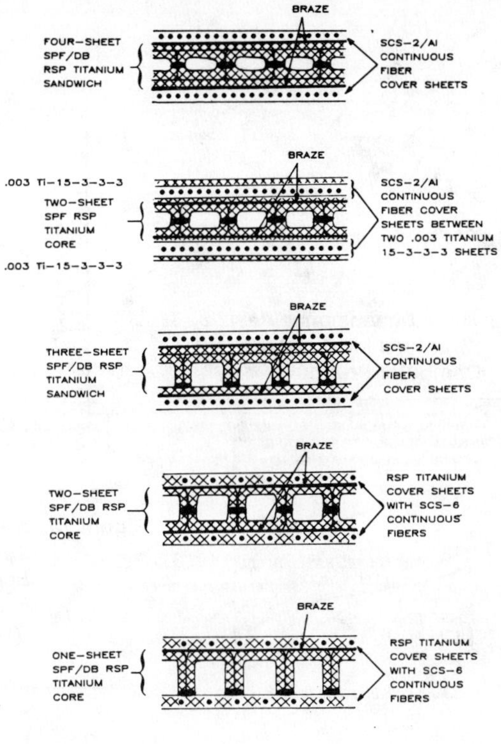

*MCDONNELL DOUGLAS PATENTED PROCESS

Figure 22. MMC Materials and Concepts with RSP Titanium Added

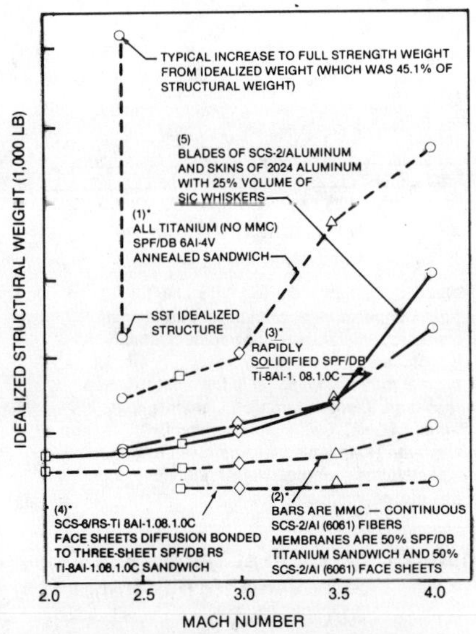

*MCDONNELL DOUGLAS PATENTED PROCESS

Figure 23. Five Structural Concepts versus Cruise Mach Number*

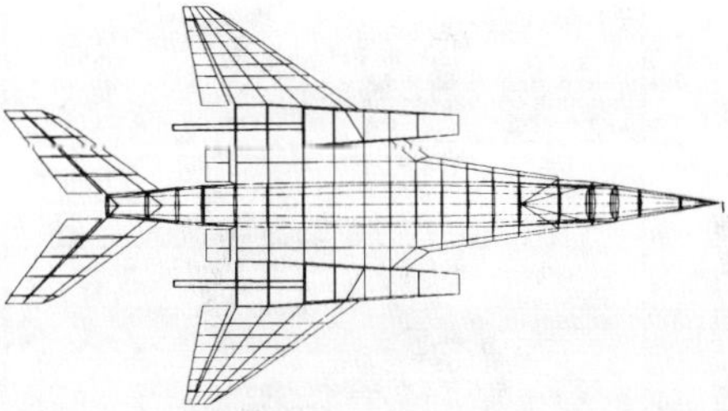

Figure 24. Plan View Structural Idealized Model — Computer Printout

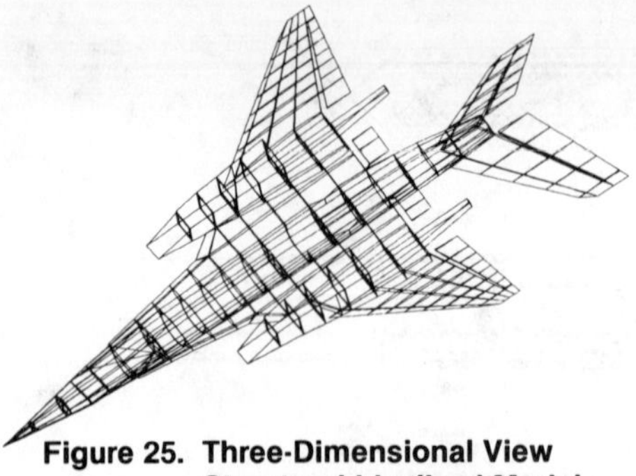

Figure 25. Three-Dimensional View Structural Idealized Model — Computer Printout

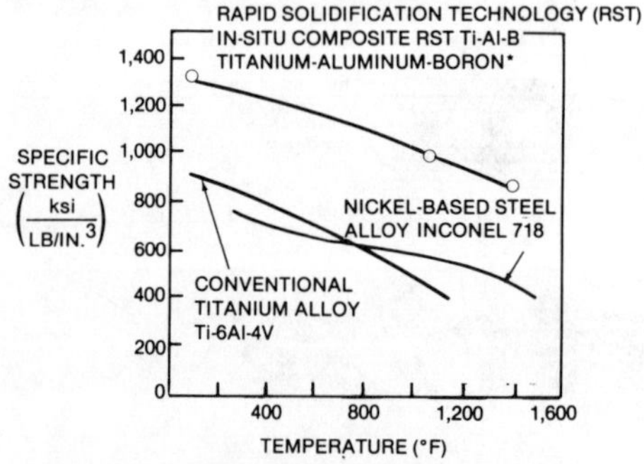

*MCDONNELL DOUGLAS PATENTED PROCESS

Figure 26. Rapid Solidification Technology (RST) Titanium Alloy for High-Temperature Applications

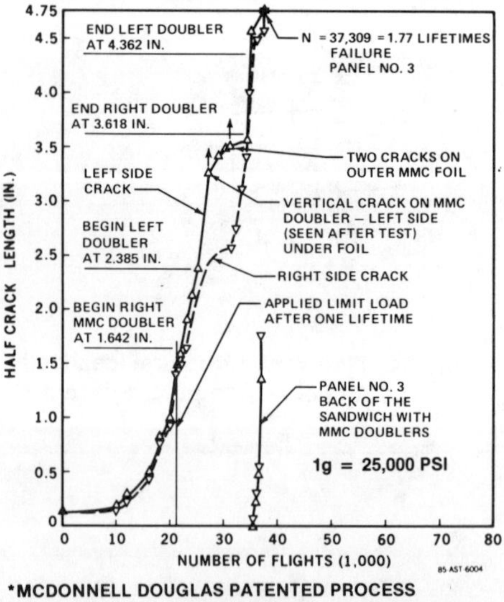

*MCDONNELL DOUGLAS PATENTED PROCESS

Figure 27. SPF/DB Titanium Sandwich with MMC Doublers*

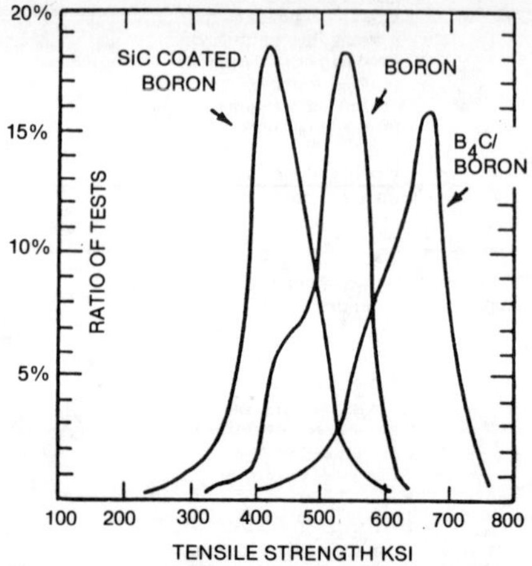

Figure 28. A Typical Histogram for the Strength of Boron, SiC Coated Boron, and B_4C/Boron Filament (Reference AVCO)

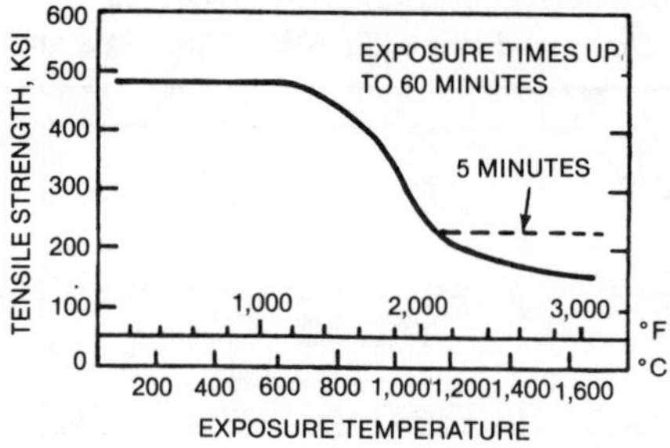

The SiC filaments are very effective for high temperature applications due to their intrinsic resistance to high temperature exposure and their lack of reaction with typical matrix alloys. Tests have established that high filament strength is maintained up to 1,200°C. SiC/Aluminum and SiC/Titanium laminate tests have demonstrated their effectiveness for high temperature structures, engines, etc.

Figure 29. Strength at Temperature for SiC Filaments (Reference AVCO)

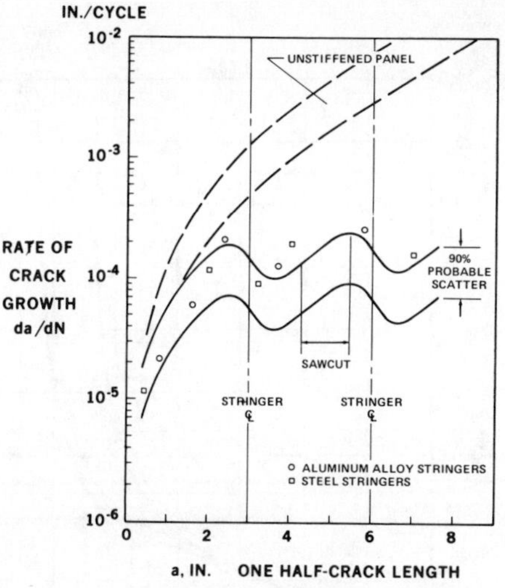

Figure 30. $\mu = 0.41$; **Aluminum Alloy and Steel Stringers Spaced at 3 Inches (Reference — POE)**

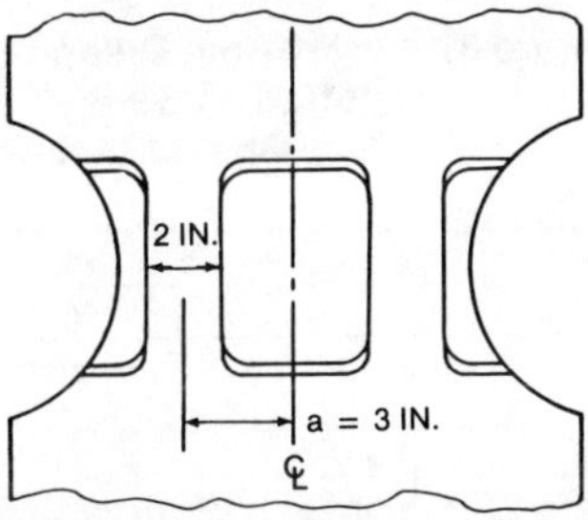

$$\mu = \frac{A_{ST}E_{ST}}{AE + A_{ST}E_{ST}}$$

$A = A_{SHEET} = 0.079 \times 6 = 0.474$

$A_{STRINGER} = 0.063 \times 2 = 0.126$

SHEET = 0.063 IN.
CORE = 0.016 IN.
Ti SPF/DB FACE SHEET THICKNESS = 0.079 IN.
(ASSUMED UNTIL MEASURED)

IF TITANIUM CRACK STOPPER $= \mu = \dfrac{0.126}{0.474 + 0.126} = \dfrac{0.126}{0.6} = 0.21$

IF MMC CRACK STOPPER $\mu = \dfrac{0.126 \times 27 \times 10^6}{0.474 \times 16.1 \times 10^6 + 0.126 \times 27 \times 10^6}$

$\mu = 0.3083$

Figure 31. Crack Stopper Stiffness Ratio for Titanium and MMC Crack Stoppers

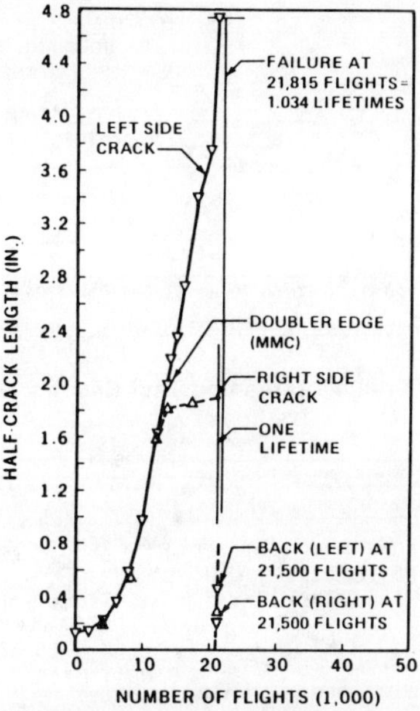

*MCDONNELL DOUGLAS PATENTED PROCESS

Figure 32. SPF/DB Titanium Sandwich with MMC Doublers*

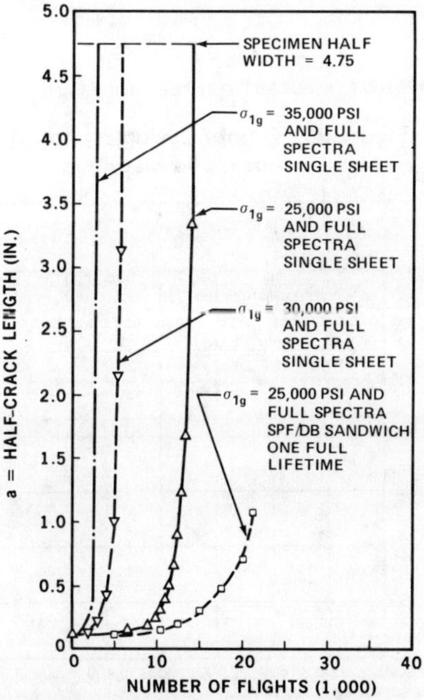

*MCDONNELL DOUGLAS PATENTED PROCESS

Figure 33. Crack Growth for Single Sheet (Cut from SPF/DB Titanium Sandwich*)

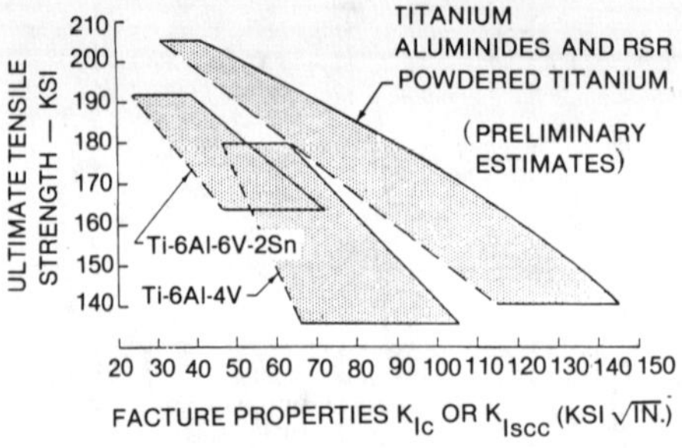

Figure 34. Screening Test Results — Ti Alloys

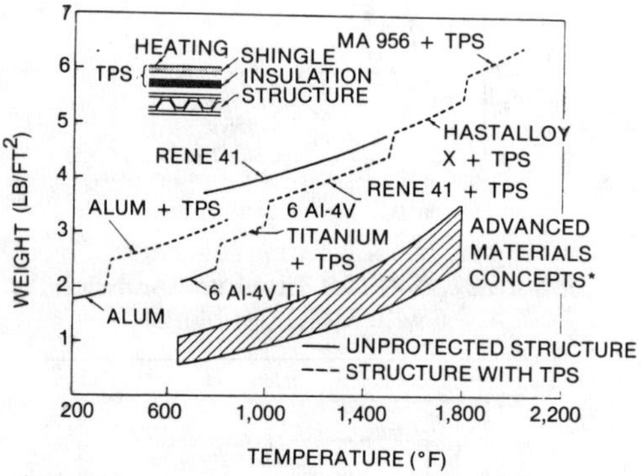

*MCDONNELL DOUGLAS PATENTED PROCESS

Figure 35. Weight Savings from Advanced Materials

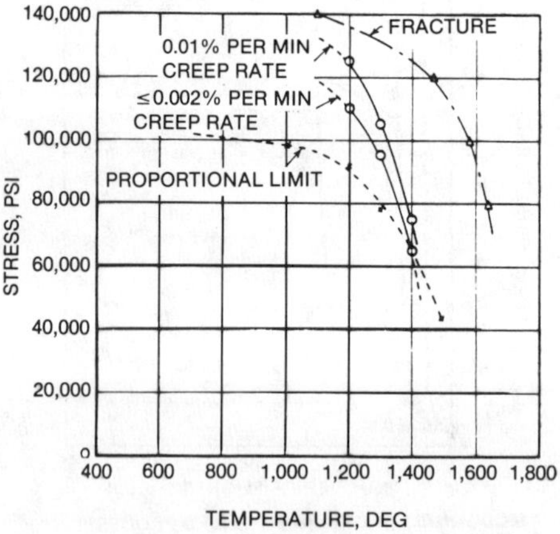

Figure 36. Approximate Lines of Constant Creep Rate

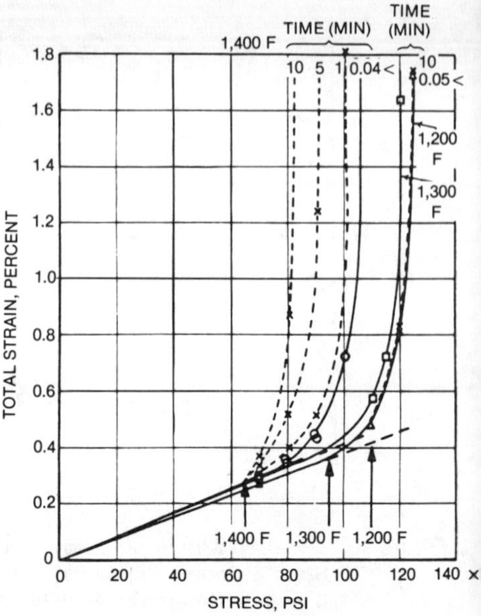

Solid lines — Strains on loading (ϵ_0) at recorded rate of 0.21 in. per in. per min.
Dotted lines — Isochronous stress-strain curves as indicated

Figure 37. Short-Time Creep Data Plotted on Stress-Strain Diagrams for Inconel 718

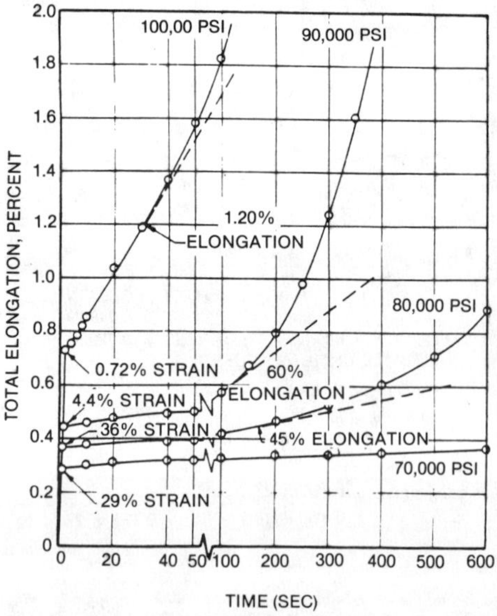

Figure 38. Short-Time Tension Creep Deformation of Inconel 718

Table 1
Idealized Structural Model versus Mach Number

NO.	CONCEPT	M=2.0	M=2.4	M=2.7	M=3.0	M=3.5	M=4.0
(1)*	ALL TITANIUM (NO MMC) SPF/DB 6Ai–4V ANNEALED TITANIUM SANDWICH	–	48,429	54,077	58,826	94,696	112,956
(2)*	BARS ARE MMC – CONTINUOUS FIBERS OF SCS–2/Ai (6061) IN AN ALUMINUM MATRIX. THE PANEL IS A 50% BY WEIGHT SPF/DB TITANIUM SANDWICH WITH FACE SHEETS OF SCS–2/Ai (6061) BRAZED TO THE SPF/DB TITANIUM SANDWICH	29,112	29,860	29,968	31,158	33,159	40,991
(3)*	RAPIDLY SOLIDIFIED SPF/DB Ti–8Ai–1.0B–1.0C SANDWICH	–	36,214	–	41,642	47,460	66,354
(4)*	SCS–6/RS Ti–8Ai–1.0B–1.0C FACE SHEET DIFFUSION BONDED TO THREE SHEET SPF/DB RS Ti–8Ai–1.0B–1.0C SANDWICH	–	–	25,028	25,248	25,658	26,366
(5)	BLADES OF SCS–2/Al AND SKINS OF 2024 Ai WITH 25% VOLUME OF SiC WHISKERS	33,668	34,606	36,730	39,554	47,690	80,698

*MCDONNELL DOUGLAS PATENTED PROCESS

Table 2
Total Structural Weights and Costs of Wing, Fuselage, and Tail for Concepts (1) and (2)

NO.	CONCEPT	M=2.0	M=2.4	M=2.7	M=3.0	M=3.5	M=4.0
(1)	(a) ALL TITANIUM (NO MMC) SPF/DB 6Ai–4V ANNEALED TITANIUM SANDWICH, TABLE I ÷ 0.451	–	107,381	119,905	130,435	209,969	250,457
	(b) COST IN DOLLARS PER POUND	281 →					
	(c) (a) × (b) = DOLLARS PER AIRCRAFT × 10^{-6} = CONCEPT (1)	–	30.17	33.69	36.65	59.0	70.38
(2) *	(a) BARS ARE MMC. CONTINUOUS FIBERS OF SCS–2/Ai (6061) IN AN ALUMINUM MATRIX. THE PANEL IS A 50% BY WEIGHT SPF/DB TITANIUM SANDWICH WITH FACE SHEETS OF SCS–2/Ai (6061) BRAZED TO THE SPF/DB TITANIUM SANDWICH. TABLE I ÷ 0.451	64,550	66,208	66,448	69,086	73,523	90,889
	(b) COST IN DOLLARS PER POUND	395 →					
	(c) (a) × (b) = DOLLARS PER AIRCRAFT × 10^{-6} = CONCEPT (2)	25.50	26.15	26.25	27.29	29.04	35.90
	(d) CONCEPT (2) – WEIGHT SAVING BASIS	WING, FUSELAGE, VERTICAL AND HORIZONTAL TAIL STRUCTURAL WEIGHT					
	(e) WEIGHT SAVING = (1)(a) – (2)(a)	–	41,073	53,457	61,349	136,446	159,568
	(f) VALUE OF A POUND SAVED ($/LB)	1240 →					
	(g) (e) × (f) = DOLLAR VALUE OF WEIGHT SAVED × 10^{-6}	–	50.9	66.29	76.07	169.19	197.9
	(h) INITIAL COST SAVING OF CONCEPT (2), DOLLARS PER AIRCRAFT AND (%) ADVANTAGE = (1)(c) – (2)(c) × 10^{-6} COMPARED TO TABLE II, (1)(c)	–	4.02 (13.3%)	7.44 (22.1%)	9.36 (25.6%)	29.96 (50.8%)	34.48 (52.4%)
	(I) TOTAL COST SAVING OF CONCEPT (2), DOLLARS PER AIRCRAFT × 10^{-6} = (g) + (h) = (LOWER INITIAL COST + LOWER OPERATING COST FOR ECONOMIC LIFETIME/AIRCRAFT)	–	54.92	73.73	85.43	199.15	232.38
	(J) TOTAL COST SAVING OF CONCEPT (2) COMPARED TO CONCEPT (1) BASELINE IN PERCENTS (2)(1) ÷ 1184.3 × 10^6 = DOC COST IN DOLLARS PER 3600 HOURS AND 20 YEARS – ECONOMIC LIFE	–	4.64%	6.23%	7.21%	16.82%	19.62%

*MCDONNELL DOUGLAS PATENTED PROCESS

Table 3
Total Structural Weights and Costs of Wing, Fuselage, and Tail for Concept (3)

RAPIDLY SOLIDIFIED SPF/DB Ti–8Al–1.0B–1.0C SANDWICH

NO.	CONCEPT	M=2.0	M=2.4	M=2.7	M=3.0	M=3.5	M=4.0
(3) *	(a) RAPIDLY SOLIDIFIED SPF/DB Ti–8Al–1.0B–1.0C SANDWICH = TABLE I ÷ 0.451	–	80,297	–	92,332	105,233	147,126
	(b) COST IN DOLLARS PER POUND	344 →					
	(c) (a) x (b) = DOLLARS PER AIRCRAFT x 10^{-6} = CONCEPT (3)	–	27.62	–	31.76	36.2	50.6
	(d) CONCEPT (3) – WEIGHT SAVING BASIS	WING, FUSELAGE, VERTICAL AND HORIZONTAL TAIL STRUCTURAL WEIGHT					
	(e) WEIGHT SAVING = (1)* – (3) *(1) FROM TABLE II	–	27,084	–	38,103	104,736	103,331
	(f) VALUE OF A POUND SAVED ($/LB)	1240 →					
	(g) (e) x (f) = DOLLAR VALUE OF WEIGHT SAVED x 10^{-6}	–	33.58	–	47.25	129.87	128.13
	(h) INITIAL COST SAVING OF CONCEPT (3), DOLLARS PER AIRCRAFT AND (%) ADVANTAGE = (1)(c) – (3)(c) x 10^{-6} COMPARED TO TABLE II, (1)(c)	–	2.55 (8.5%)	–	4.89 (13.3%)	22.8 (38.6%)	19.78 (28.1%)
	(i) TOTAL COST SAVING OF CONCEPT (3), DOLLARS PER AIRCRAFT x 10^{-6} = (g) + (h) = (LOWER INITIAL COST + LOWER OPERATING COST FOR ECONOMIC LIFETIME/AIRCRAFT)	–	36.13	–	52.14	152.67	147.91
	(j) TOTAL COST SAVING OF CONCEPT (3) COMPARED TO CONCEPT (1) BASELINE IN PERCENTS (3)(i) ÷ 1184.3 x 10^6 = DOC COST IN DOLLARS, PER 3600 HOURS AND 20 YEARS – ECONOMIC LIFE	–	3.05%	–	4.4%	12.9%	12.49%

*MCDONNELL DOUGLAS PATENTED PROCESS

Table 4
Total Structural Weights and Costs of Wing, Fuselage, and Tail for Concept (4)

SCS–6/RS Ti–8Al–1.0B–1.0C FACE SHEETS ON TITANIUM SANDWICH

NO.	CONCEPT	M=2.0	M=2.4	M=2.7	M=3.0	M=3.5	M=4.0
(4) *	(a) SCS–6/RS Ti–8Al–1.0B–1.0C FACE SHEETS DIFFUSION BONDED TO THREE SHEET SPF/DB RS Ti–8Al–1.0B–1.0C SANDWICH TABLE I ÷ 0.451	–	–	55,494	55,982	56,891	58,461
	(b) COST IN DOLLARS PER POUND	404 →					
	(c) (a) x (b) = DOLLARS PER AIRCRAFT x 10^{-6} = CONCEPT (4)	–	–	22.42	22.62	22.98	23.62
	(d) CONCEPT (4) – WEIGHT SAVING BASIS	WING, FUSELAGE, VERTICAL AND HORIZONTAL TAIL STRUCTURAL WEIGHT					
	(e) WEIGHT SAVING = (1)* – (4) *(1) FROM TABLE II	–	–	64,411	74,453	153,078	191,996
	(f) VALUE OF A POUND SAVED ($/LB)	1240 →					
	(g) (e) x (f) = DOLLAR VALUE OF WEIGHT SAVED x 10^{-6}	–	–	79.87	92.32	189.82	238.08
	(h) INITIAL COST SAVING OF CONCEPT (4), DOLLARS PER AIRCRAFT AND (%) ADVANTAGE = (1)(c) – (4)(c) x 10^{-6} COMPARED TO TABLE II, (1)(c)	–	–	3.83 (12.7%)	14.03 (41.6%)	36.02 (61.1%)	46.76 (66.4%)
	(i) TOTAL COST SAVING OF CONCEPT (4), DOLLARS PER AIRCRAFT x 10^{-6} = (g) + (h) = (LOWER INITIAL COST + LOWER OPERATING COST FOR ECONOMIC LIFETIME/AIRCRAFT)	–	–	83.7	106.4	225.84	284.84
	(j) TOTAL COST SAVING OF CONCEPT (4) COMPARED TO CONCEPT (1) BASELINE IN PERCENTS (4)(i) ÷ 1184.3 x 10^6 = DOC COST IN DOLLARS, PER 3600 HOURS AND 20 YEARS – ECONOMIC LIFE	–	–	7.07%	9.0%	19.07%	24.05%

*MCDONNELL DOUGLAS PATENTED PROCESS

Table 5
Total Structural Weights and Costs of Wing, Fuselage, and Tail for Concept (5)

BLADES OF SCS–2/Al WITH SiC WHISKERS, SKINS

NO.	CONCEPT	M=2.0	M=2.4	M=2.7	M=3.0	M=3.5	M=4.0
(5)	(a) BLADES OF SCS–2/Al AND SKINS OF 2024 Al WITH 25% VOLUME OF SiC WHISKERS TABLE I ÷ 0.451	74,652	76,732	81,441	87,703	105,743	178,931
	(b) COST IN DOLLARS PER POUND	200 →					
	(c) (a) x (b) = DOLLARS PER AIRCRAFT x 10^{-6} = CONCEPT (5)	14.93	15.35	16.29	17.54	21.15	35.786
	(d) CONCEPT (5) – WEIGHT SAVING BASIS	WING, FUSELAGE, VERTICAL AND HORIZONTAL TAIL STRUCTURAL WEIGHT					
	(e) WEIGHT SAVING = (1)* – (5). *(1) FROM TABLE II	–	30,649	38,464	42,732	104,226	71,526
	(f) VALUE OF A POUND SAVED ($/LB)	1240 →					
	(g) (e) x (f) = DOLLAR VALUE OF WEIGHT SAVED x 10^{-6}	–	38.0	47.70	52.99	129.24	88.69
	(h) INITIAL COST SAVING OF CONCEPT (5), DOLLARS PER AIRCRAFT AND (%) ADVANTAGE = (1)(c) – (5)(c) x 10^{-6} COMPARED TO TABLE II, (1)(c)	–	14.82 (49.1%)	17.4 (51.16%)	19.11 (52.14%)	37.85 (64.2%)	34.59 (49.14%)
	(i) TOTAL COST SAVING OF CONCEPT (5), DOLLARS PER AIRCRAFT x 10^{-6} = (g) + (h) = (LOWER INITIAL COST + LOWER OPERATING COST FOR ECONOMIC LIFETIME/AIRCRAFT)	–	52.85	65.1	72.1	167.09	123.28
	(j) TOTAL COST SAVING OF CONCEPT (5) COMPARED TO CONCEPT (1) BASELINE IN PERCENTS (5)(I) ÷ 1184.3 x 10^6 = DOC COST IN DOLLARS, PER 3,600 HOURS AND 20 YEARS – ECONOMIC LIFE	–	4.46%	5.50%	6.1%	14.1%	10.4%

1st International SAMPE Metals Conference
August 18-20, 1987

ELECTRON BEAM WELDING AND REPAIR OF CRITICAL STRUCTURES

R. H. Witt and A. C. Lofsten
Grumman Aircraft Systems
An Operating Division of Grumman Corporation
Bethpage, N.Y. 11714-3582

Abstract

Electron-beam (EB) welding provides unique capabilities for welding large, heavy-section structural parts in a single, full-penetration pass to produce near 100 percent structural properties in most alloys. In line with the Conference theme, this paper will present the reasons for and where EB welding should be considered competitive for past, current, and future applications in the aerospace and related industries.

1. INTRODUCTION

EB welding has been used at Grumman in production of F-14A titanium structures for more than 15 years, and over 25 miles of titanium EB welds have been accepted for flight on aircraft. Beside the F-14A applications, others developed at Grumman include titanium parts for swash plates on helicopters, missile shrouds, flap slat tracks, engine fan frame structures, Toronado wing-center-section boxes, Dassault 8A wing outer panels, and X-29 forward wing spars and ribs. Aluminum structures include a near-full-scale Army Personnel Carrier body and an ion chamber for a satellite application. Selected applications are highlighted in Sections 4 and 5. This paper describes the nature and characteristics of the EB process for welding titanium, compares EB with Gas-Tungsten-Arc (GTA) and other welding processes and discusses the Grumman facility and unique experiences with EB welding during selected applications. Inspection considerations and the significance of various joint configurations are assessed. Repair of defects and the effects on performance, overall economics of the process, and a summary of future potential based on past performance and new anticipated developments are presented.

2. THE NATURE AND UNIQUENESS OF EB WELDING

The EB gun (Fig. 1) essentially consists of an electrode, an anode, and a focusing coil. Magnetic focusing, oscillation, and deflection are provided. Welding parameters relating to the gun are:

- Voltage and Current - directly affects penetration in proportion of the voltage variation
- Focus - inversely affects penetration in proportion to the power density variation
- Welding Speed - distributes power over a given distance in a given time and, generally, inversely affects penetration.

Now, what then is EB welding? The Welding Handbook states that it "is a fusion-joining process in which the workpiece is bombarded with a dense stream of high-velocity electrons, and virtually all of the kinetic energy (energy of motion) of the electrons is transformed into heat on impact." The velocity of the electrons is about half the speed of light. This welding mechanism is used to join two or more pieces of metal by local melting which results in coalescence. Welds can be produced with or without the addition of filler metal.

How does an electron beam differ from an arc as far as the heating mode is concerned? In arc welding, as in most other conventional welding methods, a spot of heat is on the surface being welded; the heat spreads by thermal conductivity in every direction from that point.

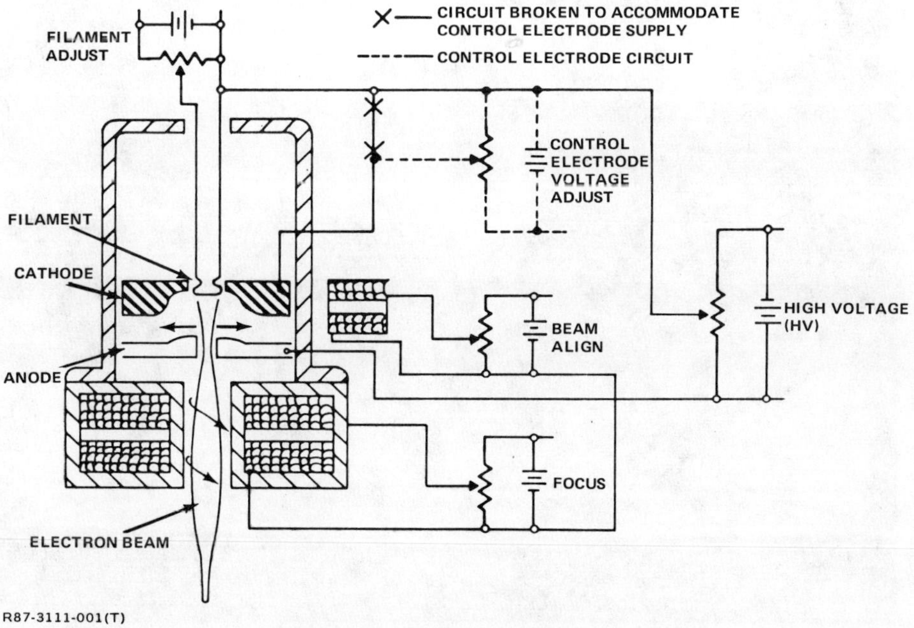

Fig. 1 Electron-Beam Gun and Associated Controls

The weld created under these conditions tends to be shallow and wide. Very low-power electron beams also produce such welds.

In a normal EB weld, a deep-fused area of high depth-to-width ratio is thought to result from the electron beam drilling a hole ("keyholing") by melting and vaporizing the metal bombarded by the electrons (fast-moving). The metal moves along the weld seam, the molten metal flows together by capillary action and freezes to form the weld. This explains why square, single-pass butt joints can be made in titanium thicknesses above 0.125 in., the potential limit for simple butt welding with most arc processes in titanium. Figure 2 depicts this rather strikingly. The 2-in.-thick GTA weld shown required over 50 weld passes to complete and a V-groove had to be machined in which to deposit the weld metal, whereas the EB weld in the same thickness was made in one pass with no filler metal on a simple butt joint with no groove machined in it. Note: When welding titanium alloys, the special filler metal required is very costly. Excessive distortion and hardness increases were also noted in the GTA weld, whereas these were absent in the EB weld and the EBW heat-affected zone rose very small.

Now that we have discussed some of the primary advantages of EB welding, we can proceed to look at the potential of the process in more detail, particularly in relation to joint design.

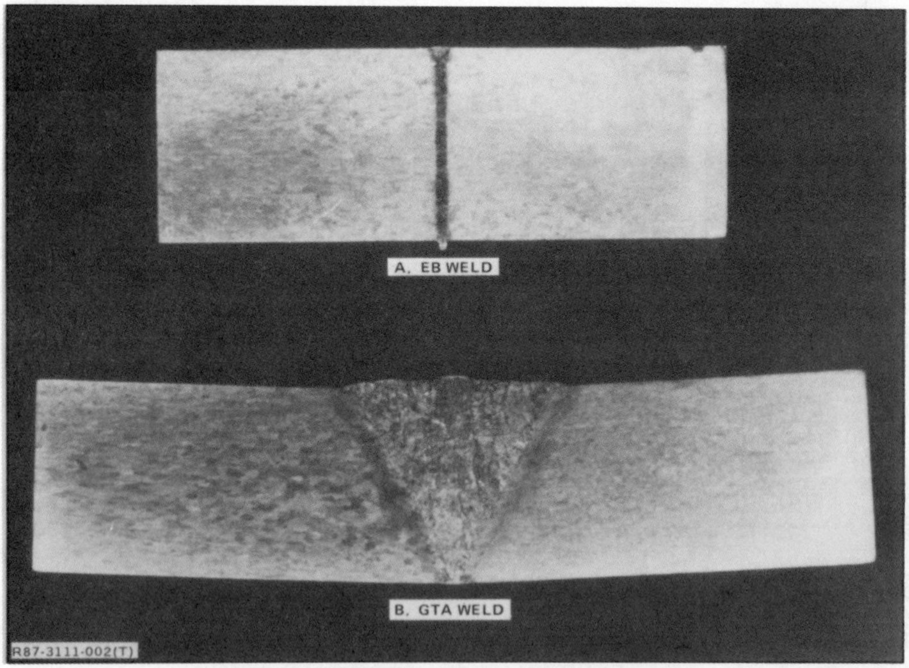

Fig. 2 Comparison of Welds in 2-In.-Thick Titanium Plate

EB welding can produce a weld with the highest depth-to-width ratio of all welding processes commercially available. It is also the only vacuum welding process which provides an ultrapure atmosphere in which to weld. Figure 3 shows the effect of pressure level on the purity of atmosphere and the relative frequency of gas molecule collisions. In a "hard" vacuum (10^{-4} to 10^{-5} torr); the electron beam is not affected much by gas molecules and can be focused over a distance of 2 feet or greater depending on gun design.

PRESSURE, TORR	GASES, PPM	NUMBER OF MOLECULES*	REL FREQUENCY OF COLLISIONS
10^{-5}	0.01	5,800	1
10^{-3}	1.3	580,000	100
10^{-1}	132	58,000,000	10,000
4×10^{-1}	500	---	---
760	---	4.4×10^{11}	100,000,000

*IN A CUBE 0.001 IN. (0.0254 MM) ON A SIDE (A VOLUME OF 10^{-6} CU IN. OR 16.4×10^{-6} CU MM)

R87-3111-003(T)

Fig. 3 Effect of Pressure Level on Purity of Atmosphere and Relative Frequency of Gas Molecule Collisions

The frequency of collisions (directly related to the concentration of gas molecules in the chamber) is still not significantly increased, even at these large distances, to noticeably reduce weld penetration at 0.01 to 0.1 torr. Above 100 torr, however, penetration rapidly deteriorates until, at 760 torr, gun-to-work distances are reduced to less than 0.5 in. to penetrate sheetmetal steels. This is very similar to most arc-welding-process arc length requirements. It is also generally necessary to employ high voltages of 100 to 175 kv to weld at atmospheric pressure. On the other hand, at levels of 100 kv in a vacuum, metals 10 or 12 in. thick have been penetrated.

The excellent purity levels of the "hard" vacuum assure that interstitial pickup of oxygen, nitrogen, and hydrogen in the weld is virtually eliminated. Hydrogen has been generally blamed for most of the problems experienced with titanium and steel weldments in the past, and only a high concentration of hydrogen in the base material or that resulting from improper joint preparation can cause problems.

Although purity of atmosphere has been generally cited as the main advantage of using EB welding, other advantages of vacuum welding have been ignored in the past.

Consider a weld joint that is greater than 1 in. thick and heavier on one side of the joint than on the other. How would you weld it to produce minimum weight, cost, shrinkage, and distortion? Figure 4 shows the design choices you would have between EB welding and the major arc welding processes presently available for production welding of heavy sections with a minimum number of weld passes. EB welding is the only current process that can produce the weld in the configuration denoted by the solid outline in Figure 4 without adding additional met-

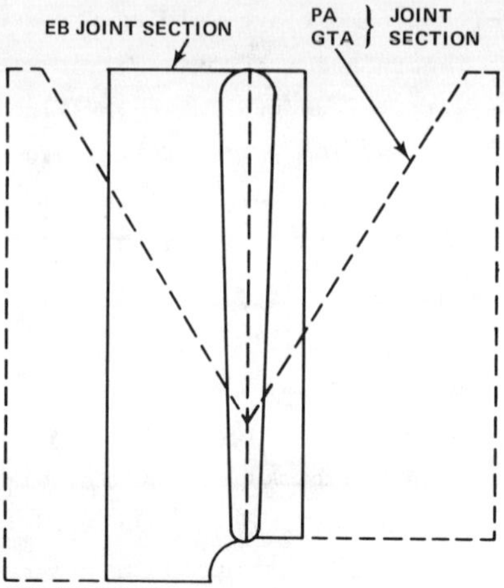

Fig. 4 Schematic Comparison of EBW Straight Butt Joint with V-Groove Joint Necessary for Arc Welding

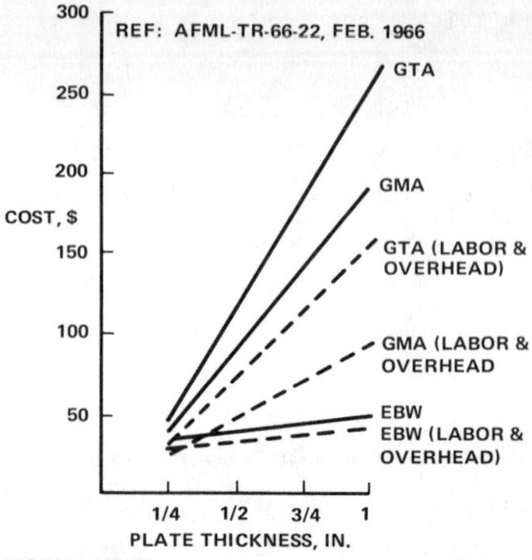

Fig. 5 Process Cost for Welding Maraging Steel (18 Ni-8Co-5Mo) 24-In.-Diameter Cylinder – Two Heads Per Cylinder

al in the joint area. A typical requirement of other processes for a welding groove is depicted by the dashed lines in Fig. 4, which show that the cross-section of the arc weld exceeds the area of the EBW by a factor of 2.5. Costs are also increased significantly when filler metal and shielding gases are required for welding such joints. Contamination is also a danger when welding out-of-chamber. The EB process is self-correcting in this respect, since the EB gun will automatically shut off when the vacuum pressure in the chamber exceeds almost 10^{-3} torr. Another great advantage of using EB butt welds is that the resultant joint is best for nondestructive inspection, particularly radiography.

Figure 5 shows that the major cost factors in most welding processes are labor and overhead. The filler metal and protective gas requirements, however, can also add very significant cost factors, particularly when welding joints greater than 1/4 in. thick.

The advantage of single-pass EB welding without filler metal is clearly demonstrated when welding thick sections in long lengths; for example, 2 in. thick by 15 ft long steel (Fig. 6). The EB welding time is 20 times faster than the best mechanized arc process and over 100 times faster than the manual arc welding process. Repeatability is also very consistent.

3. THE GRUMMAN EB WELDING FACILITY

For more than the last 17 years, Grumman has been using EB welding as a production method on the F-14A Naval Air Superiority Fighter Aircraft. Figures 7 and 8 show the wing center box and its location in the aircraft. Over 600 of these

MATERIAL: STEEL, 2" THICK X 15'-0" LONG			
PROCESS		JOINT DESIGN	H/FT
• MANUAL SHIELDED METAL ARC 5/32 ϕ ELECTRODE		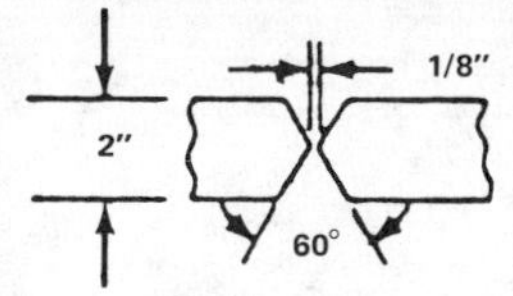	7.
• GAS METAL PULSED ARC VERTICAL POSITION	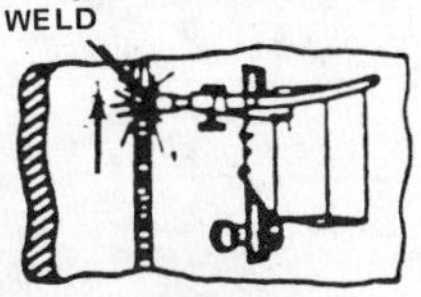	SAME AS ABOVE	2.67
• GAS METAL SPRAY ARC HORIZONTAL POSITION	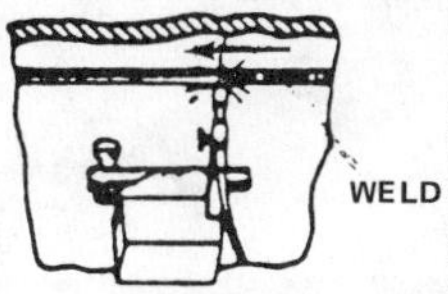	SAME AS ABOVE	1.33
• ELECTRON BEAM WELDING SLIDING SEAL VERTICAL - UP & HORIZONTAL POSITION	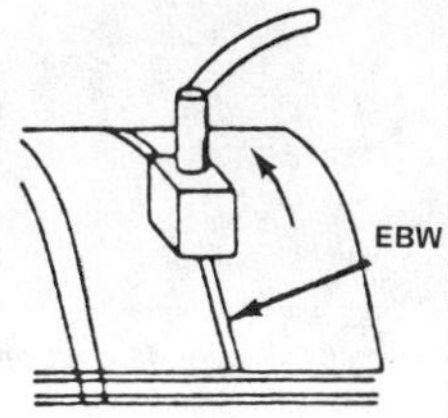	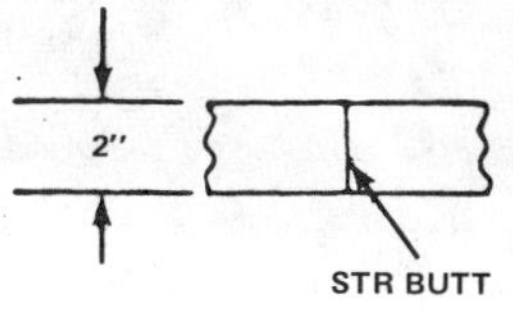	0.067

R87-3111-006(T)

Fig. 6 Comparative Welding Times for Various Processes

Fig. 7 F-14A Wing Center Section – EB Welding Top Covers

Fig. 8 F-14A Fuselage Assembly Showing Location of Wing Box

structures have been produced and accepted by Quality Assurance for actual use. This structure uses annealed titanium (Ti-6Al-4V), which has resulted in improved material utilization when coupled with EB welding to minimize weight and cost. It has been reported [(1)] that "the 2065-lb wing box is 19 lb lighter than its target weight and has become the second least expensive structure (in terms of dollars/lb) in the F-14A." In addition, the dynamic difference between the EB-welded and mechanically fastened structure is highlighted by estimates showing that bolting the upper and lower covers on the box (thereby eliminating welds) would result in a 20 percent increase in weight.

A decision in 1968 to purchase vacuum EB-welding production equipment and establish an EB Welding Center resulted from a consideration of various processes for producing titanium parts which comprise 24 percent by weight of the F-14A airframe structure. Design studies showed that titanium alloys were the most efficient on a strength-to-weight basis and provide best results on a life-cycle basis. The wing structure comprises the majority of the titanium used.

(1) Materials Engineering, Oct. 1973, p. 51

Originally, two large chambers were purchased, installed, and made operational in 1969. The facility investment was increased to over $10 million in 1971 when a third large chamber was purchased, installed, and made operational toward the end of the year. Later, the value increased even further with the addition of two medium-sized production chambers. This self-contained production facility includes controlled areas for welding, quality assurance, machining, heat treating, chemical cleaning, and forming operations.

The Bethpage, New York welding facility is located in an enclosed, positive-pressure, clean area (80,250 sq ft) in the southeast corner of a major production plant. Figure 9 shows the large EB welding chambers in the facility. This area includes a large production welding zone containing two tunnel welders, two side-loading chambers and one clamshell EB welder, additional space for loading and unloading each welder, and another large area where details and assemblies are stored and loaded into fixtures for cleaning and pre-fitting prior to welding.

Figure 10 lists the EB welding chamber capacities. All welding guns for the clamshell chamber and the side-loading S-1 and S-2 chambers are rated at 30 kW (60 kv, 500 ma). The T-1 and T-2 tunnel chambers are rated at 45 kW (60 kv, 750 ma). These chambers are being refitted with computerized Numerical Control (NC) equipment for special welding operations.

Tunnel Welder T-1 consists of a flow-through chamber, a tooling mo-

Fig. 9 Grumman EB Welding Facility Showing EB Welding Chambers Ranging from 25 Ft to 32 Ft in Length. Two 11.5-Ft Chambers Not Shown

EBW CHAMBER	CHAMBER SIZE, FT			APPROX CHAMBER VOLUME, CU FT	YEAR OPERATIONAL
	LENGTH	HEIGHT	WIDTH		
CLAMSHELL	32	8	10.5	2680	1969
T-1	25	11	9	2480	1969
T-2	30	11	9	2980	1972
S-1 & S-2	11.5	9	9	930	1975

ALL WELDING GUNS FOR THE CLAMSHELL, S-1 & S-2 CHAMBERS ARE RATED AT 30 kW (60 kV, 500 mA). The T-1 & T-2 CHAMBERS ARE RATED AT 45 kW (60 kV, 750 mA). THE T-1 CHAMBER IS BEING REFITTED WITH COMPUTERIZED NC EQUIPMENT FOR SPECIAL CONTROLLED WELDS.

R87-3111-011(T)

Fig. 10 EB Welding Chamber Capacities

bile carrier and track system, roughing and diffusion pumps, and an electron-beam gun mounted on a gantry. Access to the welding chamber area is provided through two end-opening doors. Loaded fixtures are mounted on the tooling carrier for transfer into and out of the chamber. Chambers T-1 and T-2 are separated by a continuous track; tracks are also extended out of the opposite ends of each chamber for loading and/or unloading or witness-line inspection. Provision was originally made at the exit end of the T-1 chamber for dimensional checks; however, this requirement was eliminated due to the exacting tolerances that are maintained.

4. TITANIUM WELDING FOR THE F-14A AIRCRAFT

Early tests had shown that EB welds in titanium alloys had excellent fatigue properties when all weld reinforcement was removed by machining, yielding essentially minimum-weight configurations (Fig. 11). It was then decided to design, fab-

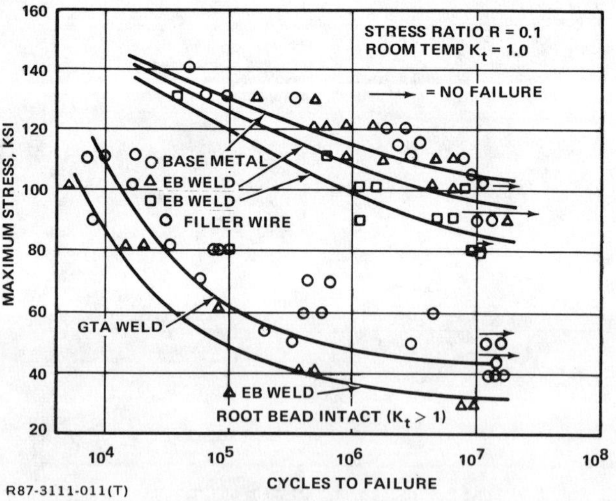

R87-3111-011(T)

Fig. 11 EB Weld and GTA Weld Fatigue Data

ricate, and test a torque box structure that was welded in a small development chamber. This structure was composed of 13 machined parts that were welded into two 26-in.-long modules; the latter were finally welded together at the centerline. Thickness ranged from 0.25 to 1.7 in. All single-pass welds were employed.

Figure 12 shows details of welding bulkheads (webs) into the modules at relatively long gun-to-work distances and the method of welding into corners where crossover defects are likely to occur. Relief holes were provided at the crossover weld areas, and end tabs were employed to prevent defects at start and stop areas. The 1.7-in.-thick girth weld joint and crossover is detailed in Fig. 12.

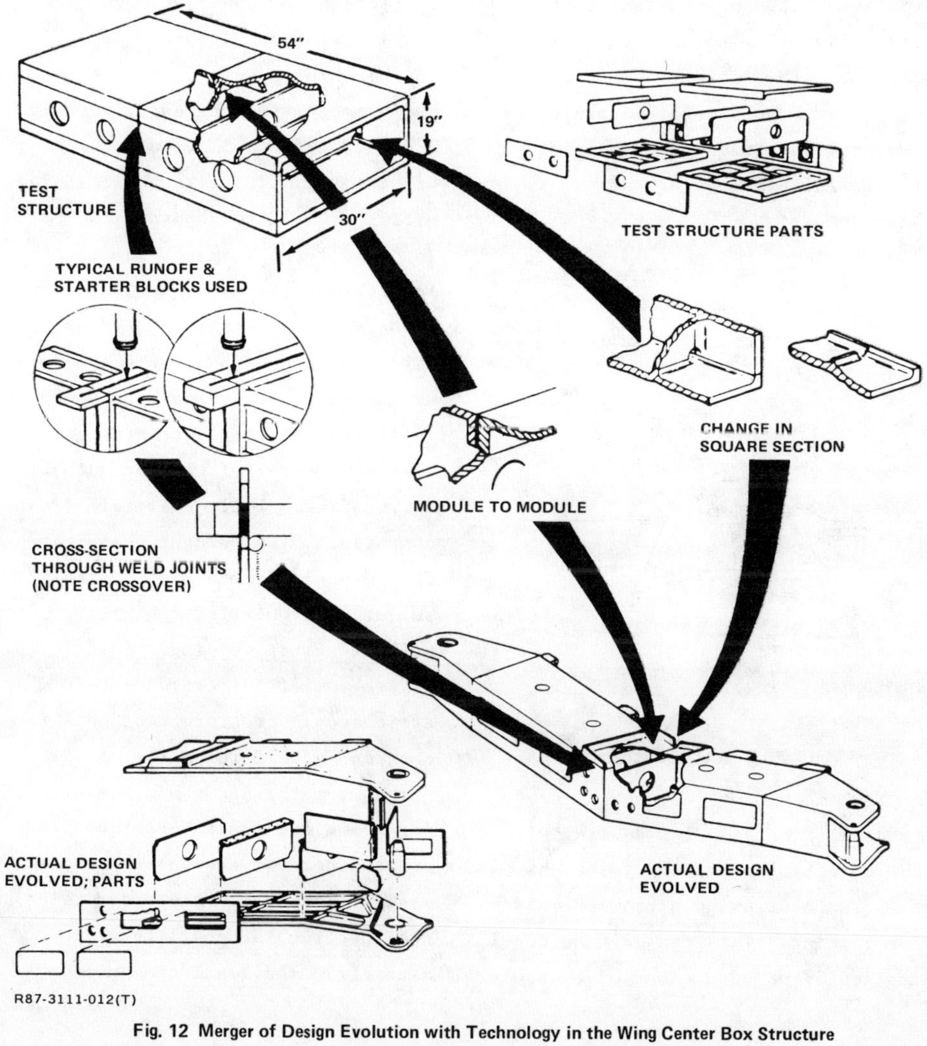

Fig. 12 Merger of Design Evolution with Technology in the Wing Center Box Structure

Following completion of joining and inspection, the torque box was fatigue-tested using a severe spectrum. After surviving the required life, loads were increased and the box ultimately failed at a location distant from any of the EB welds. The welds were re-inspected; the durability of the joint configurations was confirmed.

The actual wing center-section shown in Fig. 7 is more complex and essentially employs a weld land configuration similar to that shown in Fig. 13. This land provides the necessary material for machining and/or repair requirements prior to final finishing operations.

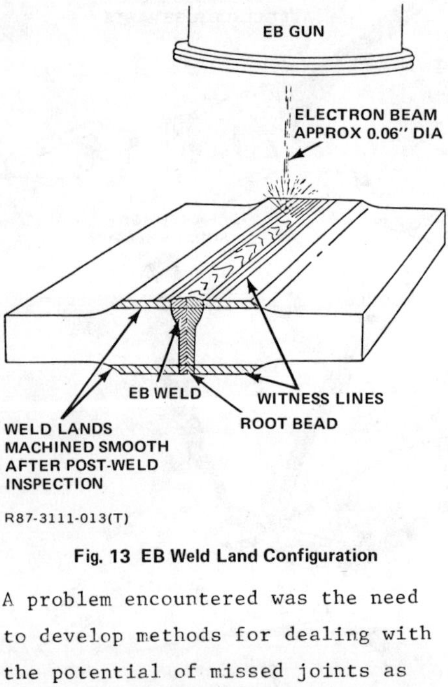

Fig. 13 EB Weld Land Configuration

A problem encountered was the need to develop methods for dealing with the potential of missed joints as the result of using narrow EB welds for joining. This can only be controlled by using production control specimens to confirm weld parameters and parallel-sideness of the EB shapes metallographically prior to welding production parts. Extensive work on this and the use of witness lines scribed on the top and root sides of the joint have virtually eliminated missed joints. It is required that at least 0.015 in. of weld be measurable on each side of the joint in all locations for acceptance. Joints not meeting visual inspection for undercut requirements or witness lines are rewelded prior to machining. Reweld studies were performed during the early work to assure the feasibility of rewelding without sacrificing mechanical properties or weld toughness.

The results of fatigue and fracture toughness tests were established to finalize the box design. Investigation showed that fatigue properties were reduced when the weld reinforcement was left intact or when stress relief was not used after welding. Fatigue efficiency is essentially 100 percent when reinforcement is removed and a stress relief at 1200°F for 4 hours is utilized after welding. Lower stress relief temperatures reduce efficiency. Fracture toughness requirements call for a K_{Ic} of 70.0 for base-metal and a minimum of 37.5 for EB-welded plate. Fatigue flaw growth tests were also run to establish critical flaw sizes for twice the spectrum fatigue life of the aircraft. The final design compensated for the reduced fracture toughness (Fig. 14).

The center wing box is essentially manufactured in modular construction from four basic boxes. Starting at the left end of the box, there is a LH outboard module, LH inboard module, RH inboard module, and RH outboard module. Each module is initially fabricated from its own bottom plate, side plates, end plates, and rib stiffeners. The top of the box is welded on last so that initial module joining includes inboard-to-outboard modules. The left side of the box is then joined to the right side at the midpoint transverse centerline.

All electron beam welds employ start and stop tabs integrally attached to the part, where possible, to avoid defects associated with starting and stopping of the weld. Integral tabs are essential, particularly for deep welds, to assure proper solidification and prevention of flaws near the finish end of the welds.

Upper and lower wing covers are fabricated using essentially the same procedure. However, the wing plank is hot-formed to a moderate curvature prior to welding. The initial weld in a typical wing cover is a relatively short one joining the pivot section to an actuator piece on the inboard cover. After removal

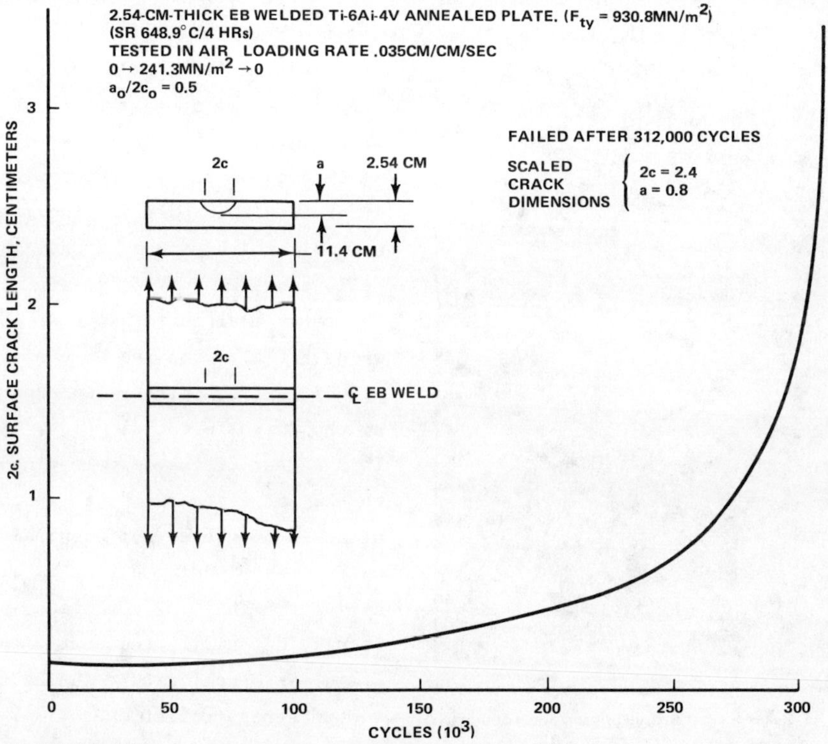

R87-3111-014(T)

Fig. 14 Constant-Amplitude Surface Flaw Growth Data

of end tabs and final machining, the inboard cover is welded to the outboard cover. A proximity control is required in making this weld to maintain constant gun-to-work distance. Figure 16 shows this weld being made and the related tooling required.

5. OTHER EB-WELDED COMPONENTS

In contrast to F-14A welding, fabrication of the Peace Keeper (MX) Missile Shroud (Fig. 15) represents EB-welding techniques closer to the other end of the spectrum. This assembly involves the joining of relatively thin sheetmetal titanium (6-4) skins and substructure in the range of 0.050 to 0.115 in. thick. The basic sequence of manufacture starts with hot-forming of the skins using male and female dies in a hot press. The skins are then rough-trimmed and then final-machined to the theoretical diameter of the conical sections.

Fig. 15 Peace Keeper Missile Shroud and Tooling

EB welding is done in the vertical-up position with the assembly and weld fixture mounted on a turntable. The welds are subsequently given visual, radiographic, and penetrant inspection to the requirements of AMS 2681.

The next operation is hot sizing of the conical skins by differential thermal expansion using a 300 series stainless steel mandrel. The cones are sized at elevated temperatures in a circulating-air oven for specified amounts of time. This sizing operation removes any slight distortion from the welding and produces a true cone.

The stiffener and mating rings are then EB-welded in place by welding through both skin and rings with a melt-through joint configuration. After completing all welding, the assemblies are vacuum stress-relieved. This not only improves the structural performance of the welds but also relaxes some of the pinching-effect distortion produced by the circumferential welds.

In the case of the lower mating ring on the aft (largest) cone, which is attached by two circumferential welds, the pinching effect is significant. This is overcome by again locally hot sizing the part with a stainless steel fixture during the vacuum stress relief operation.

Compiling and assembling the part involves the joining of 49 detail parts with 49 EB welds. The total length of weld per assembly is 430 ft.

Somewhere between the welding of heavy sections and sheetmetal, there are a number of applications where the structural efficiency of EB welding is apparent. One example is a helicopter swash plate (blade pitch control mechanism). In this case, EB welding has provided a means to produce a hollow ring forging with the required stiffness at much reduced weight.

Another example is the EB welding of the Fan Frame of the CF6 80C2 Engine (Fig. 16). This application requires the welding of 12 struts of

Fig 16 CF6 BOC2 Engine Fan Frame

the frame hub. The welds are made in the vertical-up position, again with the assembly and fixture mounted on a turntable. The detail parts are titanium 6-4 HIP castings. The most prominent feature of the assembly is that both sides of the strut are welded in one pass. This is primarily done for structural reasons because it minimizes distortion in the assembly but also has an obvious economic advantage. This part is inspected to the requirements of AMS-2631 and then vacuum stress-relieved. The stress-relief operation is accomplished in a fixture where minor changes in angular orientation of the struts and in the 8-ft diameter frame are corrected if required.

The HEAO ion chamber weld assembly (Fig. 17) was constructed of four major machined parts made of 2219 aluminum and a diaphragm of 0.040 in.-thick 2219 aluminum alloy sheet, chem milled to a thickness of 0.020, in a picture-frame configuration. No weld fixturing was required for the welding of the four corner joints. These 2-in-thick joints were made through false webs in the machined parts, where the details were bolted together with ½ in.-diameter 2219 aluminum alloy bolts. The EB weld was made through the joint (bolts and center components). After welding the false webs were machined away to blueprint configuration, leaving a vacuum-tight joint of reduced wall thickness (½ in.) The bladder weld

Fig. 17 HEAO Ion Chamber

involved a melt-through joint, welding through the 0.040-in. picture frame to the heavy frame.

The X-29A is primarily a materials, aerodynamics and avionics demonstrator. Although constructed mainly of graphite/epoxy composite skins, a role remains for metal in the structure. The forward wing beam is an EB-welded part made from five machined details and welded at the forward "kick". EB welds are made through solid sections 3½ in. thick and subsequently machined to the cross-section of the beam. Two of the main wing ribs are also of EB welded construction.

6. REPAIR WELDING

In addition to its primary use in the fabrication of aerospace structure, EB welding has complemented other welding processes in the repair of discrepant detail parts. This is particularly significant in titanium machined parts because of the inherently high material cost and the subsequent manufacturing costs. The same low distortion and high structural efficiency utilized in the design of welded structure are equally useful in salvaging discrepant parts, provided that good engineering information relative to fracture toughness and fatigue are available and the service requirements of the part are predictable. Even at relatively low production rates, thousands and even millions of dollars can be saved in weld repair. Added to this is the cost of production/schedule effects in keeping critical parts in normal manufacturing flow. EB welding is also suited for replacing metal by welding on another piece of metal, plugging mislocated holes or replacing a portion of a forging that has been severed by a misguided cutter in a machining operation.

EB welding is, however limited in instances where it is desirable to simply build up an undersize surface. Manual Plasma-Arc Welding (PAW) can be a versatile tool in repair welding. When used in repair of heavy titanium machined parts, it offers the flexibility of GTA welding with an advantage of less tendency to produce porosity and tungsten inclusions.

PAW repairs require the use of an inert gas chamber. If the parts are large, local chambers can be constructed to enclose only the damaged area (Fig 18). In all cases, it is essential to weld on a test block in the chamber and examine it for discoloration to ensure that the atmosphere is sufficiently pure to prevent contamination of the weld.

It can be estimated that nine out of ten discrepant titanium parts can be repair welded; however, the limiting factor is not welding but inspection. Accessibility of the repair weld to radiography and ultrasonic inspection are generally the controlling factor in deciding how to proceed with a repair action.

Fig. 18 PAW Repair of Titanium Component

7. CONCLUSION

It can be seen from the above examples that EB welding can be a very competitive process, particularly when welding expensive materials such as titanium alloys and advanced materials now being developed for use in the 1990's and the 21st century.

Plasma-arc repair welding can also save millions of dollars in salvaging discrepant machined parts or defective welds.

8. BIOGRAPHIES

Mr. R.H. Witt is a Principal Engineer in Manufacturing and Materials in Grumman's Advanced Manufacturing and Materials Engineering Department, where he has been employed for the last 24 years. He has contributed extensively to manufacturing and process development in the metallics regime. Mr. Witt's diversified experience includes fabrication of aluminum, titanium, steel, and high-temperature and refractory alloys by brazing, electron beam and laser welding, powder metallurgy, arc welding, rolling, and diffusion bonding. In addition, Mr. Witt has participated in the following projects at Grumman: Electron Beam Welded HIP Structures, Major Repair of Structure in an Orbital Environment, EB Welding of Titanium Wing Box Structure, Pulsed Arc Tube Welding of Stainless Steel and Titanium Alloys for Permanent Hydraulic Connections, Hot Isostatic Pressing (HIP) of Pre-Alloyed Powders, Weld Quality Definition, Sliding-Seal Electron Beam Welding of Aerospace Structures, Tooling for SPF/DB Titanium Fabrication, Laser Welding, and Plasma-Arc Repair Welding of Discrepant Machined Titanium Parts. Prior to joining Grumman, Mr. Witt was employed by Republic Aviation Corporation and Sylvania Electric Products Company on metallurgy, welding and brazing of high strength, high temperature materials. He is a member of SAMPE, ASM, AWS and Sigma Xi, and has presented more than 60 technical papers dealing with metallurgy, powder metallurgy and welding technology. He holds B.Met.E. and M.Met.E. degrees conferred by the Polytechnic Institute of New York.

Mr. A.C. Lofsten is Group Head of Welding Engineering in Grumman's Manufacturing and Materials Engineering Department where he has been employed the last 26 years. His experience is in welding of aerospace, marine, energy and transportation systems requiring the joining of low alloy high strength steels, titanium, aluminum and high nickel alloys. He has particular process experience with gas tungsten arc, electron beam, plasma arc welding, brazing and related non-destructive inspection techniques. Pertinent program experience includes the Lunar Model, F-14 aircraft, M-151/161 Hydrofoil and the Peace Keeper (MX) missile.

Mr. Lofsten is a member of ASM, SME and AWS. He served as Chairman of the Long Island Chapter of AWS in 1963 and received the AWS District Meritorious Service Award in 1983. He is a past member of the Aerospace Welding Committee and was principal contributor to two SAE Recommended Practices issued on electron beam welding and the inspection of welds. He also has chaired a Welding Workshop in an I.I.E. Conference on Plant Engineering and presented a paper on Electron Beam Welding of Structural Aircraft Components to a Conference on Small Accelerators at the University of W. Texas.

1st International SAMPE Metals Conference
August 18-20, 1987

JOINING PROCESSES USED IN THE FABRICATION OF TITANIUM AND INCONEL HONEYCOMB SANDWICH STRUCTURES

Brian Norris and Frank Gojny
Rohr Industries, Inc.
Chula Vista
California 92012-0878

Abstract

Lightweight metallic honeycomb sandwich structures have been used in high temperature aerospace applications for well over a decade. Sandwich structures typically have excellent strength and stiffness to weight ratios. In addition, these structures can be fabricated into sound attenuating ducts for use in today's quieter aircraft.

Titanium and Inconel 625 honeycomb sandwich structures play a very large part in Rohr's high temperature components product line. The titanium structures are fabricated using Liquid Interface Diffusion (LID) bonding, and the superalloy products are brazed using AMS 4777 Metglas foil.

LID bonding is a transient liquid phase bonding method combining the best features of both brazing and diffusion bonding. These features are exploited to produce titanium honeycomb sandwich products. A thin interlayer is used at the faying surface. This layer, when heated, reacts with the titanium to produce a liquid eutectic. The liquid bridges the components and allows metal diffusion to occur without the heavy pressure and rigid surface requirements of diffusion bonding. Eventual dilution of the interlayer component into the titanium finally results in isothermal solidification. High remelt temperatures and near base metal properties are produced with this technique.

Superalloy brazing is now being carried out at Rohr using braze foil which has replaced the old braze powder and binder method.

This paper will discuss the respective joining of titanium and superalloy sandwich structures. In addition, a recent process improvement, LID foil, which has greatly increased the strength and reduced the cost of the product, shall also be described.

1. INTRODUCTION

RohrBond® components are sandwich structures that comprise of titanium - 6Al-4V face sheets LID bonded to titanium honeycomb core, Figure 1. These structures allow the designer to combine the weight effective advantages of honeycomb with the high temperature strength and corrosion resistance of an all titanium structure. LID bonding combines the best features of both brazing and diffusion bonding. This technology has been successfully applied in the last 16 years to produce over 5,000 Pratt & Whitney F-100 RohrBond engine ducts, Figure 2, and over 350 large fan exit cases and nozzles for the Rolls-Royce RB211-535E4 engine. Furthermore, LID bonding has been used to fabricate various other titanium honeycomb sandwich structures for several other companies.

Over the last two years, the entire RohrBond production line has been improved through the incorporation of LID foil. Research into process improvement, however, is continually ongoing. This research has developed RADS (Rohr Activated Diffusion Spray) which is currently being used for some prototype components.

In addition to the progress made with titanium, considerable advances have been made in the brazing of Inconel 625 using Metglas braze foil instead of the conventional braze powder and organic binder. These developments in the fabrication of honeycomb sandwich structures will be described in the following sections.

2. THE DEVELOPMENT OF LID BONDING TECHNOLOGIES

LID bonding may be used to join most types of titanium structures. The prime reason, however, for developing LID was to allow thin gauge titanium honeycomb core to be bonded to titanium face sheets in order to produce a weight efficient titanium sandwich structure. Silver brazing, aluminum brazing and solid state diffusion bonding may also be used to join titanium structures. These techniques, however, suffer from one or more of the following disadvantages:

a. Weak bond strengths
b. Low remelt temperatures
c. Corrosion problems
d. High bonding pressure requirements

Contrasting to these disadvantages, LID has the low bonding pressure requirements needed for thin gauge honeycomb components coupled with a high remelt temperature, good bond strength and an absence of corrosion problems. LID is a eutectic bonding method (also known as transient liquid phase bonding, activated diffusion bonding and diffusion brazing) which relies on a thin copper-nickel interlayer at the faying surfaces. This layer, when heated in vacuum, reacts with the titanium to produce a Ti-Cu-Ni ternary liquid eutectic at 900°C (1650°F). This liquid bridges the components and allows metal diffusion to occur without the heavy pressure and rigid surface requirements of solid state diffusion bonding. Eventual dilution of the interlayer component into the titanium finally results in isothermal solidification. Solid state diffusion dilutes the copper-nickel further which results in near base metal-like properties and high remelt temperatures. A transformed β Widmanstätten structure is observed in the joint area due to the beta forming nature of the copper and nickel, Figure 3.

In the past, the copper-nickel interlayer was electro-deposited on the edge of the core in a lamellar fashion. Currently, RohrBond products are produced using a homogeneous, thin copper-nickel foil as the interlayer. The foil method has the advantage of allowing precise in-process quality controls of the interlayer thickness and composition. In addition, the foil method eliminates several complicated steps that are required in the plating process. Finally, as the foil is a homogeneous layer, it produces its available liquid all at once as soon as the ternary eutectic point is reached. This situation is an improvement over the stepwise formation of liquid which is produced from the lamellar electroplated method. Figure 3 shows the superior fillet produced by the LID foil method as opposed to the conventional electroplated method.

Apart from the LID foil advancement mentioned above, RADS is currently being used for some prototype components. RADS involves the application of the copper-nickel interlayer using a plasma-spray deposition method. This method has the advantage of eliminating the time consuming LID foil application. RADS is well suited for automation and is very versatile insofar as it allows the interlayer thickness to be easily varied depending on the application. RADS has been found to be very applicable for components incorporating very thin titanium gauge materials. RADS allows extremely thin copper-nickel interlayers to be deposited.

3. FABRICATION AND APPLICATIONS

A core sub-assembly is prepared complete with core and inserts, Figure 4. LID foil is tack welded to the face skins. The face skins are then assembled on either side of the honeycomb core.

The LID bond process depends on intimate contact between the core and face skins at the ternary liquid eutectic temperature. This requires stable tools and close tolerance parts. Delta-alpha tooling is used to provide bonding pressure. The differences in the coefficients of thermal expansion (delta-alpha) between the part and the tool are used to provide intimate contact at the liquid interface temperature. Bonding is carried out in a vacuum furnace at temperatures in the range of 900 to 980°C (1650 to 1800°F). Ultrasonic "C" scanning is used to nondestructively evaluate the part.

Actual and potential applications for the high strength, light weight, high temperature RohrBond sandwich structures are listed below:

. Engine Cases	. Bulkheads
. Thrust Reversers	. Exhaust Nozzles
. Fan and Compressor Vanes and Blades	. Hot Air Ducts
	. Sound Suppression Components
. Wing Sections	
. Insulation Panels	. Firewalls

4. THE DEVELOPMENT OF BRAZING SUPERALLOY HONEYCOMB SANDWICH STRUCTURES

Inconel 625 sandwich structures are used for hot end sound attenuating purposes. The structures usually comprise of an Inconel 625 honeycomb core brazed to solid and perforated Inconel 625 face sheets. In order to produce the correct acoustic attenuating properties, it is essential that the perforations do not become clogged with braze alloy. Figures 5 through 8 are typical of the brazed exhaust plugs, tailpipes and nozzles manufactured for the three major aero-engine manufacturers (Pratt & Whitney, General Electric and Rolls-Royce).

In the past, these components were brazed using a combination of braze alloy powder and organic binder. This combination was sprayed onto the honeycomb core.

During the braze cycle, the liquid braze would flow from the core and onto the face sheets in order to produce a braze fillet. This old method had the following disadvantages:

a. The distribution of the braze powder was hard to control. This produced poorly bonded joints in brazed-starved areas and blocked perforations in over-sprayed areas.

b. The capillary flow was blocked in areas of zero clearance producing poorly bonded areas.

c. The binder, which was baked-off during the braze cycle, deposited on the cold walls of the vacuum furnace. The resulting baked-off binder increased the maintenance of the vacuum chamber and the diffusion pumps.

All the above problems were eliminated entirely by changing over from raze powder to Metglas brazing foil MBF-20 (also specified as AWS BNi 2 and AMS - 4777B). This nickel based brazing alloy contains silicon and boron as melting point depressants. Normally, these elements embrittle the braze alloy through the formation of borides and silicides. This in turn prevents the production of braze foil using conventional rolling methods. Metglas Products have circumvented this problem by directly producing a metallic glass foil through the rapid quenching of a stream of liquid metal into a highly ductile ribbon. This supercooling bypasses the nucleation and grain growth stages to produce a very useful braze foil. The foil can be easily cut and tack welded to the face sheets.

During brazing, capillary action draws the melted braze foil to the honeycomb, thereby keeping the braze out of the perforations. This retains the required noise reduction qualities and provides highly consistent mechanical strengths. The quality of the RohrBraze product line has increased dramatically because of the introduction of the braze foil method.

5/ SUMMARY

Titanium and Inconel 625 sandwich structures have excellent high temperature strength and stiffness to weight ratios. Over the last 16 years, this has promoted the use of these structures in jet engines. During this time the technology has been continually upgraded.

These structures are ideally suited for hypersonic vehicles and their incorporation in such vehicles is envisaged in the future.

BIOGRAPHIES

Brian Norris is currently an Engineering Specialist in the Metals Research section at Rohr Industries. During his eight years at Rohr, he has concentrated on brazing and advanced metallics. He has also taught Titanium Metallurgy at Southwestern College. Previously, he worked at Rolls-Royce and at the General Electric Company. He received his batchelor's degree in Metallurgy at the University College of Wales, Swansea in 1973 and his M.S. degree in Materials Engineering at the University of Southern California in 1975.

Frank Gojny is a Senior Research Engineer in the Metals Research section of Rohr Industries. During the early part of his nine years at Rohr, he was assigned to a variety of areas, primarily in the field of metals fabrication. His work now concentrates on the fabrication of high-temperature structures for hypersonic vehicles and the development and implementation of novel structures and manufacturing methods. Previously, he was a Project Director at National Steel. Prior to this, he managed his own company.

He received his batchelor's degree at San Francisco State University in 1972.

Figure 1 RohrBond Titanium Structure

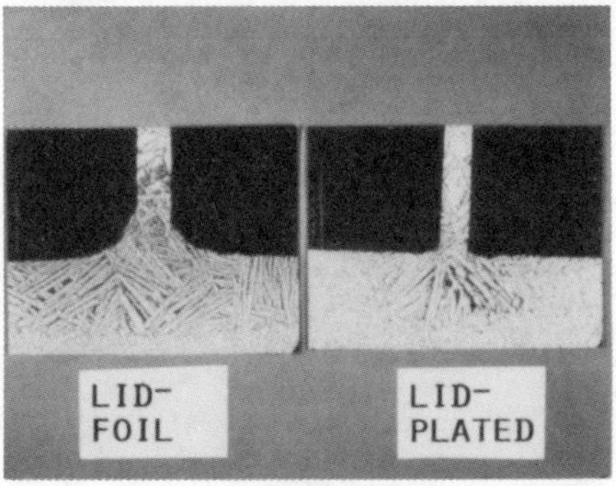

Figure 3 Micrographs of Fillets produced using the LID Electroplated and the LID Foil Methods.

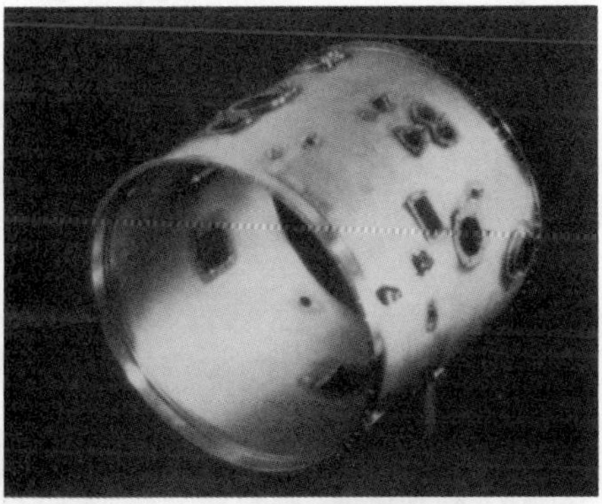

Figure 2 Pratt & Whitney RohrBonded F100 Engine Duct

Figure 4 Honeycomb Core Sub-Assembly

Figure 5 Exhaust Plug for Pratt & Whitney JT9D-7R4 Jet Engine

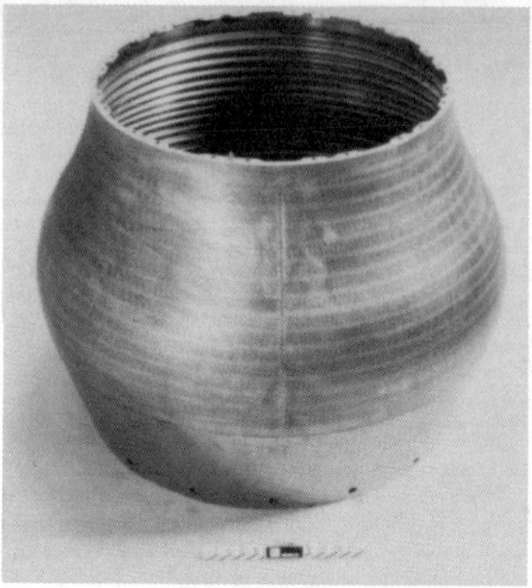

Figure 7 Exhaust Plug for CF6-80 Jet Engine

Figure 6 Tailpipe for Douglas DC-9-50 Aircraft

Figure 8 Outer Nozzle Ring for Rolls-Royce RB-211 Jet Engine

1st International SAMPE Metals Conference
August 18-20, 1987

SUPERPLASTIC FORMING/WELD-BRAZING OF Ti-6Al-4V PANELS FOR ENHANCED STRUCTURAL EFFICIENCY

Dick M. Royster
Thomas T. Bales
Randall C. Davis
NASA Langley Research Center
Hampton,Virginia

Abstract

Recent advances in the state of the art of superplastic forming (SPF) of certain metals have made it possible to fabricate design shapes that depart from past experience. Parts with intersecting compound contour surfaces can be designed and fabricated that would be impossible to fabricate with more conventional methods. With the SPF process, the designer also has considerably more freedom to design mass-efficient structures. The objectives of the NASA studies are to exploit the processing advantages of superplastic forming and joining by weld-brazing (WB) for the fabrication of titanium skin-stiffened structural components. The buckling strength of panels with stiffeners having configurations of a conventional hat shape is compared with panels having stiffeners of advanced shapes. Single- and multiple-stiffened panels are evaluated at room temperature. The SPF/WB panels with advanced shaped stiffeners developed up to 60 percent higher buckling strength than panels with conventionally shaped stiffeners. Selective reinforcement of the stiffeners using boron/aluminum metal-matrix composite material is also evaluated. Panels with selectively reinforced stiffeners developed buckling loads up to 49 percent higher than panels without selectively reinforced stiffeners. Experimental test results are in good agreement with structural analysis predictions for the panel buckling loads. Application of these advanced stiffener shapes offers the potential to achieve substantial weight savings in aerospace vehicles.

1. INTRODUCTION

Research on metals processing is being conducted at the Langley Research Center to develop improved forming and joining methods having the potential of reducing the weight and cost of future aerospace structures. The approach followed is to assess the state-of-the-art for fabricating a given structural system, define candidate methods for improved processing, evaluate the merits of each, fabricate and test subelement components, and then scale up the process to demonstrate validity.

This paper reports on the development and the state-of-the-art of superplastic forming and weld-brazing (SPF/WB) of titanium. Recent advances in the superplastic forming (SPF) of titanium have made it possible to fabricate design shapes that depart from past experience (refs. 1 and 2). Parts with intersecting compound contour surfaces can be designed and fabricated that would be impossible to fabricate with more conventional methods. With the SPF process, the designer also has considerably more freedom to design mass-efficient structures. The objectives of the NASA studies are to exploit the processing advantages of superplastic forming and joining by weld-brazing (WB) (ref. 3) for the fabrication of titanium skin-stiffened structural components. The buckling strength of panels with stiffeners having configurations of a conventional hat shape is compared with panels having stiffeners of advanced shapes. Single- and multiple-stiffened panels are evaluated at room temperature. Selective reinforcement of the stiffeners using boron/aluminum metal-matrix composite material is also evaluated. Experimental test results are compared with structural analysis predictions for the panel buckling loads. Application of these advanced stiffener shapes offers the potential to achieve substantial weight savings in aerospace vehicles.

2. DEVELOPMENT OF SUPERPLASTIC FORMING/WELD-BRAZING

2.1 Superplastic Forming of Titanium

Superplastic forming of titanium is a technology which has the potential to revolutionize the fabrication of many titanium components. An example of the superplastic behavior exhibited by titanium is depicted in Figure 1, which shows a specimen as machined and following superplastic forming. The as-machined specimen was heated to 927°C (1700°F) in vacuum, loaded in tension to a stress of 13.8 MPa (2 ksi), and strained to a total elongation of over 1000 percent. As shown, the specimen elongated uniformly without experiencing localized necking. Since the flow stress required for SPF is low, the use of gas pressure to blow form or superplastically form the material was developed.

The use of gas pressure to blow titanium into a mold was utilized to fabricate stiffeners in the development of this technology (ref. 4). A female mold and several of the male die inserts for the stiffener configurations evaluated are shown in Figure 2. The stiffener configurations selected for evaluation were based on their potential to offer increased structural efficiency in compression compared to conventional hat stiffeners. The configurations selected also exploited the increased formability of superplastic forming. Other than the conventionally shaped stiffener, the configurations would be difficult, if not impossible, to form using conventional titanium fabrication methods. The stiffener configurations that were evaluated consisted of a conventional shaped hat, a stepped-shaped web, a ribbed-shaped web, and a beaded-shaped web.

The fabrication of the stiffeners was accomplished by placing a sheet of 1.27mm (0.050 inch) thick titanium between the cover plate and the female mold as indicated in Figure 3. The mold was first sprayed with yttrium oxide to serve as a "stop off" and a lubricant. The female mold has a raised bead around the perimeter that the titanium sheet contacts to make a seal and an outlet to facilitate evacuation and purging with argon gas. Ports are also provided for the location of thermocouples inside the mold. An inlet was provided in the cover plate to facilitate pressurization of the cavity with argon gas.

The assembled tooling was placed between heated ceramic platens mounted in a press. Insulation was placed along the perimeter of the tooling to reduce heat loss. External pressure was applied by the press to maintain a seal between the cover plate, titanium sheet, and female mold as well as to react the gas forming pressure. When the titanium reached a temperature of approximately 745°C (1375°F), it was compressed over the raised bead on the mold to establish a seal. The platen temperature was then increased until the tool and titanium reached 927°C (1700°F) at which time the female side of the mold was evacuated. Argon gas pressure was then applied thru the cover plate to form the titanium stiffener. A gas pressure of 310 kPa (45 psi) was applied and held for 30 minutes. The gas pressure force generated a flow-stress for superplastic forming and the titanium was blown into the female mold cavity. The pressure of 310 kPa (45 psi) was empirically determined to provide a strain rate of approximately 10^{-4} in/in/sec. The pressure was then increased in 138 kPa (20 psi) increments with 5 minute holds until the forming pressure reached 862 kPa (125 psi). The pressure was held at 862 kPa (125 psi) for 20 minutes. The total time at 927°C (1700°F) during stiffener forming was 70 minutes. The titanium stiffener was cooled in the mold using argon gas.

The SPF stiffener was removed from the mold, cleaned, trimmed, and inspected (fig. 4). All stiffeners were formed by this method using the appropriate male die insert. These stiffeners were formed to the desired configuration and required no additional sizing or shaping following removal from the mold.

2.2 Weld-Brazing

Weld-brazing of the titanium compression panels consisted of spot-welding the SPF stiffener to the 1.27mm (0.050 inch) thick titanium skin and then brazing. The panel components and the 0.406mm (0.016 inch) thick 3003 aluminum braze alloy are shown on Figure 5. Following chemical cleaning, two rows of four spotwelds each were used to attach each flange of the stiffener to the skin. Welding parameters were developed so that weld nugget expansion established a faying surface gap of 50.8μm to 76.2μm (0.002 to 0.003 inches) between the flanges of the stiffener and the skin. Braze alloy strips were placed adjacent to the joints to be brazed and the panel then placed in the brazing furnace as shown in Figure 6. The spotwelds were sufficient to maintain alignment and no tooling was required for panel brazing.

Brazing of the panel was accomplished in a vacuum furnace at a pressure of 1.33 mPa (1×10^{-5} torr) and a temperature of 677°C (1250°F). Upon melting, the 3003 aluminum braze alloy was drawn into the faying surface gap by capillary action. The temperature of 677°C (1250°F) was maintained for 5 minutes. The panel was allowed to cool to room temperature before being removed from the furnace. A superplastically formed/weld-brazed (SPF/WB) panel following brazing is shown in Figure 7. Visual inspection and radiography verified good wetting between the stiffener and skin and establishment of a sound, integral joint.

A typical cross-section of a weld-braze joint is shown in Figure 8. The braze is shown to have drawn through the gap to form a fillet between the face sheet and the inner radius of the stiffener. The photo-micrograph shows good integrity of the joint in the vicinity of the weld nugget. The resulting weld-braze joint produced a hermetically sealed, high strength joint.

3. STRUCTURAL DESIGN AND PROCESS VERIFICATION

3.1 Panel Test Procedure

Following brazing the SPF/WB panel was trimmed on the edges and ends using a conventional band saw. After trimming the panel was potted on both ends using an epoxy potting compound, and a finished panel is shown in Figure 9. Potting material was used to facilitate grinding the ends of the panel flat, parallel to each other, and perpendicular to the skin. The finished panel was 254mm (10 inches) long and 121mm (4.75 inches) wide. The potting also served to prevent local failure of the ends during compression testing.

The SPF/WB skin-stiffened structural panels were tested in end compression using a 1.3MN (300 kip) hydraulic testing machine. The edges of the specimens were supported with knife edges positioned 6.35mm (0.25 inch) from the edge. Relative motion between the upper and lower heads of the testing machine was measured using linear variable differential transformers (LVDT). Foil strain gages were attached to the stiffener and skin to measure local strains. Strain reversal was used to define onset of elastic buckling. Data were recorded every two seconds until two-thirds of the predicted local instability load was achieved and every second from there to maximum load. All tests were conducted at a load rate of 890 N/S (12,000 lb/minute).

3.2 Panel Performance

Three panels were fabricated and tested at room temperature from each of the four stiffener configurations (conventional shaped hat, stepped-shaped web, ribbed-shaped web, and beaded-shaped web). The buckling strengths of the complex shaped stiffener configurations were compared with those of the conventional shaped hat stiffener. The data in Figure 10 show that the panels with the complex shaped stiffener configurations were capable of carrying 50 to 60 percent more load than the panels with the conventional shaped hat stiffener. All of the panels tested were of equal weight and cross section. The improvement in structural efficiency was attributed to the advanced structural shapes that were formed by the superplastic forming process.

3.3 Selective Reinforcement of Single Stiffener Panels

The structural efficiency of titanium skin-stiffened compression panels has been increased significantly through the use of advanced forming and joining technology and by selective reinforcement using metal-matrix composites. Metal-matrix composites such as boron/aluminum have a high specific strength and stiffness in the direction of the filaments and can be attached to titanium panels by weld-brazing to improve the load-carrying ability of compression members. Each of these technologies was incorporated into the fabrication of Ti-6Al-4V skin-stiffened compression panels with complex shaped stiffeners. The panels were tested and the results compared with the results from unreinforced panels.

The specific compression buckling strengths of superplastically formed and weld-brazed (SPF/WB) hat-stiffened panels which were selectively reinforced with boron/aluminum are compared with unreinforced hat-stiffened panels in Figure 11. The use of selective reinforcement increased the strength of the panel with the conventionally shaped hat stiffener by 35 percent. Combining the selective reinforcement and the more efficient superplastically formed beaded-shaped web stiffener or ribbed-shaped web stiffener increased the strength of the panels by 49 percent and 46 percent respectively over the strength of the unreinforced panels of the same stiffener configurations. These results suggest a possible approach to improve the load carrying capability of titanium structure.

3.4 Multiple Stiffener Panel Fabrication and Testing

To further assess the potential of SPF/WB for fabricating titanium compression panels, the SPF/WB process was scaled up to fabricate and test multiple stiffener panels (ref. 5). A tool having multiple inserts was fabricated to superplastically form a multiple stiffener or corrugated sheet from a single piece of titanium using the same parameters and processes established for the small single stiffener panels described previously. Four titanium multi-stiffener panels were successfully fabricated using the SPF/WB technique. Three panels had conventionally shaped hat stiffeners and one panel had beaded-shaped web stiffeners. The SPF/WB panel [559 mm x 762mm (22 in. x 30 in.)] fabricated having nine conventional shaped hat stiffeners is shown in Figure 12. Following inspection the panels were trimmed, the ends were potted, and the panels were machined to the proper length. The panels were instrumented with strain gages and tested at room temperature to failure in end compression in a hydraulic test machine. None of the weld-braze joints in the panels failed or separated during the tests (fig. 13).

The beaded-shaped web stiffened panel buckled at a load of 1.19MN (268.5 kips). The buckling loads from the tests showed the beaded-shaped web stiffened panel had an increase of 13 percent in local buckling load over the conventionally shaped hat stiffened panels. Maximum loads for the panels were limited by the yield strength of the material and the conventional shaped hat stiffened panels failed at a 6 percent higher load than the beaded-shaped web panel.

3.5 Analysis and Fabrication of Advanced Structural Concepts

The recent phase of the LaRC superplastic forming program was begun by conducting an analytical study on structural efficiency of advanced panel design concepts. The results of analysis on these configurations are shown on Figure 14. In the mass coefficient, $\bar{t}/L$, $\bar{t}$ is the mass-equivalent thickness for the corrugated cross section and L is the panel length. In the load coefficient, Nx/EL, Nx is the compressive load per unit width, E is the elastic modulus and L is the panel length. The mass-strength data are dimensionless and are, therefore, independent of the materials used to make the panels. The mass-strength data for 1 percent honeycomb sandwich is a standard of comparison for low mass structures. It represents the desired goal in mass efficiency and is considered just beyond the current state-of-the-art in manufacturing technology. As Figure 14 shows, some advanced configurations show mass-strength efficiencies better than 1 percent honeycomb. These configurations can be fabricated with current state-of-the-art superplastic forming techniques. Processing of panels having light weight, sine wave webs with minimized thinning in the flanges is an important fabrication goal. This goal has lead to the development of new tooling for superplastic forming and a new processing procedure (ref. 6) to explore the possibility of fabricating panels similar to those shown in Figure 14.

Two tooling concepts for superplastic forming stiffeners are shown in Figure 15. The full-hat male-die concept was utilized to form the stiffeners discussed previously in this report. After forming a stiffener, the thickness of the titanium in the crown of the stiffener was equal to the approximate thickness of the titanium sheet prior to forming. The resulting webs and flanges of a stiffener had variable thicknesses. The half-hat die concept, which required lower forming strains, was developed to reduce metal thinning across the web and to develop equal thicknesses in the flanges. In the half-hat die concept only half of a hat segment was superplastically formed in one operation. Figure 16 shows a superplastically formed part before and after trimming. During superplastic forming of the half-hat segments, the thickness of the titanium material was reduced 16 percent from the original thickness in the highest strained regions.

The half-hat process was utilized to fabricate both single-stiffener panels and the multi-stiffener panel shown in Figure 17. The half-hat segments with beaded-shaped webs were joined to caps by the weld-braze process. The process offers the opportunity to fabricate full size multi-stiffener panels limited only by the size of the brazing furnace.

Three panels of each configuration were tested in compression to failure at room temperature. The experimental buckling load was within 3 percent of the calculated theoretical load for both panel configurations. They failed at compressive loads approaching the yield strength of the titanium materials. The tests verify that the superplastic forming process can be used to achieve panel components having advanced shapes and that the weld-brazing process was a highly effective technique for joining panel elements.

4. CONCLUDING REMARKS

The two titanium processing procedures, superplastic forming and weld-brazing were successfully combined to fabricate titanium skin-stiffened structural panels. Stiffeners with complex shapes were superplastically formed using simple tooling. These stiffeners were formed to the desired configuration and required no additional sizing or shaping following removal from the mold. The weld-brazing process by which the stiffeners were attached to the skins utilized spot-welds to maintain alignment with no additional tooling required for brazing. The superplastically formed/weld-brazed panels having complex shaped stiffeners developed up to 60 percent higher buckling strengths than panels with conventional shaped hat stiffeners.

The superplastic forming/weld-brazing process was successfully scaled up to fabricate full size panels having multiple stiffeners. The superplastic forming/weld-brazing process was also successfully refined to show its potential for fabricating multiple-stiffener compression panels employing unique stiffener configurations for improved structural efficiency. Combining selective reinforcement with the complex shaped stiffeners increased the strength of the resulting panels up to 49 percent over unreinforced panels of the same stiffener configurations. These results suggest a possible approach to improve the load carrying capability of titanium structure.

5. REFERENCES

1. Bales, Thomas T., ed.: SPF/DB Titanium Technology. NASA CP-2160, 1980.

2. Evaluation of Superplastic Forming and Co-Diffusion Bonding of Ti-6Al-4V Titanium Alloy for Supersonic Cruise Aircraft Research. Douglas Aircraft Company, NASA CR 178006, 1985.

3. Bales, Thomas T.; Royster, Dick M.; and Arnold, Winfrey E., Jr.: Development of the Weld-Braze Joining Process. NASA TND-7281, 1973.

4. Royster, Dick M.; Bales, Thomas T.; and Wiant, H. Ross: Superplastic Forming/Weld-Brazing of Titanium Skin-Stiffened Compression Panels. Materials overview for 1982, Volume 27 of National SAMPE Symposium and Exhibition, Soc. Advance Mater. and Process Eng., 1982, pp. 569-582.

5. Davis, Randall C.; Royster, Dick M.; and Bales, Thomas T.: Analysis and Test of Superplastically Formed Titanium Hat-Stiffened Panels Under Compression. NASA TM 88989, 1987.

6. Royster, Dick M.; Davis, Randall C.; Shinn, Joseph M., Jr.; Bales, Thomas T.; and Wiant, H. Ross: Fabrication and Evaluation of Superplastically Formed/Weld-Brazed Corrugated Compression Panels with Beaded Webs. NASA TP 2512, 1985.

6. BIOGRAPHIES

Dick M. Royster joined NASA in 1959 and is a senior materials research engineer responsible for processing and testing of materials for high temperature applications on advanced aircraft. He has worked on forming and joining methods to fabricate structures from titanium, aluminum and nickel alloys and metal-matrix composites. Mr. Royster has a Bachelor's Degree in Civil Engineering from Davis and Elkins College and is a member and past chairman of the Hampton Roads Chapter of the American Society for Metals.

Mr. Thomas T. Bales is a senior research engineer in the Metallic Materials Branch of the Materials Division at the NASA Langley Research Center. He is a research team leader in advanced metals processing and during his 24 years at NASA, has authored over 20 publications on this topic. Based on his accomplishments, he is recognized as an expert in advanced processing and fabrication methods for metallic airframe structures. He is recognized as the inventor of the weldbrazing processing for joining titanium and has been issued several patents on processing and structural concepts for airframe applications. He is currently conducting research on the development of joining and forming processes for the fabrication of efficient high temperature structures for applications on future high speed aircraft. He obtained a M.S. degree from Virginia Tech (VPI&SU) in 1965.

Randall C. Davis received his Ph.D. from Rensselaer Polytechnic Institute, Troy, New York in 1970. Dr. Davis joined NASA in 1970 and has been involved in design and testing of advanced structures for supersonic and hypersonic transports.

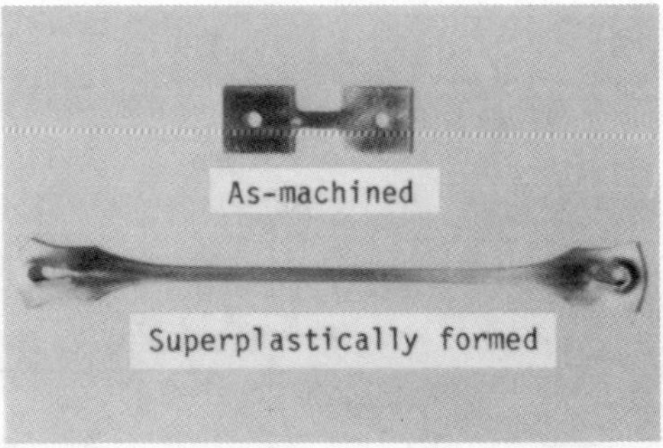

Figure 1. Ti-6Al-4V titanium alloy tensile specimen in the as-machined condition after being superplastically formed.

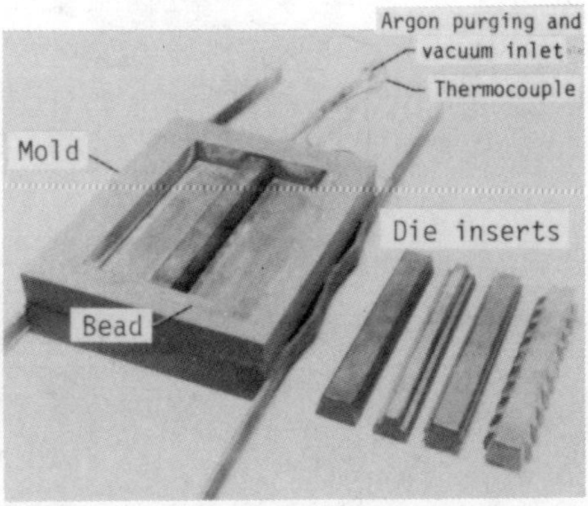

Figure 2. Tooling for superplastic forming of titanium stiffeners.

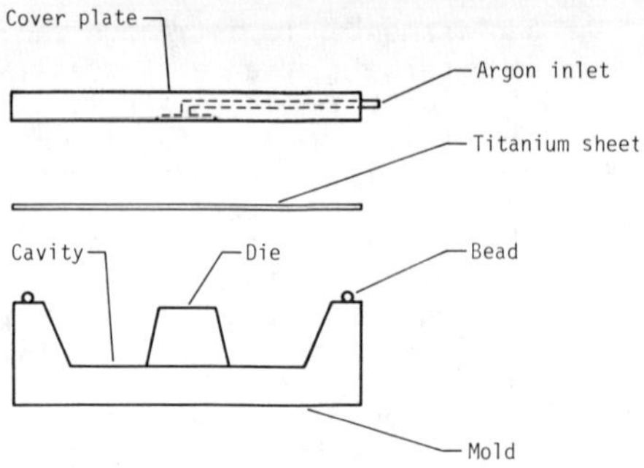

Figure 3. Cross-section of tooling for superplastic forming.

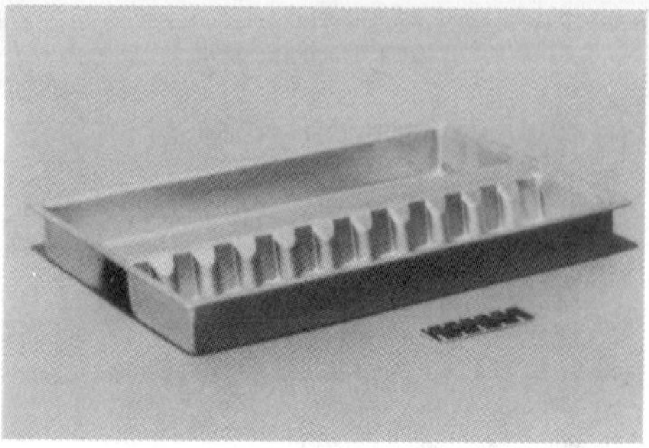

Figure 4. As-formed beaded-shaped web stiffener.

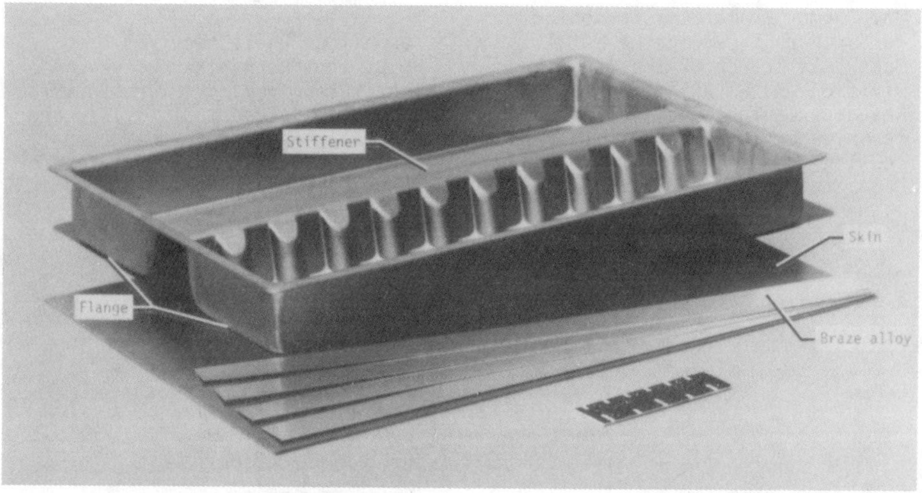

Figure 5. Panel components for joining by weld-brazing.

Figure 6. Panel assembly in vacuum furnace prior to brazing.

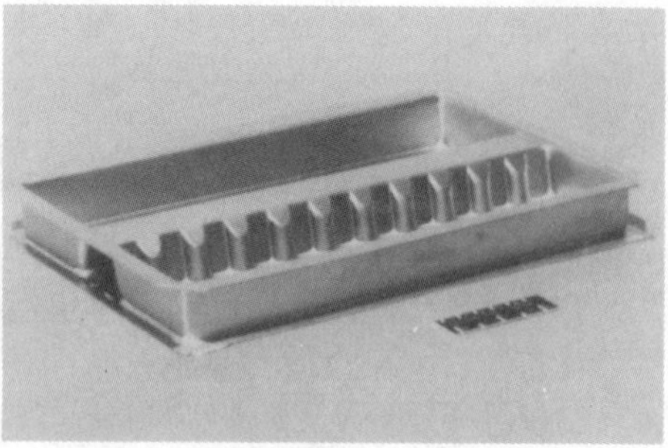

Figure 7. Superplastically formed/ weld-brazed panel section.

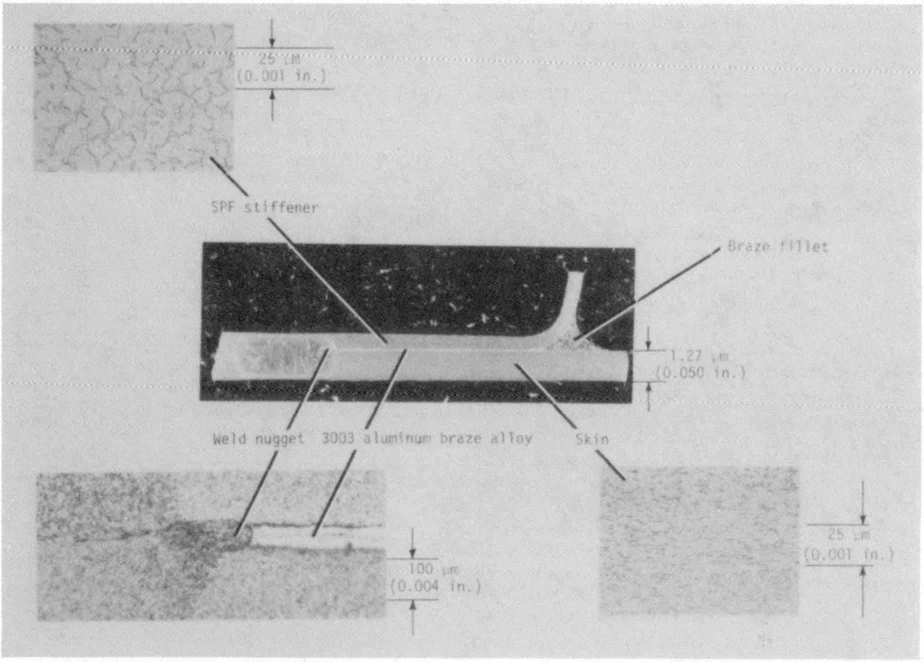

Figure 8. Photomicrograph of a superplastically formed/weld-brazed panel section.

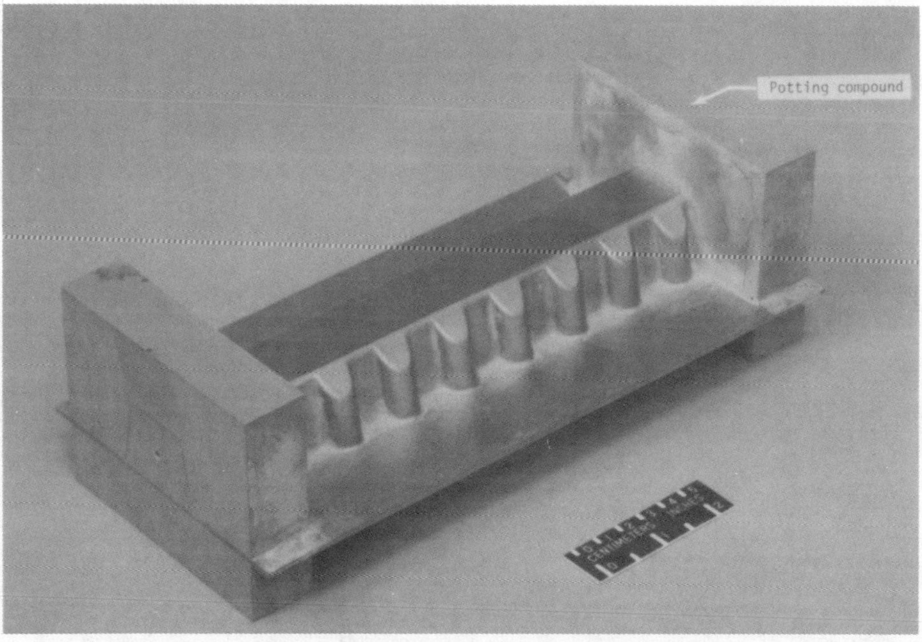

Figure 9. Skin-stiffened test panel with beaded webs.

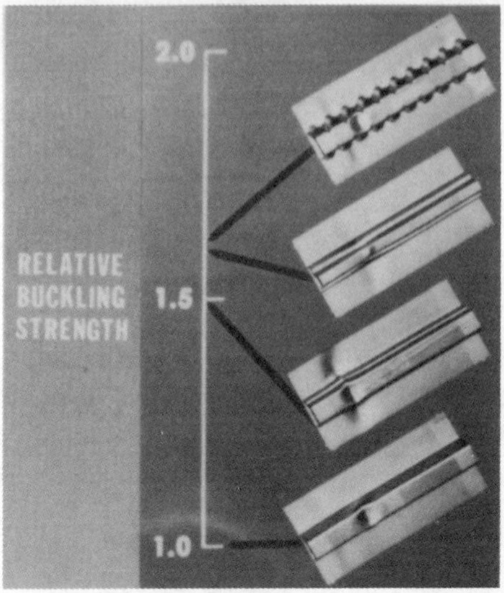

Figure 10. Structural efficiency of superplastically formed skin-stiffened compression panels.

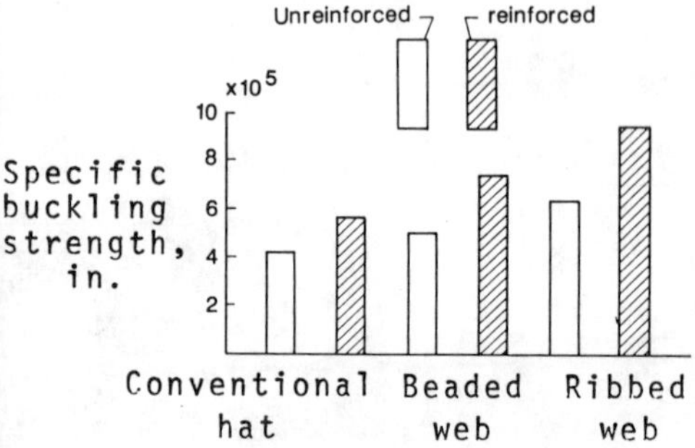

Figure 11. Specific buckling strength of superplastically formed and weld-brazed compression panels with selective reinforcement.

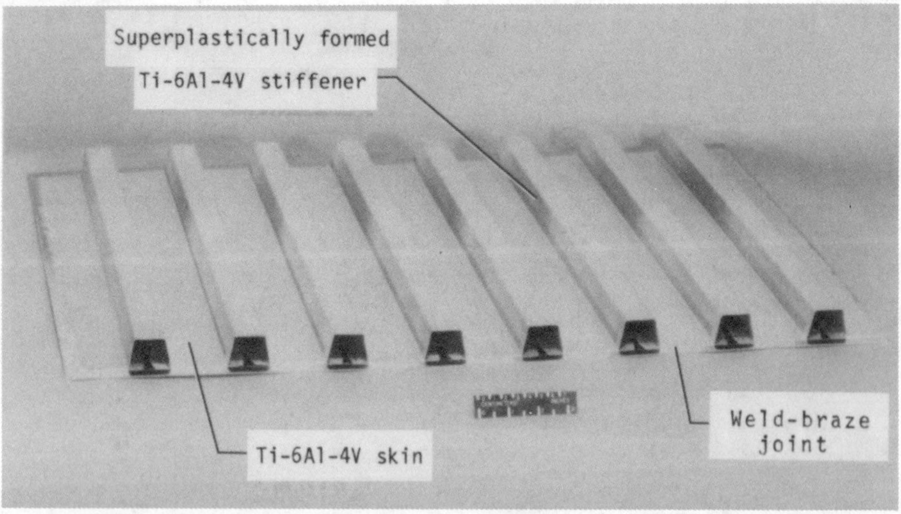

Figure 12. Multiple stiffener compression panel with conventionally shaped hat stiffeners.

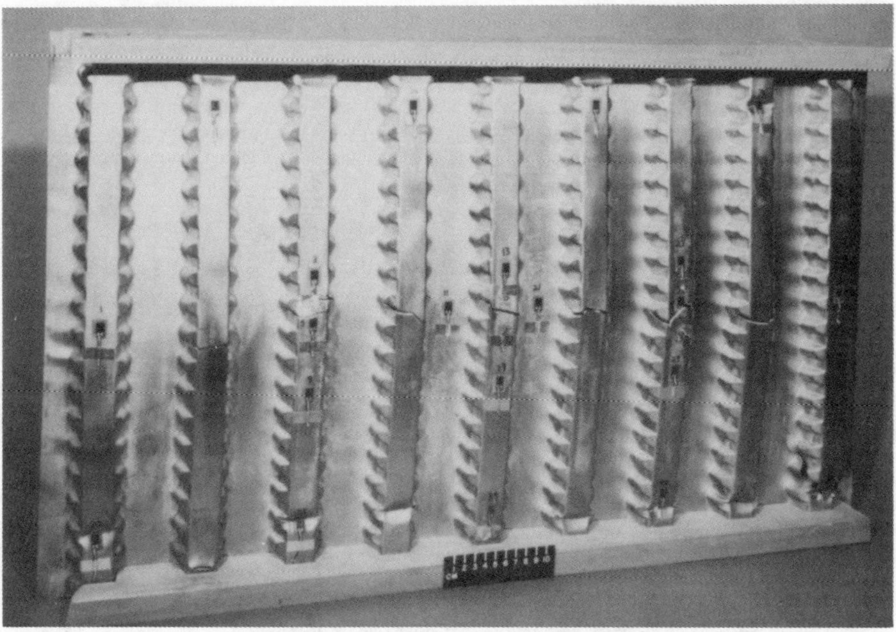

Figure 13. Failed panel with beaded-shaped web stiffeners.

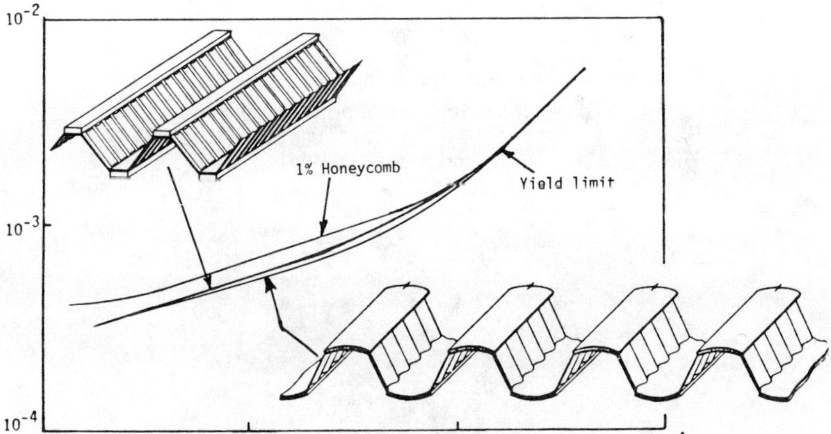

Figure 14. Structural efficiency of advanced panel concepts.

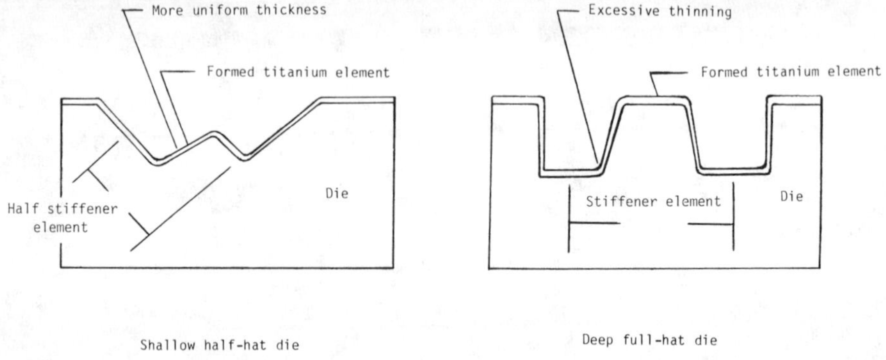

Figure 15. Cross-sections of tooling configurations for superplastic forming hat-shaped elements.

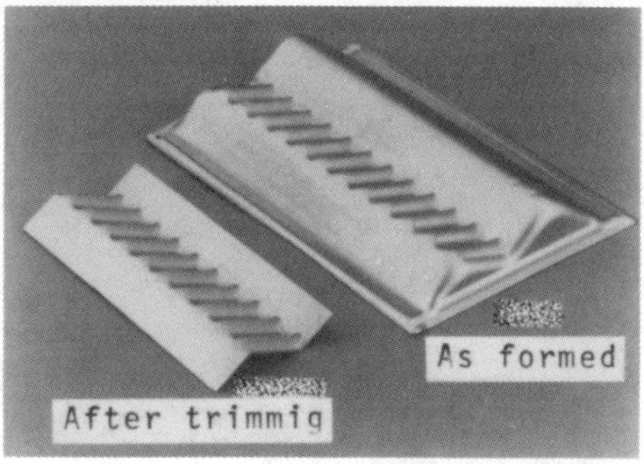

Figure 16. Superplastically formed Ti-6Al-4V titanium alloy half-hat element.

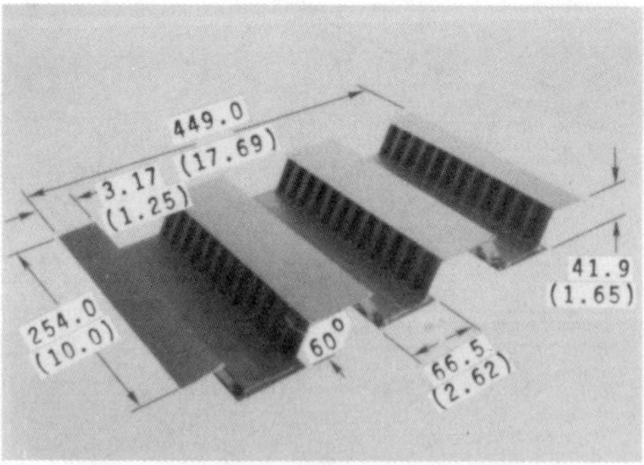

Figure 17. Multiple corrugation compression panel. Dimensions are in mm (in.).

1st International SAMPE Metals Conference
August 18-20, 1987

EVOLUTION OF ROTOR BLADE ABRASION STRIPS AT BELL HELICOPTER TEXTRON

Joe B. Dunham
Senior Engineering Specialist, Producibility
Bell Helicopter Textron, Inc.
Fort Worth, Texas

Abstract

Designs, Materials, Tooling, and manufacturing concepts for rotor blade abrasion strips have all changed during the past 35 years at Bell Helicopter Textron. Early rotor blades utilized .508mm (.0.020-inch) thickness stainless steel for abrasion strips. The customer wanted better wear resistance due to rain and sand erosion. Thus, various materials were evaluated such as urethane coatings, nickel plating, cobalt sheet and titanium sheet. Engineering had to compile field service and wear data for the various materials and operating conditions.

Today, different materials are selected to meet the operating specifications of the customer. As a result, manufacturing methods vary. Some stainless steel abrasion strips are nickel plated after stretch forming without difficulty. Others require tool developmental work to compensate for the nickel plating thickness distribution peculiarities. Also, in past years, titanium abrasion strips have been formed using matched die tooling and then trimmed by machining. Today, the titanium abrasion strips for Bell's most recent model employ a new, modified superplastic forming technique and are trimmed with an abrasive water jet.

1. INTRODUCTION

Helicopters operate in a variety of commercial and military environments such as hot and cold temperatures, rain, and sand. Consequently, the rotor blade abrasion strips must meet many requirements simultaneously. They must protect the blades from erosion and foreign object impact, not impair the operation of deicing system, and contain aerodynamic contour. They must have a predetermined useful life and be designed so that they can be replaced as they wear. To meet these requirements, the abrasion strips are fabricated from a variety of materials, in several geometries, and by many manufacturing methods.

In general, blade geometry has evolved from blades having a constant cross section and a uniform rate of twist to tapered blades with curved leading edges and nonuniform twist rates. The abrasion strips must be fabricated to conform to the blade to which they are attached. Generally, Bell's abrasion strips are designed so that both trailing edges of the part are congruent in the untwisted condition so that the edges can be profiled simultaneously to reduce cost. Special mill fixtures must be created to hold the abrasion strips in the untwisted condition for the profiling operation.

2. ENGINEERING DESIGN CONSIDERATIONS, EQUIPMENT AND TOOLING DEVELOPMENTS

As flight performance requirements change, the geometry and materials for the blades and abrasion strips change. The following depicts the evolution of engineering requirements, equipment and tooling, model by model, over the years at Bell Helicopter Textron.

Model 47 helicopters, our first production model, had wooden main rotor blades with .305 mm (0.012-inch) thick stainless steel abrasion strips. These were satisfactory at that time because the blade tip speed was relatively slow. Consequently, rain and sand eroision were not a problem.

These abrasion strips were formed by stretch forming. Stretch forming is accomplished by gripping the ends of a press brake formed preform and stretching the work piece between a punch and die which contain the desired geometry. When the work piece is stretched just beyond its yield strength, it will retain the desired geometry. Special grip jaws had to be developed for stainless steel materials. Conventional jaws designed for aluminum were not satisfactory. Grip jaws for these early abrasion strips were custom made for each different shape and the cost was several times that of conventional jaws for aluminum.

The next generation, UH-1 helicopters, had increased blade tip speed and the abrasion strip thickness was resultingly increased to .508mm (0.020-inch) to compensate for erosion problems. These stainless steel abrasion strips wore away too quickly to satisfy the Army, so the outboard four feet of the stainless steel was subsequently replaced cobalt. The cobalt details were simply press brake formed.

This two piece abrasion strip was satisfactory for their application until cobalt became in short supply. The cobalt then had to be replaced with stainless steel once again, but thicker. The stainless steel abrasion strips were stretched formed.

Early 214 models used abrasion strips which were fabricated from 1.778mm (0.070-inch) thick stainless steel to prolong their useful life and to extend the time between blade repairs. The abrasion strips tapered linearly from 1.778mm (0.070-inch) at the leading edge to .635mm (0.025-inch) thick at the trailing edge. This was simply accomplished by chem milling the stretch formed shape.

Next came the Model 222 main rotor blade. It has a constant cross-sectional shape from the root end to the tip. The abrasion strip is roll formed from 2.54mm (0.100-inch) thick stainless steel so that the outside surface of the part is to the blade contour. Roll forming was selected because the abrasion strip was a constant cross section and Bell felt that the production quantities justified the additional tooling expense. The leading edge of the abrasion strip remains 2.54mm (0.100-inch) thick and the sides are chem milled down to 1.016mm (0.040-inch) thick. Conventional progressive contour rolls were utilized to roll form the 2.54mm (0.100-inch) thick material to the desired shape. Fourteen pairs of progressive rolls were required. Progressive roll forming is an art, not a science, and a development effort was required to establish the geometry of the roll.

These roll formed work pieces contained residual stresses which had to be neutralized by stretching. When the parts were chem milled without being stretched, they would wrinkle and warp so severely that they could not be used.

The Model 406 helicopter blade uses .508mm (0.020-inch) thick stainless steel abrasion strip substrate which is nickel plated to a combined thickness of 2.235mm ().088-inch). As in the past, the stainless steel substrate is stretch formed.

As with all blade development progress, the abrasion strip changes as the blade geometry progresses. Consequently, a standard design for the stretch form punch and die evolved reducing turn-around time and cost for reworked tools.

The steel punches are now N/C machined to the inside shape of the part. Steel boxes for the die component are constructed so that plastic and fiberglass inserts can be cast to the desired shape from the punch which has been waxed up to represent the work piece thickness.

When the stainless steel abrasion strips are to be nickel plated, the stretch forming tooling and the nickel plating mandrel must usually be developed. Compensation for the nickel thickness distribution peculiarities due to the plating process and the "clam-shelling" effect or spring back tendency of the abrasion strip due to compressive forces in the outer surfaces from plating must be established. Personnel from Bell Helicopter and the plating vendor must work together as a team to accomplish the desired result.

In addition to developing a standard design for the stretch form punch and die, special jaw designs were also developed for the stretch forming operation. This was done so that a complete jaw assembly would not have to be created for each new abrasion strip shape or replaced when a repair was required. Today, only the peculiar inserts need to be created and installed into a universal jaw holder. This concept resulted in very significant tooling cost savings.

The Model 205 composite blade has a urethane abrasion strip which is superior in sand but not in heavy rain. A nickel electroform strip is available to add over the outboard leading edge of the urethane if the helicopter is expected to operate in a rainy environment. This strip is bonded to the urethane abrasion strip in a secondary operation.

The Model 412 composite blade has a stainless steel tapered cross section outboard strip and a titanium constant thickness inboard strip. The stainless piece is stretch formed and the titanium piece is hot formed with matched dies.

The titanium abrasion strips are formed by placing the work piece between a punch and die under pressure which are heated to 649°C (1200°F). The matched dies are of conventional design except that they are designed to accept a work piece which contains press brake formed flanges along both long edges. These flanges help to prevent the hot work piece from bending under its own weight when it is being removed from the hot forming press.

The Model 214ST composite blade, one of the largest Bell blades, has two titanium abrasion strips. The outboard strip is 2.032mm (0.080-inch) thick at the leading edge and tapers to .635mm (0.025-inch) thick at the trailing edge. The inboard strip is a constant 1.016mm (0.040-inch) thick. Both abrasion strips are not formed with matched punch and die sets.

The most recent blade design specified a titanium sheath to

which a nickel electroform is bonded. This sheath will be super plastic formed at 899°C (1650°F). Super plastic forming tooling for titanium consists of only one tooling component, which is the punch. Argon gas serves as a fluid die and forces the work piece onto the punch.

3. SUMMARY

Abrasion strips are a necessity for the protection of the rotor blades from erosion and impact damage. These strips must meet close tolerance requirements so that they contain correct aerodynamic configuration and so that they will bond properly to the blade. They must weigh within a small tolerance requirement so that the blades can be properly balanced. Abrasion strips can be costly to produce. Bell Helicopter is constantly seeking ways to reduce costs by considering new materials, designs, and manufacturing methods.

BIOGRAPHY

DUNHAM, J. B. - Bell Helicopter Textron

Mr. Joe B. Dunham, Senior Engineering Specialist, Manufacturing Development, Producibility, has been selected as the Bell Helicopter principal investigator for this program. Mr. Dunham has an extensive background in the field of sheet metal forming, developing most of the tooling for forming sheet metal at Bell Helicopter.

Mr. Dunham is the sheet metal fabrication consultant to Planning, Tool Design, Numerical Control, Loft, Equipment Engineering, Tool Manufacturing, Design Engineering, Engineering Labs, Purchasing and Industrial Engineering Departments. He also provides liaison and technical support to outside machine and tool companies engaged in the design and manufacture of tools and equipment for Bell Helicopter.

A summary of Mr. Dunham's accomplishments includes:

- Hot forming of titanium in a heated platen press with matched tooling.
- Stretch forming stainless steel sheet metal and extruded shapes using stretch and wrap equipment.
- Deep drawing sheet metal pan shape parts with heated and cold draw dies.
- Designed 214ST firewall configuration and developed manfacturing method.
- Manufacturing tools and methods fabricate infrared suppressors AH-1G, UH-1D, and UH-1H helicopters.
- Tools and manufacturing methods for in-house fabrication of leading edge abrasion strips.
- Tooling and manufacturing methods for the firewall ejector system.
- Superplastic forming of aluminum test articles.

1st International SAMPE Metals Conference
August 18-20, 1987

SUPERPLASTIC FORMING - A NEW PRODUCTION TECHNOLOGY

P.N. Comley
Manager Technical Services
Murdock, Inc., Compton, California

1. ABSTRACT

Superplastic forming, or SPF, is a technology that offers a new way to form sheet metal. It utilizes a peculiar phenomenon that is observed in certain metal alloys that allows the alloy to plastically deform to large elongations without rupture. The process is done hot, and at controlled strain rate, to form complex parts in sheet metal in a method similar to vacuum forming in sheet thermoplastics. The process offers advantages in lower die costs, replacement of assemblies of parts by a single part, and no spring-back of the formed metal, and has found extensive use in the aircraft industry. Metals such as titanium, aluminum and stainless steel can be SPF´d. This paper outlines the method of the technology, the alloys most suitable and the equipment used, and shows parts that have been made by superplastic forming on a variety of aircraft.

2. INTRODUCTION

If a piece of sheet metal is pulled in tension, it will stretch. If the tension is released, the sheet will revert to its original size. This stretching is called elastic deformation. If a lot of tension is applied the metal may stretch beyond its elastic limit and when the tension is released the sheet will acquire a small amount of permanent stretch. This permanent stretch is called plastic deformation, and the amount of plastic deformation that can be put into a metal varies between around zero for a brittle case iron to about 20% for a soft ductile aluminum.

There is, however, a small group of special metal alloys that, when subjected to the right conditions of temperature and strain rate, will stretch to extraordinary lengths. The amount of plastic deformation that can be induced in these alloys varies between 400% up to nearly 5000%; hence the metallurgical term super-plastic is used to describe their behavior.

3. HISTORY

Superplastic forming or SPF had its origins back in the 1930´s when Pearson (1) discovered during experiments in extrusion, that a Bismuth-Tin alloy exhibited nearly

2000% of plastic elongation when stretched at a slow speed. The phenomenen remained a laboratory curiosity, however, until the mid 1960's when workers in the USA and USSR reported other alloy systems could be made to exhibit superplastic deformation when subjected to a controlled strain rate. The first alloy system to be commercially used was a zinc-aluminum alloy and IBM and others used it for making covers, bezels, etc. In the early 1970's, interest spread to other alloy systems, including aluminum, stainless steel and titanium. At the present time superplastic forming is finding the majority of applications in the aerospace industry with SPF aluminum and titanium being applied in many military and commercial aircraft.

4. THE SHEET FORMING PROCESS

The dominant process application has been in the forming of sheet material. In the sheet forming process a sheet of superplastic material is clamped over a heated die and a heated upper chamber or caul plate is placed over the sheet (Fig 1). A continuous indent around the die or caul plate presses into the sheet to form a gas tight seal around its periphery. Gas pressure is applied through the caul plate to blow the sheet downward and into the die. Exhaust holes in the die enable trapped gas below the sheet to escape. The sheet finally fills all the contours of the die, the forming gas pressure is released and the part is removed. In some cases a tool box is used instead of a die. In this case, die inserts are placed in the tool box and the sheet wraps around the inserts. The die must generally be clamped in a heated platen press to withstand the opening forces of the gas pressure and to maintain the correct temperature.

5. SPF ALLOYS

The explanation of the phenomenen of superplastic forming is still a matter of debate among metallurgists and there have been many papers over the last 20 years outlining the theories and mechanisms of superplastic forming. The results of their work are outside the scope of this paper, but there is general agreement on the basic requirements of any alloy system to be superplastic. Firstly, the alloy needs to have a very fine grain size, preferably under 5-8 micron, which is preferably equi-axed. Secondly, there must exist a way of keeping this grain size small at elevated temperatures of up to 0.9 of the melting temperature and for long periods of time, up to 2 hours or more.

Commercially available alloys that meet these requirements and exhibit superplastic properties are:

1) Aluminum. Two 2000 series alloys, designated 2004 and 2419. 7475 alloy that has been thermomechanically processed to make it superplastic. The new aluminum -lithium alloys such as 2090, 2091, 8090. Forming temperatues are generally over 500 degrees C.

2) Stainless steel. Most of the microduplex alloys, where there is a 50/50 balance of austenite and ferrite. Examples are

the generic 2205 series. Forming temperature is 1000 degrees C.

3) Titanium. All of the alpha-beta alloys, such as the industry standard Ti-6Al-4V and 6-2-4-2.

The alpha-beta titanium alloys are unusual in that standard mill run sheet can be superplastically formed, whereas most of the other alloy systems have to be specially manufactured and processed.

The property that makes these alloys superplastic is that they exhibit a high degree of strain rate sensitivity. To effect a higher strain rate in the material requires a disproportionatly higher stress. In addition, the material strain hardens, so that a strained piece of material requires a greater stress to deform it than an unstrained piece. These properties result in the material being able to resist local necking. While being stretched, the material deforms uniformly all over instead of suddenly developing a neck that would lead to premature failure.

To superplastically form, for example, Ti-6Al-4V, it is necessary to heat the material up to around 900 degrees C, and impose a sheet stress of 15 MPa. This results in the material straining at a rate of 3% per minute. Most parts requiring 50 to 200% elongation can therefore be formed in a few tens of minutes.

It is worth noting that the forming stress to superplastically form SPF alloys is less than 5% of their room temperature annealed yield stress, which offers advantages in the consideration of making parts.

6. THE ADVANTAGES OF SPF AND THE AREAS OF APPLICATION

Titanium and high strength aluminum alloys in sheet form have many attractions for the aerospace industry due to their high strength, low density, heat resistance, etc. However, many of the sheet metal structures in airframes have complex shapes, tight bends, joggles, and compound curvature, and aerospace metal alloy sheets cannot be generally worked into complex shapes due to their limited formability. Their high strength and modest modulus means that springback and distortion are problems, and wrinkling and cracking of sheet can also occur. The standard method of making titanium sheet parts, for example, is to first cold form a preform by hydro-forming, press braking, etc. and then hot sizing or creep forming to final shape. The forming stress while making the preform is high, above the yield stress, and the maximum allowable elongation of the material is low, less than 10%.

The promise of superplastic forming can offer various advantages over the existing method of forming sheet metal parts.

Firstly, the material is capable of high deformation. Parts can be designed with up to 200% plastic strain in them. What this means is that very complicated parts can be made. Items with compound curvature are easily formed with no wrinkling (Fig 2). A single sheet can be formed into a complex arrangement of ribs and stiffeners and hence replace an assembly of parts, together with their associated rivets, etc (Fig 3). The single sheet is lighter than its equivalent assembled structure due to the absence of rivets, and the more even distribution of stress means that a thinner sheet gauge can be specified. Also the cost of the part is very much less than the equivalent assembly due to the absence of assembly time, the fact that there are no assembly jigs and fixtures, no subpart inventories and tracking, and fewer drawings. (Generally, the cost of each drawing in the aerospace industry is very high.)

Secondly, the high forming temperature effectively stress relieves the part while being formed. After the part is removed from the press and cooled down, there is little or no residual stress. Hence the finished part sits where it is supposed to sit, and there is excellent repeatability between pressings. Post formed springback is not a problem.

Thirdly, the method of manufacture leads to inexpensive tooling. Conventional forming methods require a matched pair of tooling and often additional pairs of tooling for multiple blows or for making the preform as well. Superplastic forming requires only a single die, whether it is male or female, and the expense of making a matched pair of dies is avoided. In the aerospace industry, with its short runs and limited production quantities, the cost of tooling is often very significant in the total cost of the produced part.

Another consideration is the size of the equipment needed to form parts. The parts shown in Figure 2 are approximately 700mm diameter and 300mm deep, are made from 2.5mm starting gauge Ti-6Al-4V and yet require a press of only 30 tonne capacity to form them. Superplastic forming presses can therefore be of low tonnage in relation to their platen area.

7. CONSIDERATIONS IN SUPERPLASTIC FORMING

The application of SPF technology to the manufacture of a particular item has to take into consideration a number of factors in order to implement a successful program.

As is always the case when a new technology is introduced, there is a lack of field data on the performance of that technology. SPF titanium underwent an extensive R&D examination by many of the major aerospace companies during the mid 1970's and there are now many SPF parts finding their way onto aircraft structures. Most of the companies have specifications covering the forming and inspection of parts and so superplastically formed titanium has now become an accepted manufacturing process, which was a major hurdle to overcome for the introduction of the technology. SPF aluminum is now coming on stream.

Design conservation is another hurdle. Designers are only just becoming aware of the possibilities of superplastic forming and in order for the application area of SPF to grow, they must be informed to a greater level than at present. The current generation of aircraft being built today was designed a few years ago, before Superplastic forming was established, and it is a credit to the process and to the aircraft builders that many of the SPF parts now flying are redesigned ones that were originally made by conventional methods, but have proven to be cost and weight effective as a superplastically formed part. The introduction of an SPF part to replace a conventional part has two obstacles to overcome; firstly, the cost of change and secondly the fact that the part is not designed with SPF in mind and so is not using its optimum SPF properties. Historically a desire to change to SPF has also occurred because of manufacturing difficulties with the current design, so SPF has been looked at as a sort of panacea for problems. Unfortunately, that is not the best way to implement superplastic forming. Often the structure around the part to be changed could also benefit by being changed as well. Recently however, new aircraft are being made with SPF as an integral part of the design in mind and it now appears that SPF primary and secondary structures will be prominent in future airplanes.

8. DESIGN CRITERIA IN SUPERPLASTIC FORMING

In the design of parts to be superplastically formed, some criteria have been established to help both the designer and the manufacturer of parts. The first concerns sheet thinout across the formed piece. During forming the sheet is clamped around the edge of its die and blown into it. Thus all the material forming the sides and base of the part comes from the projected area of the initial blank over the die. This compares with, for example, deep drawing, where material is drawn in from outside the die cavity to help form the walls. It follows that an SPF´d part has thinner wall thickness than a deep drawn part (from the same starting gauge) although, of course, the starting blank is also smaller. Techniques have been perfected for the superplastic forming of aluminum alloys that create a uniform wall thickness in the part by a "maleform" process; however, this process is technically very difficult to use in titanium and stainless steel SPF forming due to the higher forming temperature. To keep from excessive thinning, the depth of a formed part should, as a guideline, be kept less than half the width. Radii should be kept above 2 times material thickness, as male radii below this give rise to a high degree of thinning in and adjacent to the bend and female radii require high pressure and long time to form below this figure. Generally, radii of 5 times sheet thickness are acceptable. A second consideration as far as aluminum alloys are concerned is that small voids, or cavities, appear in the material after large amounts of superplastic deformation. There is considerable debate as to the detrimental effects of cavitation on properties such as fatigue life. At the present time it is argued that the small amount of cavitation, about 2-3% after 150% superplastic deformation, is not detrimental to secondary structure. However, aircraft manufacturers are unhappy about putting SPF aluminum parts that has cavities into primary structure. Recently a process known as the back pressure technique has been introduced where an isostatic gas pressure equal to the forming stress is applied to the part during forming which effectively suppresses cavitation. Cavitation is not a problem with titanium alloys.

The best applications of superplastic forming fall into two categories. Firstly, where a deep shape with compound curvature is required (Fig 2). In this instance, an SPF part can be made with no wrinkling. Parts are either female or drape formed (as with male inserts in a tool box). Dependent on inside radii, sheet thicknesses up to 6mm have been used. Joggles, details, etc. can all be designed into the part. To enable the part to be removed from the die, a 3 degree draft angle per side is recommended.

The second category is in the application of flat or curved pan shapes with complex detail (Fig 3). These parts generally replace an assembly of parts, where all the longerons, hat sections, gussets and plates are made in one piece. Again, joggles and compound curves are easily accomodated in the design. There are many examples of enclosures, door stiffeners, access panels, and equipment covers that are readily and simply made by superplastic forming.

9. TOOLING

Tooling for superplastic forming is different from that required for other forming processes. As previously mentioned, only one die, either male or female, has to be made. The pressure or load on the die is not great, being of the same magnitude as the gas forming pressure. There is no ironing or squeezing of material between die halves. However, the operating temperature is very much higher than with other processes, 900 degrees C for titanium and 500 degrees C for aluminum.

Ceramic tooling can often be used for prototype or short run work. A castable high temperature ceramic with stainless steel chopped fibre to minimize cracking, can be cast off a pattern as an inexpensive way to make a die. After a few pressings, the die may spall or crack, so ceramic dies are not generally used for long runs. However, there has been some success where ceramic male dies have lasted over 250 pressings. (2) (Fig 4)

For most work metal dies are used (Fig 5). The choice of alloy for titanium is important as there are three criteria that the metal must meet in order to be a candidate for tooling. Firstly, the alloy must have reasonable strength and good oxidation resistance at 900 degrees C and be thermally dimensionally stable. Many of the stainless steels satisfy the first two points but some, for example 17-4 P.H., exhibit some shrinkage after repeated thermal cycling. Secondly, the metal must not react with titanium at temperature. It has been found that alloys with medium to high nickel contents contaminate the titanium and only those with less than 7% Ni appear non reactive. Thirdly, the alloy needs to be machinable, as the SPF´d parts will often be of complex form. Again, many of the high temperature alloys are not easily machined.

The industry has determined from the above that wrought 22-4-9 stainless steel and cast 49M are the optimum choice of tooling material for superplastic forming of titanium. NP464 and 431 stainless are also candidates. 304 stainless has sometimes been used.

When designing in 22-4-9, a differential thermal expansion factor has to be considered, with the die being 0.007 smaller than the part. For 49M cast material, an additional shrinkage factor of 3% has also to be allowed during casting. The design of the top of the die or toolbox has a sealing ring to bite into the titanium sheet to effect a gas tight seal during forming. Sometimes the sealing ring is on the toolbox lid or on both items.

For aluminum forming ordinary 304 stainless, type 4340, 4130 steel and even a simple 1020 or 1040 mild steel have all been used with success. Cast aluminum alloys can be used for the 2004 alloy, but the forming temperature of the other alloys is too high for aluminum tooling.

10. PROCESS PARAMETERS

In order to achieve high superplastic elongation and good distribution of sheet thickness, it is necessary to form the sheet within certain bands of temperature and strain rate. The optimum temperature for SPF of Ti-6Al-4V, for example, is 900 degrees C., but on large dies it can be difficult to hold that temperature and a tolerance band of +/- 15 degrees C is specified. The stress versus strain rate curve at 885 degrees C is different to that at 915 degrees C and so, in the production of parts, some deviation from the theoretical curve can be expected.

The stress is applied to the sheet by gas pressure and the correct pressure to apply can be calculated knowing the local radius, sheet thickness, and desired stress. For forming an unconstrained sphere or a long channel, the stress/pressure relationship is well known ($S = Pr/2t$ and $S = Pr/t$) but for most SPF shapes, the equations are more complex. In addition, the radius and thickness are constantly changing so the pressure has to vary with time in order to keep the stress constant. As local radii and thickness vary across the forming sheet, only one part of the sheet has the correct stress applied to it, all the other areas are being stressed at some level other than optimum. A further factor in the calculations is that the alloys strain harden as they superplastically form so that the applied stress has to increase in order to maintain constant strain rate.

All these different parameters can be handled by a computer program and each company involved in SPF has developed models to predict the correct pressure/time relationship to form a part. The models vary from simple approximations to full finite element or finite difference simulations. A combination of computer prediction and practical experience is generally sufficient to make successful parts.

When the sheet meets the die surface it can either stick or slide, depending on the friction. Lubricants are applied to enable the sheet to slip and achieve more uniform thickness distribution (3). Boron Nitride is usually used for titanium and graphite for aluminum.

11. THE FORMING EQUIPMENT

SPF parts are made on an hydraulic press (Fig 6) with heated platens. The platens are made from a high strength oxidation resistant stainless steel and are usually heated by cartridge heaters. Platens can be segmented with up to 30 independent heating zones to provide a constant temperature at the die. Heat shields around the platens are required to retain temperature and prevent heat loss to the press frame and operators.

Integral to the SPF press is the control and gas management system. The control of the forming gas pressure, which varies with time, is crucial to the SPF process, and in a production environment automatic control is required. The pressure/time profile and other data is downloaded to a microprocessor or computer via a floppy disk etc. for each part.

The control panel for a typical SPF/DB press contains two independent gas pressure zones and a third argon low pressure zone for die purging. Sometimes a purge/vacuum system is employed to completely eliminate air from the die.

12. THE MANUFACTURE OF AN SPF PART

After a part has been designed for SPF, a computer analysis can be made to predict the thinnout of the wall in critical areas. From that basis the starting gauge for the part is selected. The die is manufactured and material ordered. After shearing to size, the blanks require cleaning to remove markings, oil, etc. They are then sprayed with a Boron Nitride lubricant which can be mixed with a binder such as Nicrobraze for adhesion. The die is brought up to temperature in the heated platen press and flooded with argon to minimize contamination of the sheet. The pressure/time program is loaded into the machine microprocessor. The blank is then loaded. After a short time to allow it to reach temperature, ram pressure is applied to the die to squeeze the lip seal into the blank to create a gas tight seal. The operator then pushes the cycle start button to form the part and argon gas is automatically introduced above the sheet to form the part. At the end of the cycle the die is opened and the part removed. For large and deep parts the die is first taken out of the press, the top plate is lifted off and the part removed. Items requiring a stabilizing heat treatment are recoated with a protective coating and heat treated.

Parts are then descaled by abrasive blasting and titanium parts chemically milled to a depth of 0.08mm on each side to remove alpha case. To verify that all of the alpha case has been removed, the parts may be given an ammonium biflouride dip. Sometimes sample coupons are removed from the trim area for mechanical testing. Parts are trimmed to final size and then penetrant inspected and visually inspected prior to I.D. marking and packing for shipment.

13. APPLICATIONS OF SPF IN THE AEROSPACE INDUSTRY

Many of the major aircraft manufacturers are becoming involved in superplastic forming and the following photographs show some of the parts being formed.

Rohr Industries have been heavily involved in superplastic forming of titanium, and have replaced or substituted many conventionally made parts by SPF parts. The left hand side of Fig 7 shows the bulkheads in a pen nib fairing as conventionally formed. Each of the three bulkheads were made in three pieces and the whole required 54 tools for their manufacture. The right hand side shows the SPF replacement, where a one piece pressing replaces the old assembly. Fig 8 shows that all three bulkheads are formed in one operation on one die. The conventionally made bump pan for the Boeing 757 (Fig 10) consists of 41 detail parts and 386 hi-lok rivets. The next photo (Fig 11) shows the SPF replacement made in just one part.

Another company that has made extensive use of the technology is McDonnell Douglas in St. Louis and Long Beach. Fig 9 is a photo of a complete landing gear door made in one operation by combined superplastic forming and diffusion bonding. MacAir have now made some 10,000 parts on the F15, F18 and AV8B aircraft and are currently developing large primary structural parts on the F15 for SPF and SPF/DB.

Rockwell International did much of the pioneering work on SPF and have made many details for the B1 and space shuttle. Shown in Fig 12 is a bay on the B1 with many SPF parts in the assembly. Some 70 SPF details go on the B1 bomber.

14. CONCLUSION

Superplastic forming is a technology that is now a tried and tested method of producing cost effective and reduced weight parts for the aerospace industry. In the recent past it has found many applications as a replacement for assemblies of parts or for parts that were difficult to form by conventional means. At the present time aircraft are being designed with Superplastic Forming as an integral part of the design and the industry can look forward to a healthy future as SPF production expands into the next generation of aircraft.

15. REFERENCES

1) Pearson, C.E. "The Viscous Properties of Extruded Eutectic Alloys of Lead-Tin and Bismuth-Tin" J. Institute of Metals, 54, 1934, pp 111-123

2) Cadwell, G., Rohr Ind., private communication

3) A.K. Ghosh, and Hamilton, C.H. "Superplastic Forming of a Long Rectangular Box Section - Analysis and Experiment," ASM Materials & Processing Congress, Nov 1979

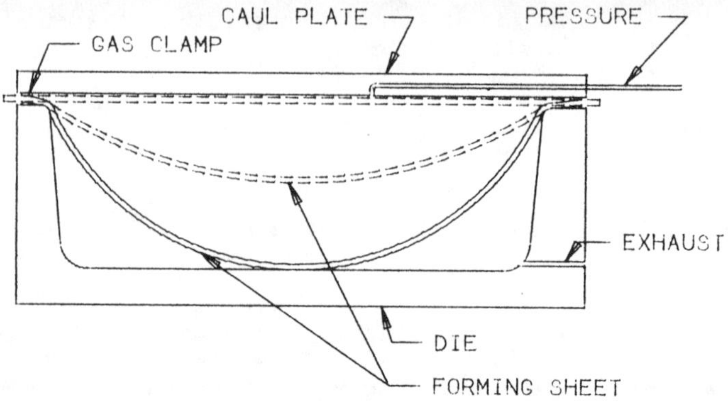

Fig 1

Fig 2

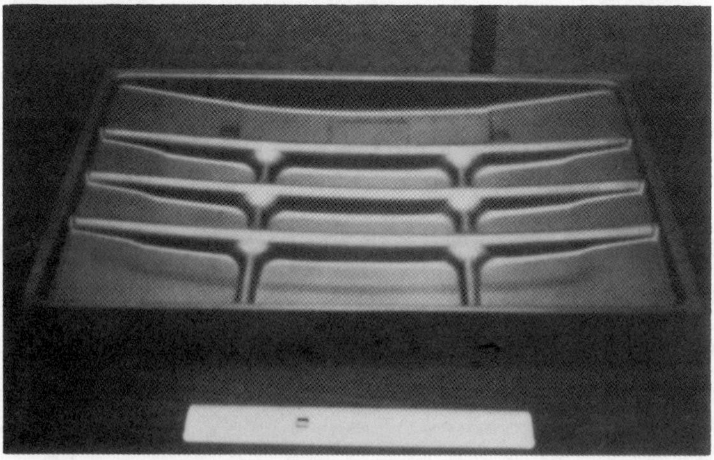
Fig 3

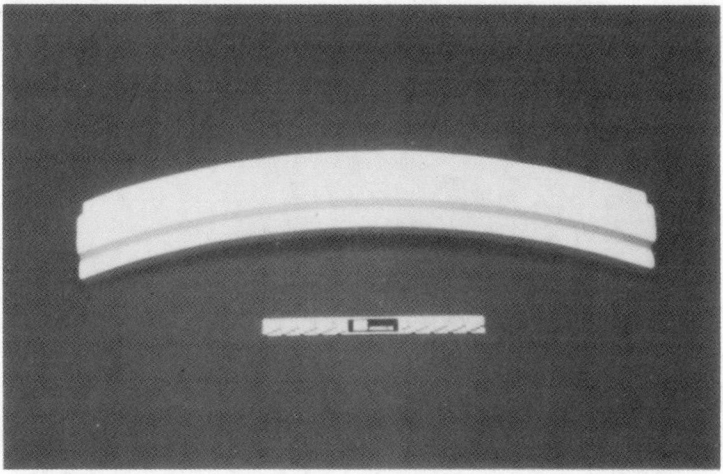

Fig 4

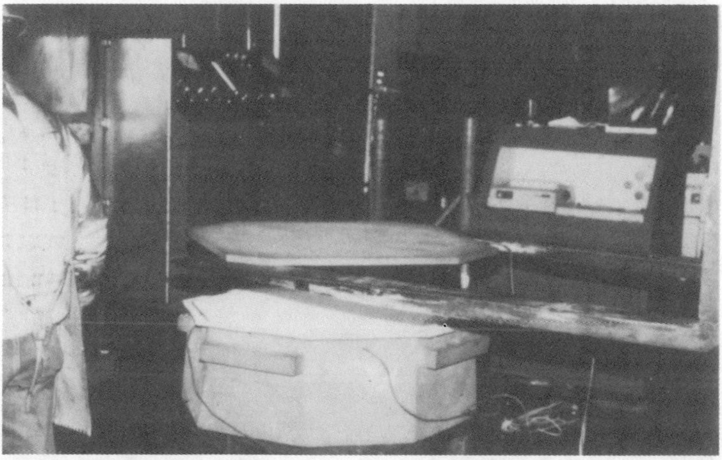

Fig 5

Fig 6

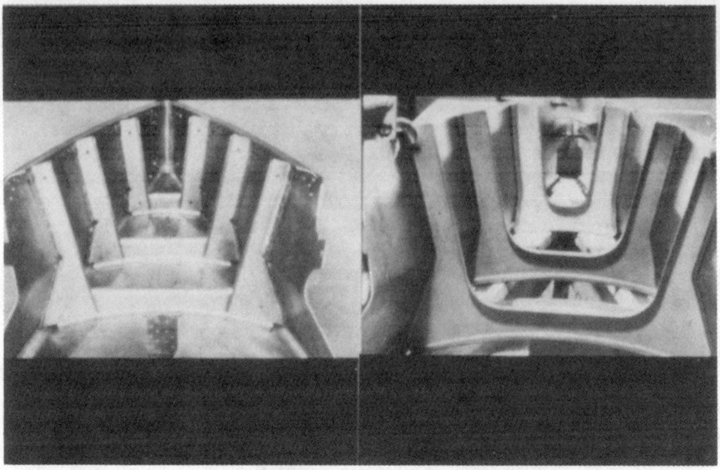

Fig 7

Fig 8

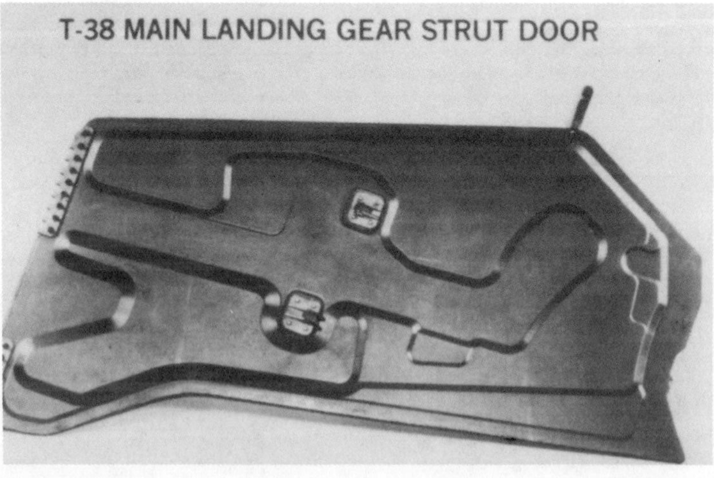

Fig 9

Fig 10

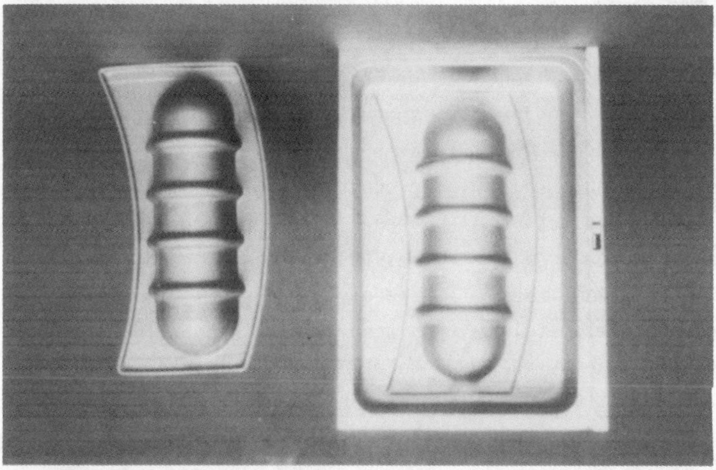

Fig 11

Fig 12

1st International SAMPE Metals Conference
August 18-20, 1987

WHITHER TITANIUM POWDER METALLURGY?

Gerald Friedman
P/M Prototype Manufacturing
PCC Airfoils, Inc.
Cleveland, Ohio

Abstract

In the more than two decades since the first appearance of prealloyed titanium powder, there has been only one component, an engine mount support, manufactured and installed on an airplane; this part had a very short production run, and was used on only six aircraft. Much effort has been expended on titanium powder metallurgy by both industry and government in the U.S. and Europe. These efforts resulted in the eventual development of (1) a number of powder production processes, ranging from the original tungsten-cathode rotating electrode process to melt extraction and vacuum melt/gas atomization techniques, (2) at least three powder consolidation techniques, (3) an assortment of post-consolidation heat treatments, (4) mechanical property evaluation ranging from room temperature tensile tests, to fatigue, fracture toughness and crack growth testing, and (5) cost comparisons for small, medium and large components. These efforts resulted in the demonstrated ability to produce sound, strong, cost-effective components, which never obtained production status. The effort has been disproportionate to the results obtained, and contrasts sharply with the manufacturing history of superalloy powder metallurgy. This paper will attempt to explain the reason(s) for this phenomenon.

1. Introduction

The United States government, through the mechanism of Navy and Air Force manufacturing technology development programs, has devoted considerable time and effort promoting the concept of replacing forged titanium alloy components with one-step-consolidation P/M (powder metallurgy) parts. That the effort has largely been unsuccessful has not been due to lack of trying, or to incompetence, or to lack of suitable raw materials. This paper will attempt to analyze, through a discussion of the materials, procedures, and potential applications, the role of P/M technology in the aircraft industry.

2. Powders

The "powder" in powder metallurgy refers to either (blended) elemental powder or (pre-)alloyed powder; both have their advantages.

Elemental powder is just that, a commercially pure metallic powder, present as a single element, e.g., copper, tungsten, or titanium, or mechanically blended with other elements. Typical commercial blends are carbon blended with iron and aluminum-vanadium blended with titanium. The advantages of elemental blends are the (relative) ease with which they can be consolidated, and their low cost. Compared to their alloy counterparts, elemental powders are soft, and frequently are irregular enough in shape to permit consolidation at room temperature into useful shapes. The disadvantages of elemental blends are the difficulty in achieving homogeniety in complex compositions, and in some systems, the lower purity levels, reflecting the absence of a pyrometallurgical alloying/refining operation.

Alloy powders are homogeneous; each particle, regardless of size, can be expected to be of the same composition as the others. With respect to titanium alloys, there are two main classes of alloy powder, powder produced in a solid state reaction, and powders derived from a liquid.

HDH titanium alloy powder is produced from the solid by hydriding, crushing, and then dehydriding titanium alloy ingot, wrought bar, or scrap. Although this process is intrinsically non-contaminating, in practice a great deal of care is required if powder oxidation and contamination from the crushing and screening operations is to be avoided. HDH powder is compactible at room temperature; compacts, however, are neither so strong, nor as complex, as those made from blended elemental titanium.

Most of the titanium alloy powder in use today is made by atomization, i.e., each particle is a casting resulting from the freezing of a molten droplet in vacuum, argon, or helium.

Titanium atomization processes are either rotary, pressurized gas, or a combination of the two. In a subdivision of the first category, the end of a rotating titanium alloy bar is melted in helium by a tungsten arc or plasma torch; the droplets are thrown off by centrifugal action and freeze in flight. In a variant of this method, the end of a titanium bar or ingot is melted by an electron beam and allowed to drip upon a rapidly rotating disc, which atomizes the liquid stream. Because the EB operates in vacuum, rather than inert gas, particle cooling is less efficient; innovative engineering solutions are applied to freeze the droplets within an economical chamber size.

In another category of rotary atomization, titanium is melted in a crucible and poured onto a rotating disc; additional cooling of the freezing droplets is provided by helium jets impinging on them as they fly outward.

Recently, pressurized gas has been employed to atomize a stream of crucible-melted metal, in a manner analagous to the procedures employed for superalloy and tool steel powder manufacture.

3. Consolidation

All of the atomization processes result in quenched (hard) spherical particles that cannot be consolidated at room temperature. Although some investigators have consolidating atomized titanium powder in rigid tools, by vacuum hot pressing (VHP), this approach, employing low pressure over a relatively long time, has not been very successful, because of difficulties associated with tool strength and low throughput. More work has been done employing a combination of an expendable tool/ mold in a forging press, in which very high pressure is brought to bear over a very short time period; this approach addresses both of the shortcomings of the VHP approach. Most consolidation efforts, however, have been devoted to hot isostatic pressing, HIP, employing metal or ceramic molds, at pressures and times intermediate between the two perceding techniques.

4. Applications: Past

Almost without exception, the powders and consolidation technologies described above have been employed to demonstrate the feasibility of replacing a forging with a near net powder component. And, almost without exception, the reason for doing so was to reduce the cost of the component. Unlike the case of powder metallurgy superalloy turbine discs, which were introduced to remedy the problem of the newer alloys' poor forgability, there was no technical problem which P/M titanium was introduced to solve. Whereas P/M super-alloy discs represented "market-pull", P/M titanium has been a "technology push". The "push" has been largely unsuccessful, although in most cases the mechanical properties of the potential P/M component were equivalent to, or greater than, those of the forging to be replaced. The problem was/is that the projected savings have not been great enough to overcome the classic aerospace dictum, that "If it ain't broke, don't fix it", when indeed, "it ain't broke".

5. Applications: Future

What then, is the reason for further pursuit of P/M titanium technology (other than to maintain in interesting employment a handful of metallurgists)? On the horizon there are, or will be, problems, "market pulls", which P/M titanium technology is uniquely qualified to address. These "pulls", new alloys and complex configurations, are not at all contradictory:

5.1 New Alloys

5.1.1 Those with poor workability

This category is directly analogous to the case of the superalloy discs, and is best represented by the TiAl family of light-weight, high strength (gamma) aluminides which possess such low ductibility, that they are quite difficult to forge.

5.1.2. Dispersion-strengthened titanium

By definition, this is a powder metallurgy application, since dispersion strengthened alloys are produced by techniques--gas atomization, melt extraction--which result in particulates rather than ingots. To date, most effort has been devoted to DS titanium alloy development and testing. Ordinarily, once an alloy composition has been arrived at, component manufacture procedes from the production, in turn, of ingot, billet, and bar, followed by forging to shape.

In the case of P/M superalloys or the high strength, P/M aluminum alloys, ingot production is replaced by the atomization of powder, which is then consolidated, by HIP or hot pressing, into a billet; the P/M billet is then converted to bar and component in a manner analagous to the processing of ingot-metallurgy material. It is, however, questionable whether the fine dispersion produced during the atomization of a dispersion strengthened titanium alloy would survive such extensive thermo-mechanical processing; the (rare earth oxide) dispersoid would likely grow, and/or agglomerate, as a result of the consolidation and working energy applied to it. In this case, direct HIP of the dispersion strengthened alloy powder to the component shape would be a logical fabrication procedure.

5.2 Complex configurations

There are two categories of special configurations for which direct HIP of alloy titanium powder represents an advantageous fabrication choice: bi-metal parts, and thin-wall, hollow, structural components.

5.2.1. Bi-metal parts

This category of component is typified by the one-piece, dual alloy, integrally bladed rotor or disc, in which the hub and flange are made of a high-tensile alloy, while the blades are made of a different alloy, one which is better able to resist creep and oxidation. The concepts of bonding powder to solid, and powder to powder have already been satisfactorily demonstrated.

5.2.2. Thin, Hollow Structurals

This class of (turbine engine) component represents a point on a component-complexity continuum, at one end of which are forgings, followed by welded fabrications, and by castings. At the other end of this spectrum there is a "pull" for components which can only be made by P/M; these parts are lightweight, complex, hollow, and (preferably) monolithic. The compositions are those of existing forgings and castings, as well as the aluminides. The parts have long, narrow internal passages with webs and outer walls .25 - .50mm thick. The applications require a finer grain size than can be obtained in castings, with far more complexity, inside and out, than can be achieved with forgings. Critical surfaces are net; attachment surfaces are machined.

The preceding analysis of P/M titanium technology--what it is, where it has been, and where it is going--is the view of one who has spent some time making, consolidating, heat treating, and testing blended elemental and prealloyed titanium powders. It is felt that this technology is only now coming of age.

6. Biography

Gerald Friedman is the Manager of Powder Metallurgy at Precision Castparts Corporation. He is presently responsible for setting up and running a facility for the prototype manufacture of complex, thin-wall hollow P/M titanium, titanium aluminide, and columbium alloy structural components for aircraft gas turbine engines.

He has thirty years' experience in the powder metallurgy of reactive and refractory metals and the nickel superalloys during which time he developed encapsulation and consolidation techniques for the fabrication of complex, hollow, thinwall titanium, titanium aluminide, and columbium alloy turbine engine components.

Mr. Friedman developed fabrication techniques and produced prototype quantities of sintered, forged, and HIP titanium alloy missile gyroscope components; he also developed both CIP/sinter and HIP procedures for the manufacture of high strength, weldable, P/M columbium alloy components for the Trident missile.

He holds three patents, and co-authored the "Titanium P/M Technology" section of the Metals Handbook, and the "Powder Metallurgy" chapter in the book, The Superalloys.

Mr. Friedman received the Metal Powder Industry Federation "Part-of-the-Year" award for the fabrication of a complex, hollow P/M superalloy structural component by the filled-billet extrusion technique.

1st International SAMPE Metals Conference
August 18-20, 1987

POWDER METALLURGY MANUFACTURING PROCESSES: APPROACHES, STATUS, AND FUTURE DIRECTIONS

Alan Lawley
Department of Materials Engineering
Drexel University
Philadelphia, PA 19104

Abstract

The status of powder metallurgy processing technology in near net shape manufacturing is reviewed with particular reference to specialty alloys for the aerospace, transportation, space and basic metal industries. Steps in the powder metallurgy process are first delineated. This is followed by a description of commercial and near commercial methods of prealloyed metal powder production by atomization, including gas atomization, the rotating electrode process, soluble gas atomization, centrifugal atomization, and ultrasonic gas atomization. Commercial techniques utilized in the densification of prealloyed specialty alloy powder are then considered, namely conventional pressing and sintering, injection molding, powder preform consolidation, hot extrusion and hot isostatic pressing of canned powder, and spray deposition. It is demonstrated that powder metallurgy is a source of new and innovative manufacturing processes, with compelling advantages over conventional ingot practice.

1. INTRODUCTION

The metals industries are in an era of increasingly intense global competition, particularly within the United States. Previous approaches to manufacturing efficiency through economy of scàle must, in many cases, be replaced by flexible manufacturing methods. This new and changing economic and technical climate has resulted in intense interest world-wide, in net or near net shape manufacturing (NNSM). Powder metallurgy (P/M) is a potent and compelling technology for NNSM with several economic and technical advantages over conventional ingot metallurgy (I/M) practice. Thus, the theme of the conference 'competitive advancements in metals and metals processing' is timely, as is a consideration of the status of P/M in the context of NNSM.

The conventional P/M route of pressing and sintering has been used for several years in the manufacture of near net shapes. More recent P/M processing innovation has led to the development of several viable processes for the hot consolidation of alloy powders to fully dense net or near net shapes. P/M technology now encompasses aluminum alloys, iron-base compositions, magnesium alloys, superalloys and titanium alloys. Starting with a prealloyed powder or powder blend of elemental powders, near or net shape complex parts or wrought forms (plate, bar, sheet, tubing...) can be produced via several P/M routes. Alternatively, spray deposition offers several attractive features for the production of near net shape preforms.

This contribution is a state-of-the-art review of the technology of powder making and consolidation with particular reference to the NNSM of specialty alloys. In particular, unifying concepts in the manufacture of parts and wrought structural alloys via P/M processing for the aerospace, transportation, space and basic metal industries are identified.

2. THE POWDER METALLURGY PROCESS

P/M processing encompasses the material processing techniques of powder production and subsequent powder consolidation. Compared to I/M, P/M offers enhanced compositional flexibility, chemical homogeneity, fine-scale microstructures and a high degree of materials utilization.

The three major steps in P/M processing are illustrated in Figure 1. Pressing followed by sintering results in a semi dense part or component, while fully dense wrought alloys are fabricated by hot consolidation of the powder (1,2).

3. POWDER PRODUCTION

The most important method for making prealloyed specialty alloy powder is atomization, or melt disintegration. In its various modes, atomization offers a reasonable degree of control over powder chemistry, cleanliness, size and shape. There are three commercial forms of atomization, namely gas atomization, the rotating electrode process, and vacuum (soluble gas). Brief descriptions are given of each approach; detailed reviews of atomization are given elsewhere (3,4).

3.1 Gas Atomization

The principle of gas atomizaton or two fluid atomization, is simple (2,3,4); a continuous stream of liquid metal is broken down into droplets by a subsonic or supersonic gas stream or jet. The overall configuration of a gas atomization unit is shown in Figure 2.

The process is governed by a number of interrelated operating parameters. Controllable variables include jet distance, jet pressure, nozzle geometry, velocity of gas and metal, and metal superheat. Several empirical expressions have been proposed that relate particle size distribution to atomization conditions. Gas atomization pressures are typically in the range 14 x 10^5 Pa to 42 x 10^5 Pa at gas velocities from 50 m sec^{-1} to 150 m sec^{-1}.

Two fluid atomization accounts for greater than 95% of atomizing capacity (about 1 Mt/year) worldwide. Typically, gas atomized powder is spherical with a log normal size distribution. Mean particle size is usually in the range 10µm to 300µm.

3.2 The Rotating Electrode Process

The principle of the rotating electrode process is shown in Figure 3. Material in the form of a rod electrode is rotated at ~250 rev/sec while it is melted by a plasma arc. The molten metal is ejected centrifugally in the form of molten metal droplets which solidify before hitting the walls of the inert-gas-filled chamber. Use of a transferred-arc helium plasma torch eliminates the tungsten contamination characteristic of earlier designs (3).

Rotating electrode powders are usually spherical, and have high surface quality but a relatively large particle diameter (~150µm). For comparable size particles, cooling rates are less for the rotating electrode process than for gas atomization.

3.3 Vacuum (Soluble Gas) Atomization

This is a commercial batch process based on the principle that when a molten metal supersaturated with gas under pressure is suddenly exposed to vacuum, the gas expands, comes out of solution, and causes the liquid metal to be atomized into spherical powder (5). A schematic of the vacuum atomization process is given in Figure 4.

3.4 Near Commercial Atomization Methods

In centrifugal atomization, molten metal is ejected in the form of droplets from a rapidly spinning crucible, plate, or disk. Particles subsequently cool in the collection chambe,r or can be gas-quenched as they leave the rotating carrier (6-8). Possible configurations for the process are shown schematically in Figure 5. In each, particle shape can be varied from spherical to flake. The powders are relatively coarse, with an average diameter of ~200 μm.

In ultrasonic gas atomization, disintegration of the liquid stream is brought about by means of high intensity shock waves. The method looks attractive for the production of small diameter (<50 μm) particles with a narrow size distribution at high quench rates (~ 10^5°C/sec). Hartman shock wave tubes are used to deliver high velocity pulses of gas (up to Mach 2) at frequencies in the range of 60,000-80,000 cps (9).

4. POWDER CONSOLIDATION

The extent to which the powder is densified during consolidation is dictated by end usage. Load bearing applications call for full-density. Pores in the consolidated material act as crack sites and centers of stress concentration.

4.1 Conventional Pressing and Sintering

Typically, compaction is performed in a simple double acting die under uniaxial loading with the lubricant mixed with the powder, Figure 6. In the sintering of compacts, the aim is to achieve a high sintered density since this improves mechanical properties. Control of final part dimensions is also important and this mandates only a small shrinkage from the green condition.

Pressed and sintered material, primarily iron and low alloy steels, at about 95% theoretical density is characterized by limited ductility and low toughness and fatigue resistance (1,10). Repressing, infiltration and post heat-treatments serve to improve mechanical property levels. However, the effect is usually insufficient to warrant load- bearing applications.

4.2 Injection Molding

Injection molding, originally developed for plastics, has now been adapted to allow for the molding of metal and ceramic powders. This had led to a number of viable processing routes in the production of complex net shape close-tolerance parts which would be impossible to fabricate by conventional press and sinter or other methods. Two patented methods (11,12) have been developed for the injection molding of metal or ceramic powders. These differ primarily in relation to the type of binder(s) used, the mode of introduction of the binder, and the mode of binder removal.

The amount of shrinkage is high in injection molding when the compact is sintered. Non-uniformity in sintering is not, however, a problem and part dimensions on intricate shapes can be held to within normal engineering standards, e.g. ±0.08 to ±0.12 mm (+0.003 to 0.005"). This usually eliminates the need for post-production machining.

4.3 Powder Preform Consolidation

In powder preform consolidation, a porous sintered preform is prepared which is then worked to full density (10,13). Working modes include forging, rolling, and extrusion. Powder preform forging is considered here.

Powder preform forging is a commercially viable method of fabrication for low alloy steel high-strength parts such as automotive camshafts, connecting rods, and pinion gears. These are two regimes of powder forging namely repressing (Figure 7(a)) and upset forging (Figure 7(b)). In repressing, extensive flow of material in the lateral direction is prevented by die wall constraint. Upset forging is characterized by extensive unconstrained (free) lateral flow of preform material during forging. If, during the later stages of upset forging, lateral flow of the preform is restricted, then subsequent densification will be in the repressing mode. Metallurgically, the two forging modes give rise to differing levels of microstructural integrity.

4.4 Direct Powder Consolidation

There are several commercial methods available for the densification of a loose powder mass. Hot isostatic pressing (HIP) and hot extrusion are considered here.

HIP utilizes a combination of elevated temperature and isostatic gas pressure to consolidate the powder (14,15). It is extremely versatile and commercial applications include the fabrication of tool steels, superalloys and titanium alloys. Powder consolidation to near-net-shape provides a high materials utilization factor. Prior to HIP, the powder is sealed in a flexible container and vacuum degassed. Metal cans are utilized with simple shapes while ceramic/glass molds allow for the HIPing of complex forms, Figure 8. As a specific example superalloy powders are HIPed at pressures ~100 MPa in the temperature range 1100-1300°C (3,16). Typically, a shrinkage of about 35% accompanies densification.

The capital cost of a HIP facility is high

and cycle times are long, typically several hours. These disadvantages have stimulated interest in, and development of, pseudo HIP processes for NNSM. Examples are rapid omnidirectional compaction (ROC), and the CERACON, STAMP, and ceramic mold processes (17,18). In these processes a granular medium is utilized to transmit pressure to the canned powder (or powder preform) in a pseudo isostatic manner. Thus, conventional mechanical or hydraulic presses can be used.

Powder consolidation to full-density by hot extrusion utilizes a combination of hot compacting and hot mechanical working (1,19). Microstructural control is achieved by manipulation of the parameters of the extrusion process; in particular the extrusion ratio controls the amount of deformation in the fully densified compact. Prior to hot extrusion, the powder is either precompacted or put in a metal can under inert gas and sealed. The can serves as an extrusion billet (Figure 9) and is either extruded directly to shape, or compacted against blank tooling to give a fully dense compact which is subsequently re-extruded or forged to the required shape. Because of the shear stresses intrinsic to extrusion, there is extensive shearing and flow of the particles, particularly at high extrusion ratios. Compared to their ingot metallurgy counterpart, the P/M extruded alloys exhibit finer and more equiaxed grain structures.

4.5 Spray Deposition

It is possible to produce and consolidate particulate material in an integrated processing mode, rather than by going through discrete, separate steps. Spray forming is a promising net shape powder technology. As the name implies, the material in the form of a spray of liquid metal droplets impinges on a substrate in order to build up a thick preform.

In the Osprey process (20) a melt of the alloy is inert gas atomized and the spray of metal droplets is directed onto a substrate, Figure 10. Several preform geometries can be achieved, depending on the configuration and movement of the substrate relative to the spray of metal droplets, Figure 11.

Using compact spray chambers, rapidly solidified powders and preforms have been produced for a wide range of alloys, including carbon and low-alloy steels, bearing steels, tool steels, stainless steels, and Ni- or Co-base superalloys. Typically, preforms exhibit densities >98% of theoretical at deposition rates up to 1500 lb/hr. Melt stock can be in the form of scrap. The several Osprey modes include spray rolling, spray forging, centrifugal spray casting, spray peening and splat coating.

5. TRENDS

Evolution of the P/M technologies described above for the production and consolidation of specialty alloy powders has resulted in a new and relatively mature metalworking activity. Looking ahead, a redirection in powder processing is emerging with the focus on real-time monitoring and control to assure 'built-in' quality. Equipment revisions are directed at defect minimization and computer simulation for better control of parametric limits, with a decrease in the time frame for reduction to practice. In terms of materials, recent trends reflect the coupling of P/M and rapid solidification in the processing of new classes of aerospace specialty alloys.

6. REFERENCES

1. Lenel, F.V., Powder Metallurgy: Principles and Applications, Metal Powder Industries Federation, Princeton, NJ 1980.
2. German, R.M., Powder Metallurgy Science, Metal Powder Industries Federation, Princeton, NJ 1984.
3. Gessinger, G.H., Powder Metallurgy of Superalloys, Butterworths, London, 1984.
4. Lawley, A., Journal of Metals, 33, #1, 13 (1981).
5. Wentzell, J.M., J. Vac. Sci. and Technology, 11, #6, 169 (1974).
6. Hodkin, D.J., Sutcliffe, P.W., Mardon, P.G. and Russell, L.E., Powder Met., 16, #32, 227 (1973).
7. Dougherty, T.S., Powder Met., 11, #22, 342 (1968).
8. Cox, A.R., Moore, J.B. and Van Reuth, E.C., Proc. Third Int. Conf. on Superalloys, Kear, B.H., Muzyka, D.R., Tien, J.K. and Wlodek, S.T., Eds., Claitor's Pub. Div., Baton Rouge, LA, 1976, pp. 45-53.
9. Grant, N.J., Journal of Metals, 35, #1, 20 (1983).
10. Lawley, A., Powder Metallurgy Processing - New Techniques and Analyses, Kuhn, H.A. and Lawley, A., Eds., Academic Press, NY, 1978, pp. 139-171.
11. Rivers, R.D., U.S. Patent #4,113,480, "Method of Injection Molding Powder Metal Parts"; September 1978.
12. Wiech, R.E., Jr., British Patent #1,516,079, "Manufacture of Parts from Particulate Material"; June

1978.
13. Lawley, A., Advances in Powder Technology, Chin, G.Y., Ed., American Society for Metals, Metals Park, Ohio, 1982, pp 75-97.
14. Hanes, H.D., Siefer, D.A. and Watts, C.R., Hot Isostatic Pressing, MCIC Report #77-34, Battelle Columbus Laboratories, Columbus, Ohio, 1977.
15. Fischmeister, H., Powder Met. International, 10, #3, 119 (1978).
16. Thompson, E.R., Ann. Rev. Mater. Sci., 12, 213 (1982).
17. James, W.B., The Int. J. of Powder Met. and Powder Tech., 21, #3, 163 (1985).
18. Ferguson, B.L., The Int. J. of Powder Met. and Powder Tech., 21, #3, 201 (1985).
19. McShane, H., Tutcher, M.G. and Sheppard, T., Powder Met., 21, #2, 47 (1978).
20. Leatham, A.G., Brooks, R.G. and Yaman, M., Modern Developments in Powder Metallurgy, Aqua, E.N and Whitman, C.I., Eds., Metal Powder industries Federation, Princeton, NJ, 15, 157 (1985).

7. BIOGRAPHY

Alan Lawley is Professor of Metallurgy in the Department of Materials Engineering at Drexel University. He received B.Sc. and Ph.D. degrees from the University of Birmingham, England. He was a Post-Doctoral Fellow at the University of Pennsylvania (1958-1961), Manager of the Solid State Laboratory at the Franklin Institute (1963-1966) and Head of the Department of Materials Engineering (1969-1979) at Drexel University. Dr. Lawley is a Fellow of ASM, a former President of the Metallurgical Society, President of AIME, a member of the National Materials Advisory Board, and Editor of the International Journal of Powder Metallurgy. His interests are in the areas of powder metallurgy, composite materials, and engineering education.

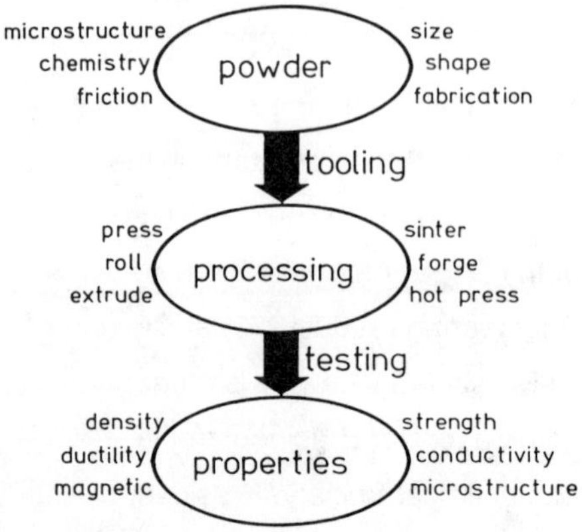

Figure 1: Schematic of powder processing steps (2).

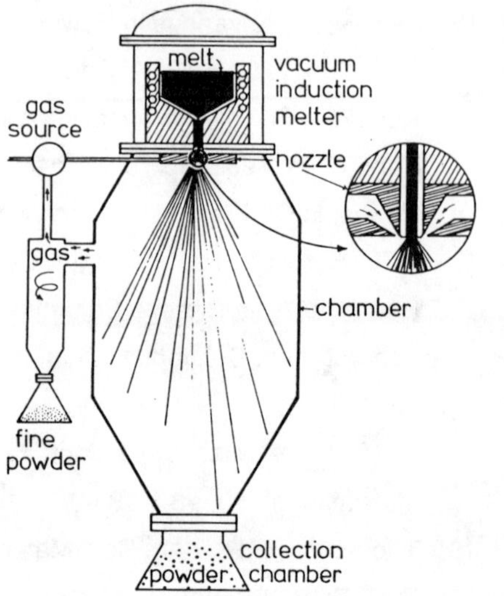

Figure 2: Schematic of gas atomization (2).

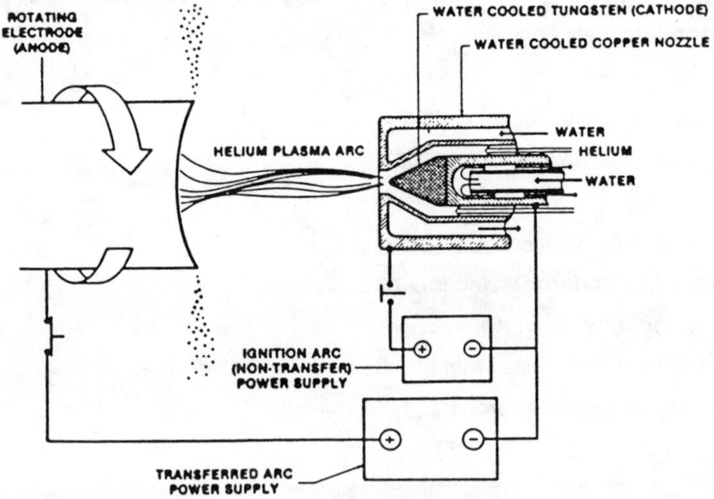

Figure 3: Schematic of the plasma rotating electrode process (3).

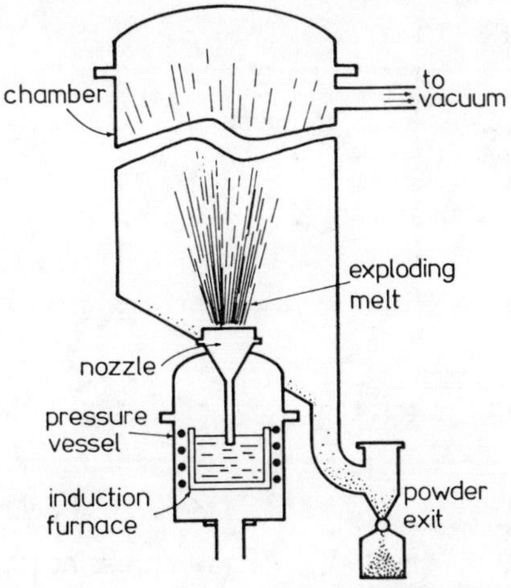

Figure 4: Vacuum (soluble gas) atomization (2).

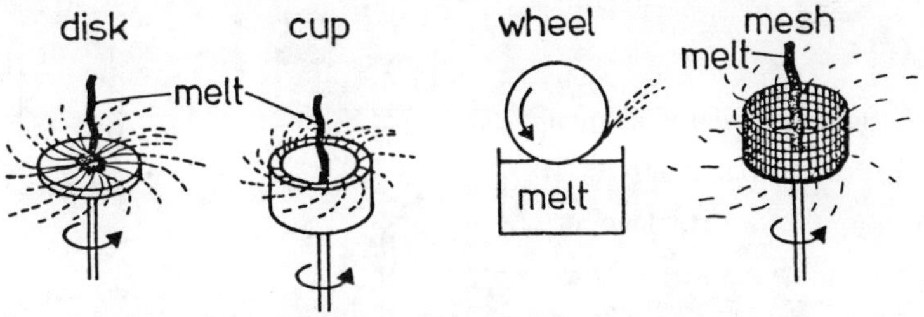

Figure 5: Centrifugal atomization configurations (2).

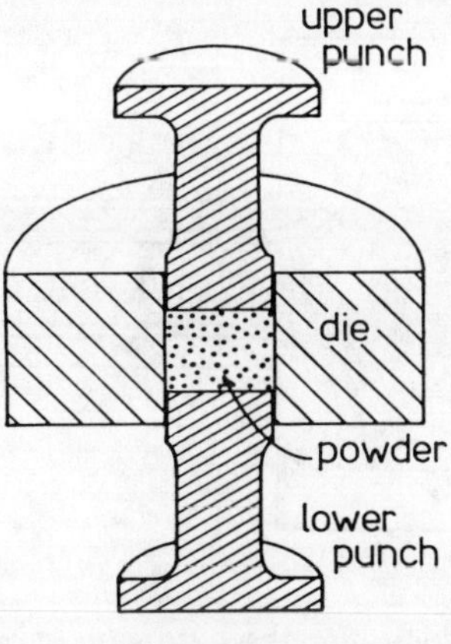

Figure 6: Double acting die for powder compaction - schematic (2).

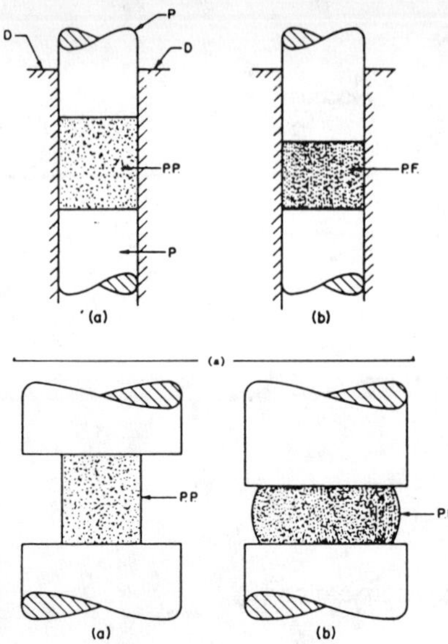

Figure 7: Powder preform forging. (a) repressing, (b) upset forging. D = die. P = punch. P.P. = powder preform. P.F. = powder forging (13).

Figure 8: Schematic illustrating HIP. (a) HIP unit. (b) simple shape (billet). (c) complex part. A = autoclave, G = gas. H = heater. P = powder. V = vacuum. M = metal can. C/G = glass/ceramic envelope (13).

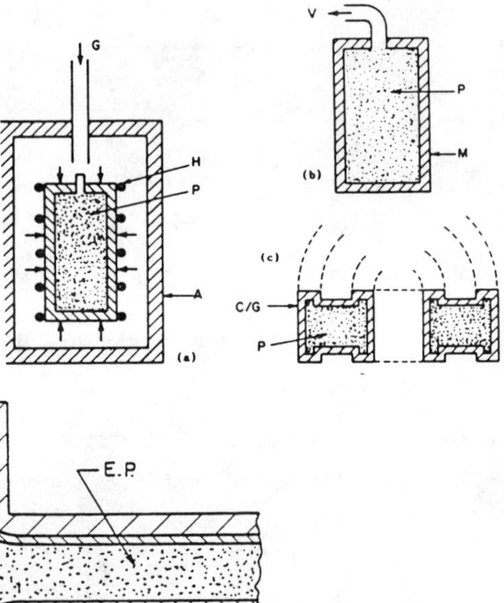

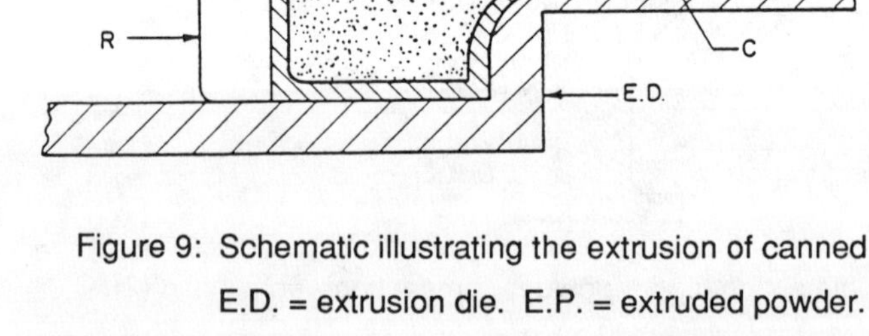

Figure 9: Schematic illustrating the extrusion of canned powder. C = can. E.D. = extrusion die. E.P. = extruded powder. R = ram. (13).

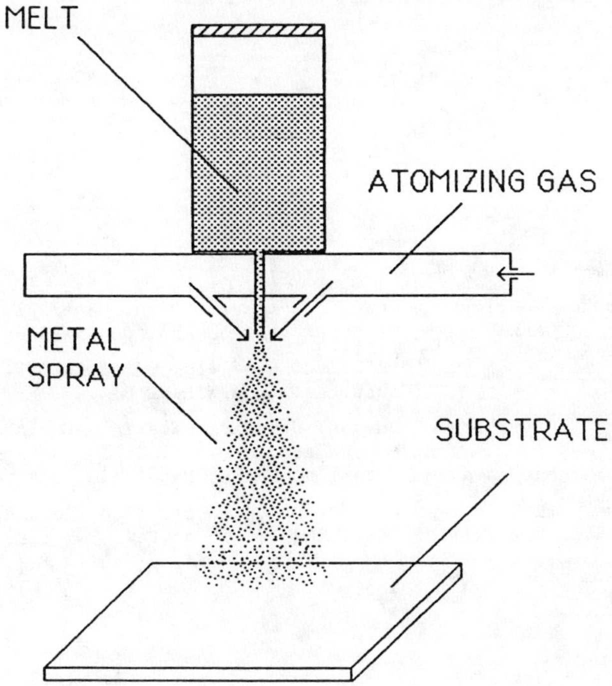

Figure 10: Schematic of the Osprey process.

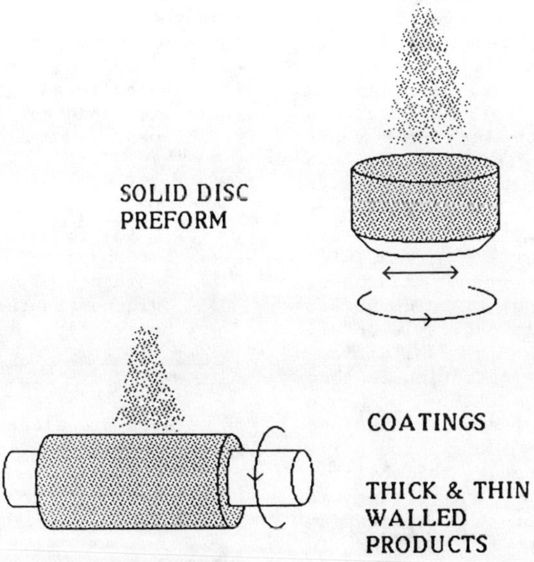

Figure 11: Preform geometries via spray deposition.

1st International SAMPE Metals Conference
August 18-20, 1987

POTENTIAL OF TITANIUM POWDER METALLURGY

F. H. Froes[a], H. B. Bomberger[b], D. Eylon[c], and R. G. Rowe[d]

[a]Air Force Wright Aeronautical Laboratories, Materials Laboratory
AFWAL/MLLS, Wright-Patterson AFB, OH 45433-6533

[b]385 Bradford Drive, Canfield, OH 44406-1004
(Consultant to AFWAL/MLLS)

[c]University of Dayton, Graduate Materials Engineering
KL262, Dayton, OH 45469-0001
(Work done as a consultant to Metcut-Materials Research Group)

[d]General Electric Company, Corporate Research and Development Center
P.O. Box 8, Bldg. K-1, 263MB, Schenectady, NY 12301-0008

Abstract

Traditionally titanium has only seen use in aerospace applications when superior performance was required. This was because of the high cost of producing titanium components, a result of the costly nature of extracting, working, and machining this metal. To decrease the cost, efforts in recent years have been directed towards net shape technologies, one of which is powder metallurgy. However, powder metallurgy parts of titanium alloys have still to receive wide acceptance. This paper discusses the three facets of titanium powder metallurgy: blended elemental, prealloyed, and rapid solidification approaches. Attention is also given to factors which influence the penetration of a new technology into the marketplace, including cost, performance, reliability, applicability, and the intangible "emotional factor." Titanium powder metallurgy is discussed in light of these factors and some projections are made on when significant use of this technology will be made for aerospace applications.

1. INTRODUCTION

The four major advantages of powder metallurgy (PM) are summarized in Table 1. These attributes are utilized in various proportions in current titanium PM technology which involves blended elemental (BE), prealloyed (PA), and most recently, rapid solidification (RS) techniques.

TABLE 1. MAJOR ADVANTAGES OF PM

1. Net shape and low cost
2. Avoidance of fusion state
3. Unique structures/properties
4. Homogeneity

BE methods use elemental and master alloy additions combined with pressing and sintering. To date little has been done to develop new alloys using this method.

The PA approach generally involves melting and atomizing, although comminuting can also be used to produce powders. When sufficient care is exercised parts have mechanical properties at least equivalent to those exhibited by conventional cast and wrought ingot metallurgy (IM) products. Greatly reduced directionality in PM compacts can result in higher design allowables because of reduced scatter of mechanical properties compared to similar IM products.

The aim of this paper is to examine current and future applications for the three PM methods. For more detail the reader is referred to recent reviews [1-5].

2. BLENDED ELEMENTAL (BE) TITANIUM POWDER METALLURGY

The BE approach is a press and sinter method [1, 2]. Since titanium alloy components are typically targeted for fracture and fatigue resistant high performance applications, efforts have been made in the past few years to produce compacts with close to 100% density for enhanced behavior. The BE material is produced from a mixture of "sponge fines" and elemental or master alloy powder [1, 2]. The sponge fines are the small (typically -100 mesh [-150 μm]) irregular shaped particles of high purity titanium produced in the reduction process. Since the commercial processes use chlorine [6], sponge fines contain about 0.15 wt% Cl and about 0.10 wt% Mg or Na. After blending, the desired alloy powders are cold compacted under vacuum up to 400MPa (60 ksi) with a mechanical press or a cold isostatic press (CIP) to 85-90% density. Relatively small components with no re-entrant angles can be produced by mechanical pressing. Larger and more complex shaped products are made by the CIP process [7, 8]. Compacts are vacuum sintered at about 1260°C (2300°F) to a density of 95-99%. Careful control can lead to 99% sintered density without secondary pressing [9]. Post-sinter hot isostatic pressing (HIP'ing) gives densities as high as 99.8%. BE sintered compacts have relatively little porosity, and can be HIP'd without containment [10, 11]. HIP'ing increases the density with about a 10% increase of the process cost. A 99.8% dense CIP+HIP compact microstructure is shown in Fig. 1a. Salts in BE material prevent 100% density on secondary pressing [12, 13] even with chlorine levels as low as 0.016 wt% [14, 15] and can lower resistance to crack initiation.

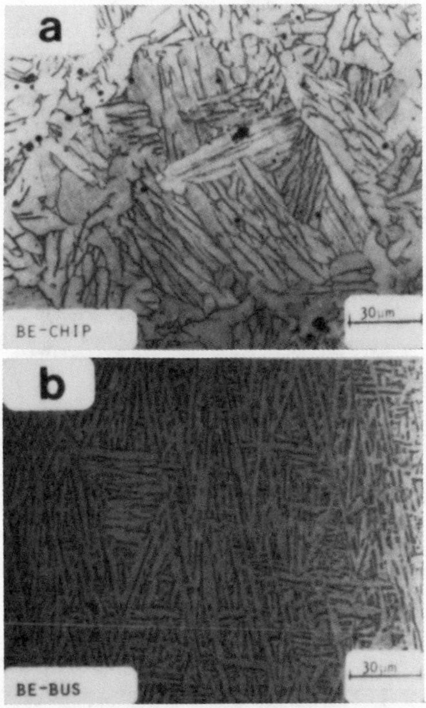

Fig. 1 - Microstructures of 99.8% dense Ti-6Al-4V produced by (a) CIP+HIP BE method and (b) after applying the BUS treatment.

Recently, salt-free (<0.010 wt% Cl) powder, produced by the hydrid-dehydride (HDH) process, became available and 100% density was achieved in CIP+HIP BE material [8, 16, 17]. Both static and dynamic mechanical properties of BE titanium compacts improve with density. The 99% dense sintered products meet minimum required IM tensile properties and fracture toughness [9, 14, 18]. However, small levels of porosity lead to degradation of crack initiation related properties such as fatigue

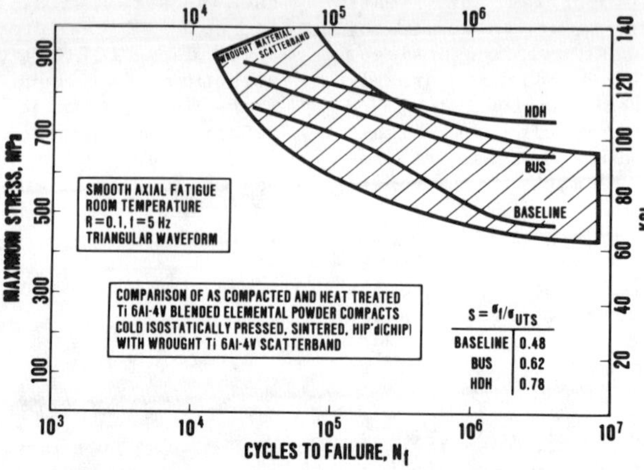

Fig. 2 - Comparison of fatigue behavior of 100% dense BE Ti-6Al-4V compacts with the wrought material scatterband.

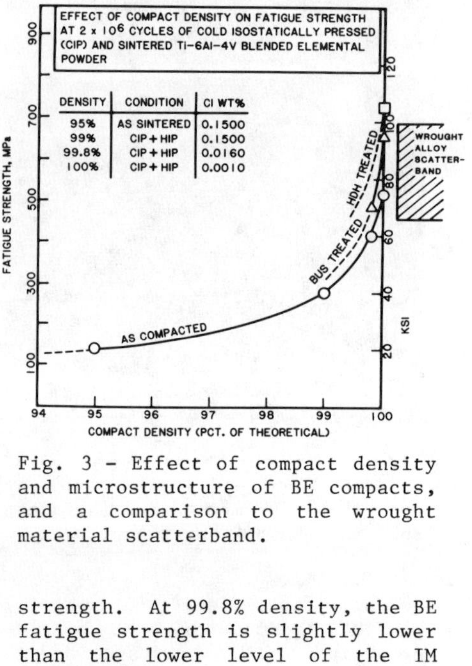

Fig. 3 - Effect of compact density and microstructure of BE compacts, and a comparison to the wrought material scatterband.

strength. At 99.8% density, the BE fatigue strength is slightly lower than the lower level of the IM scatterband [8, 14, 15]. Only 100% dense BE compacts had fatigue strengths equivalent to IM material [8, 17] (Fig. 2). Fatigue strengths of 99.8% and 100% dense materials can be improved with the "broken up structure" (BUS) treatment (Fig. 1b), thermochemical processing with the HDH treatment [17, 19], and by quenching from the beta phase field prior to HIP'ing [20]. Such improvements are shown in Figs. 2 and 3 and have opened the door for BE components in critical aerospace applications. The main benefit of BE technology is its relatively low cost as shown in Table 2.

Because of the relatively low price, several components are manufactured on a regular basis for aerospace and non-aerospace industries. The missile housing, Fig. 4a, was produced in a series of more than 10,000 parts by two manufacturers. A price savings of 60% over IM machined parts was demonstrated in the mirror hub shown in Fig. 4b. This is a net shape component with 99% as-sintered density. A BE lense housing is shown in Fig. 4c in the preform and final machined product, and an airframe part of 100% dense Ti-6Al-4V BE compact is shown in Fig. 4d. Commercially pure (CP) titanium components are shown in Fig. 5. BE parts for automobiles [21], such as a retainer cap and a connector rod, indicate substantial weight savings could result.

3. PREALLOYED (PA) TITANIUM POWDER METALLURGY

Starting stock for PA powder production contains the required alloy elements. Powder production is achieved by (a) comminuting or (b) melting, followed by atomizing, and subsequent cooling [2].

TABLE 2. COST[a] FOR TITANIUM POWDERS AND BE Ti-6Al-4V COMPONENTS[b]

Product	Cost Range, \$/lb
Sponge fines (0.15 wt% Cl)	3.5-5
HDH CP titanium powder (<0.001 wt% Cl)	20-23
Mechanically pressed compact (no HIP)	15-40
Cold isostatically pressed (CIP) compact (no HIP)	30-50
CIP+HIP compact	35-55
100% dense compact (HDH powder, CIP+HIP)[c]	60-80

[a]Based on a survey of four BE component producers.

[b]Based on components ∿1 lb in weight.

[c]Cost over conventional compact is due to higher cost of low chloride powder.

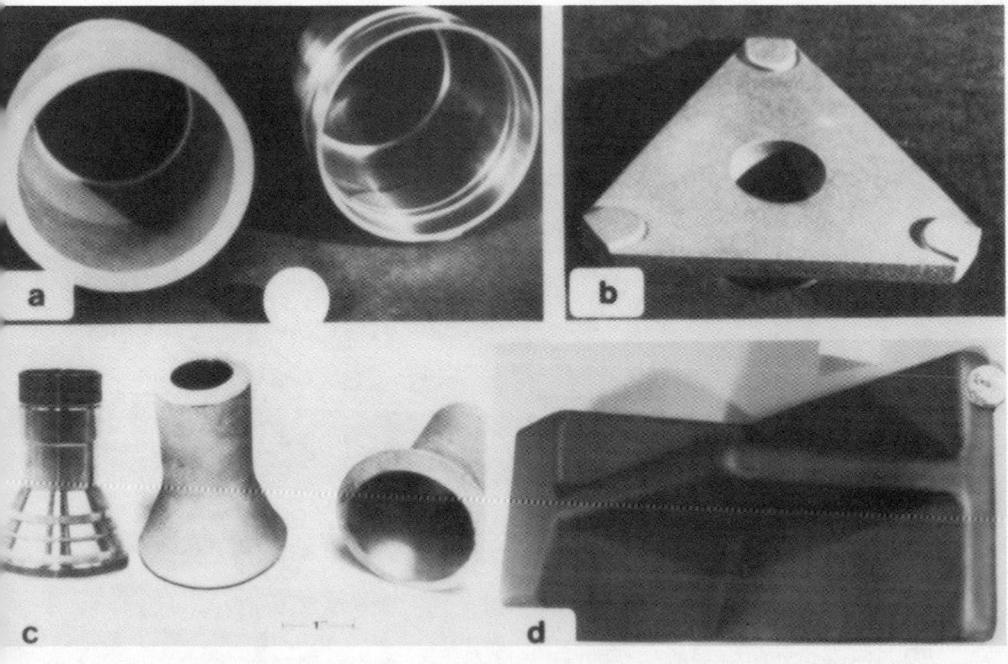

Fig. 4 - Production run of (a) BE Ti-6Al-4V missile housing (courtesy of Dynamet Technology), (b) BE Ti-6Al-4V net shape mirror hub (courtesy of Valform), (c) BE lense housing blank and finished product (courtesy of Clevite Industries), and (d) a BE Ti-6Al-4V 100% dense airframe component produced from CIP+HIP HDH powder (courtesy of Dynamet Technology).

For producing high-integrity near-net shapes, clean, spherical, PA powder is generally required [1, 2]. However, because of the reactivity, titanium powder cannot be easily produced by the processes used for less reactive metals [22, 23]. This led to processes in which only local melting occurs. Only the REP process is in production, using a plasma heat source (PREP) to yield powder free of contaminants [1, 2]. The PSV process appears promising but can suffer from aluminum segregation [1, 2]. Gas atomization appears to have potential for scale-up [24]. Attention has also been given to PA powder from chemical precursors [2, 6].

Fig. 5 - An assortment of (a) commercially-pure PM titanium parts for the chemical industry and (b) porous BE titanium filters for electrochemical processes (courtesy of Clevite Industries).

Three compaction processes are being developed: hot isostatic pressing (HIP'ing), fluid die compacting (FDC), and vacuum hot pressing (VHP'ing) [1, 2]. On HIP'ing, pressure and temperature in an autoclave result in full density. FDC [1, 2] involves production of shaped cavities or dies which are then filled with powder. A recent refinement involves Rapid Omnidirectional Compaction (ROC) [25].

Complex PM shapes have been made by one of four competing techniques: metal can, ceramic mold, fluid die [1, 2] and electroplated nickel can [26]. The ceramic mold process relies on investment casting technology [1, 2, 27].

Fatigue resistance of Ti-6Al-4V can be improved by either deforming the powder prior to compaction, by high strain rate compaction, by ROC [1, 2] or by compacting at relatively low temperatures and high pressures [2]. The microstructure of compacts (Fig. 6a) can also be modified by the BUS treatment (Fig. 6b) and by thermochemical processing (TCP) (Fig. 6c). These processes raise the high cycle fatigue strength of clean Ti-6Al-4V PREP compacts to the best levels in IM products [2]. Use of higher strength alloys results in tensile strengths and fatigue performance exceeding those of IM Ti-6Al-4V [2].

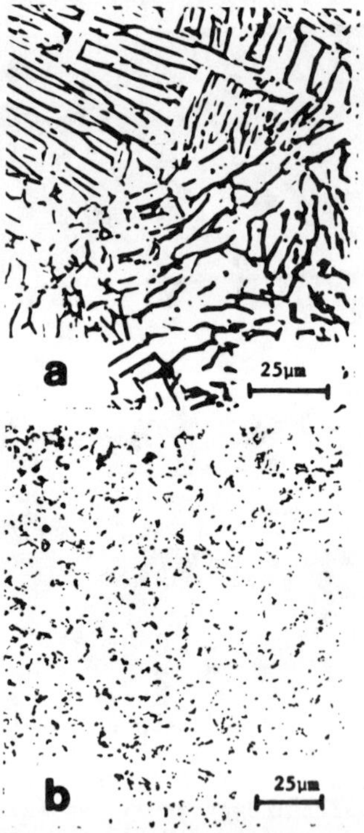

Fig. 6 - Microstructures of PA Ti-6Al-4V powder compacts (a) PREP as-compacted by HIP, (b) modified using the broken-up structure (BUS) heat treatment [1, 2], and (c) modified by thermochemical processing using hydrogen as a temporary alloying element [28, 29].

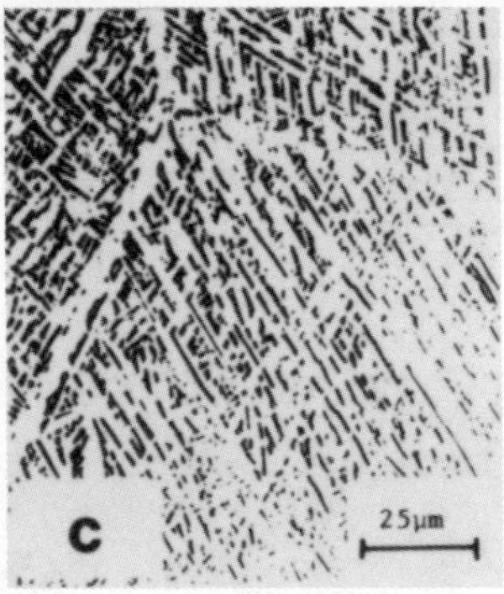

Fig. 6 - (continued)

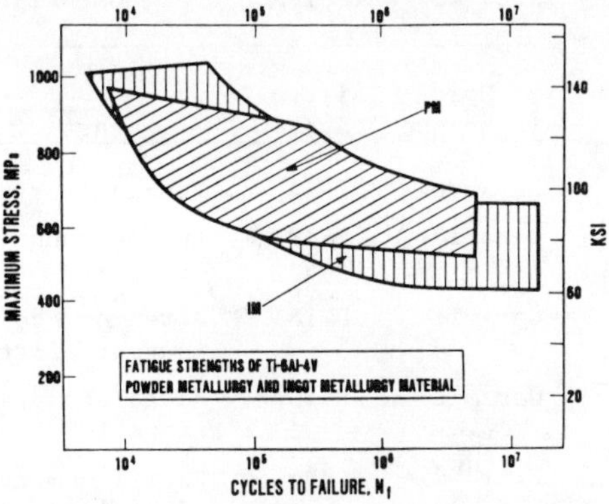

Fig. 7 - Fatigue behavior of PA compacts compared to IM material.

Properties of IM and PM products are affected by microstructures [30], foreign particles and contaminants [2, 22]. Contaminants have little effect on static properties [1, 2] and cyclic crack propagation [31], but can degrade crack initiation resistance [32, 33]. For IM material the chance of foreign particles being present is extremely small, and the very small chance of a failure in a component due to this reason is accepted on a statistical basis. Recent improvements in PM have resulted in low contaminant levels [2], so that PM should now be accepted in a similar manner to the IM route. In addition, work is in progress to improve NDT for considerably smaller inclusions [34, 35]. As a consequence, the mechanical properties of powder products are now at least equivalent to IM levels for tensile behavior, fracture toughness (Table 3) and critical fatigue behavior (Fig. 7). However, the final hurdle is to convince design engineers that they can now use powder products. This can be achieved by accumulation of data so statistically significant design curves can be developed.

The cost of titanium IM components lies mainly in material, forging, and machining [37, 38, 39]. Selection of parts for PM can be made by evaluating these factors. Generally, PM is most attractive for large, complex parts with a high buy-to-fly ratio. Presently, the largest autoclave available is 120 cm diameter x 240 cm [36]. Table 4 lists weights and anticipated cost savings for various parts produced by the PA technique [1-5].

Thus complex shapes with properties at least equivalent to those in IM products can be obtained in alloys such as Ti-6Al-4V and Ti-10V-2Fe-3Al using PM. An example is the impeller shown in Fig. 8 [40, 41]. Additional examples are given in other references [42, 43]. The remaining factor to consider is the economics of PM compared to the alternative cast and wrought methods. Recent studies by Witt [44], Moll [45], Kelto [46], Grewe [40, 41], Keinath [47], and Winkler [48] have shown that with certain

TABLE 3. TENSILE AND TOUGHNESS PROPERTIES OF PA Ti-6Al-4V COMPACTS [36]

0.2% YS, MPa (ksi)	UTS, MPa (ksi)	Elongation, Percent	RA, Percent	K_{Ic}, MPa $\sqrt{m}$ (ksi $\sqrt{in.}$)
930 (135)	992 (144)	15	33	77 (70)

TABLE 4. WEIGHT AND COST COMPARISONS OF INGOT AND POWDER PRODUCTS [1-5]

Mold Technique	Aircraft System	Part	Part wt. (lb) Forged Billet	PM	Final Part	Potential Cost Savings
Ceramic	F-14	Fuselage brace	6.2	2.5	1.7	50%
Ceramic	F-18	Engine mount support	17.0	5.5	1.1	20%
Ceramic	F-18	Arrestor hook support fitting	175.0	55.0	28.4	25%
Ceramic	F-107	Radial compressor impeller	32.0	6.2	3.6	40%
Fluid die	AH64	Radial compressor impeller	21.0	5.0	2.34	35%
Ceramic	F-14	Nacelle frame	315.0	181.0	53.2	50%

parts PM can be significantly less expensive.

Grewe and co-workers [40, 41] used ceramic dies for parts such as impellers. With this technique cost savings up to 40% (Fig. 9) could be obtained for parts as shown in Fig. 8. Conventional forging resulted in the costs shown in Table 5 [40, 41]. Equivalent figures for PM are given in the same table. Ways to lower PM costs further are to reduce powder costs (about 25% of the cost of the PM part) and to make parts closer to net-shape to decrease machining costs (about 75% of the cost of the PM impeller).

Fig. 8 - Impeller produced with PA powder [40, 41].

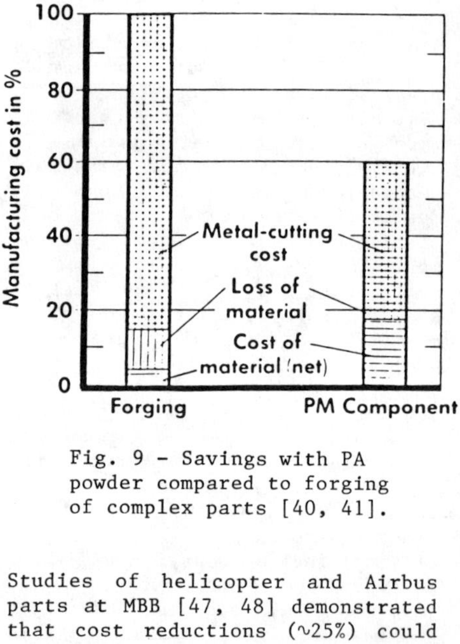

Fig. 9 - Savings with PA powder compared to forging of complex parts [40, 41].

Studies of helicopter and Airbus parts at MBB [47, 48] demonstrated that cost reductions (∿25%) could be achieved using PA powders. A helicopter fitting and a connector arm from the Airbus (Fig. 10) were produced with reductions in material use and machining cost. However, optimization of the conventional forging route resulted in a cost reduction (∿34%) and negated production of PM components.

In another analysis, Moll [45] used Crucible's ceramic mold process and an F-18 engine mount support (Fig. 11), and found that powder

TABLE 5. FORGING AND PM COSTS FOR A COMPLEX IMPELLER [40, 41]

Processing Step[a]	Conventional Forging $	Conventional Forging %	Powder Metallurgy $	Powder Metallurgy %
Forged rod[b], 19.25 lb	460	12	-	-
Centerless ground billet[c]	-	-	≅100	≅2.5
Powder making[d]	-	-	≅400	≅11
Shape making[e]	-	-	≅100	≅2.5
Machining and inspection[f]	3,330	88	1,670	44
TOTAL	3,790	100	2,270	60

[a] Using $1 = 2.1DM, 1 lb = 0.45 kg.

[b] $24/lb, final component weight 3.85 lb.

[c] ≅$25/lb.

[d] ≅$100/lb atomization.

[e] $20-40/lb, ≅$25/lb used. Includes qualification and testing.

[f] From German Aircraft Industry. PM part costs 50% of forged part.

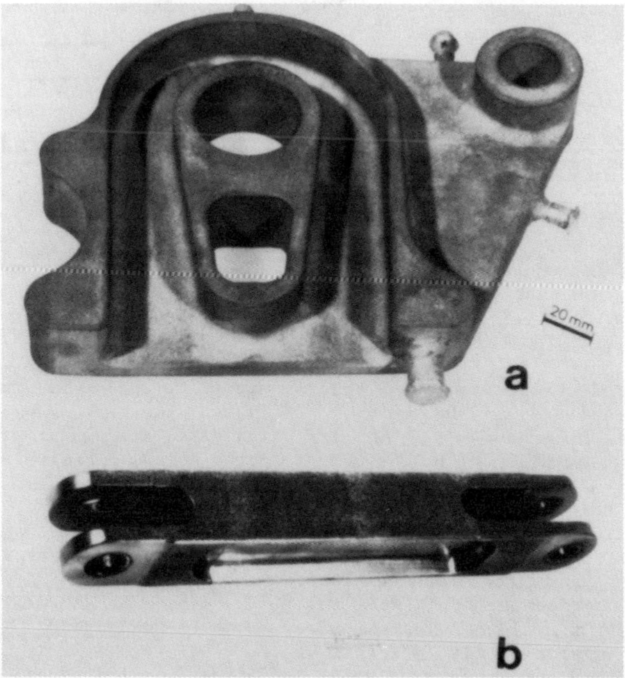

Fig. 10 - Prealloyed powder components: (a) helicopter blade connector fitting and (b) Airbus connector arm produced by MBB [47, 48].

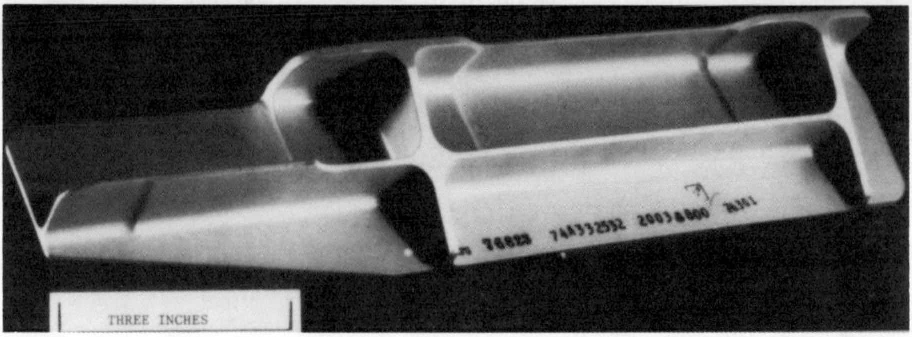

Fig. 11 - F-18 engine mount support produced with the Ceramic Mold Process (courtesy Crucible Materials Corporation).

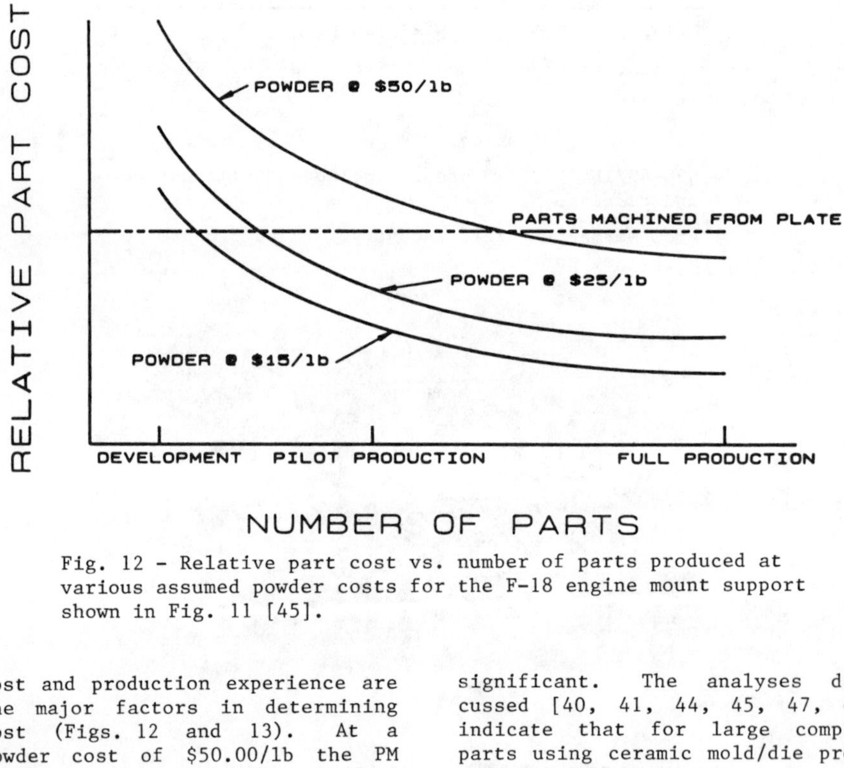

Fig. 12 - Relative part cost vs. number of parts produced at various assumed powder costs for the F-18 engine mount support shown in Fig. 11 [45].

cost and production experience are the major factors in determining cost (Figs. 12 and 13). At a powder cost of $50.00/lb the PM part cost is only reduced marginally below that of the part machined from a plate even in a full production environment. If powder cost could be reduced to $10.00/lb for comparison, Moll's analysis indicates that PM parts would cost approximately 1/3 that of conventional parts in the same full production mode. For more complex parts the savings with a $10.00/lb powder would be even more significant. The analyses discussed [40, 41, 44, 45, 47, 48] indicate that for large complex parts using ceramic mold/die processes PM can be highly competitive with conventional processing. Similar conclusions were reached by Kelto [46] using ROC.

4. RAPID SOLIDIFICATION TECHNOLOGY (RST)

Rapid solidification extends alloying and processing in a number of alloy systems [23, 49]. Requirements imposed by the reactivity of

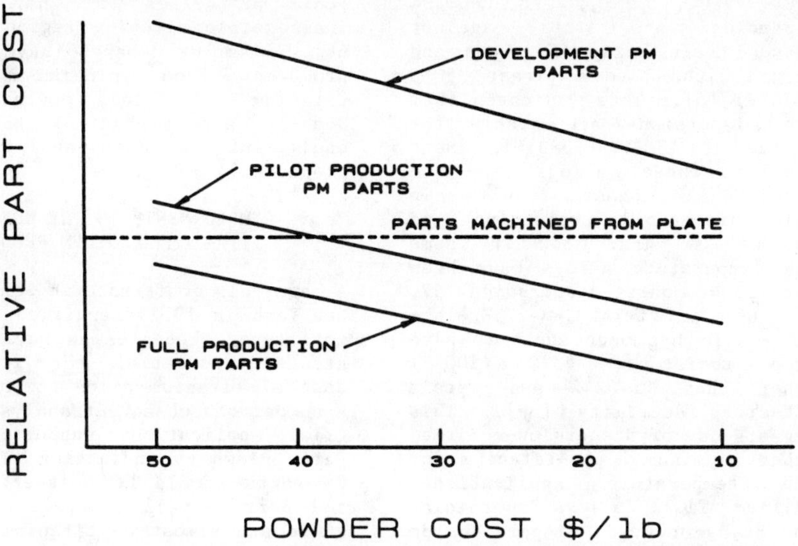

Fig. 13 - Relative part cost vs. assumed powder cost under various production environments for the F-18 engine mount support shown in Fig. 11 [45].

liquid titanium hinder such studies, but the potential benefits appear to be as great as those for less reactive metals [3]. RS has led to unique microstructures in otherwise conventional alloys, fine dispersions of stable second phases, ultra-fine grains, chemical homogeneity, and extension of solid solubility [3-5].

Several RST methods have been developed for producing contamination-free titanium alloys. One of these is the PREP process [40] described earlier. Although coarse powders made by PREP are not strictly rapidly solidified, the cooling rate for particles less than 100 microns in diameter may be as high as 10^5°C/s [50]. Gas atomization [24, 51] can produce fine powders which solidify at a comparable cooling rate. Substrate quenching, as in melt extraction and melt spinning, offers solidification rates in excess of 10^5°C/s, but has not produced titanium particulate in quantity [52]. Other processes have been explored for small quantities. These are laser spin atomization, electron beam splat quenching, laser or electron beam surface melting, and hammer-and-anvil splat quenching [53-56].

Fine dispersions of rare earth oxides, sulfides, and oxysulfides have been produced and retained after high temperature consolidation. Fig. 14 shows a dispersion of Er_2O_3 in a high temperature alpha-titanium alloy [57, 58]. The dispersoids have produced a 200 MN/m^2 (30 ksi) increase in yield strength and an order of magnitude reduction in creep rate in binary Ti-Er alloys [59].

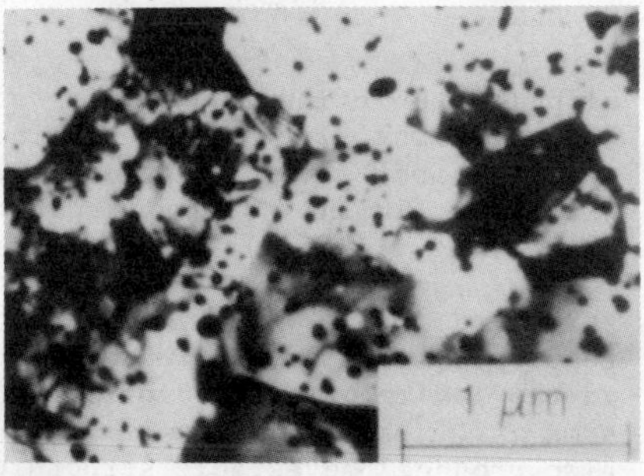

Fig. 14 - Melt extracted Ti-5Al-2.5Sn-3Zr-3.5Er after HIP, extrusion (840°C), aging (750°C/2 hr) [57]. Note fine erbia particles.

Er_2O_3, Y_2O_3, Ce_2S_3, and $Ce_4O_4S_3$ dispersions are stable against coarsening in hexagonal alpha and ordered alpha-2 to at least 950°C (1678°F) [5]. This indicates that such dispersions are stable for service at 750°C (1380°F). Heat treating these alloys in the beta-titanium phase results in rapid dispersoid coarsening and loss of strength. However, some high temperature alloys have been improved by coarse dispersoids [57, 59, 60]. Ti-7Al-3Sn-2.5Zr-1Nb-0.2Mo-2.1Er has been shown to have creep resistance 93°C (200°F) higher than Ti-6242 and retain post-creep ductility [59]. This suggests use of dispersion-modified higher aluminum alpha-titanium for high temperature applications. Modified Ti_3Al alloys containing fine Er_2O_3 or Ce_2S_3 dispersions do not exhibit greater ductility than alloys without a dispersion [61]. This suggests dispersion-strengthening rather than reduction of slip localization in alloys with aluminum content greater than 10 wt%.

Dispersion strengthening by metalloids B, C, and Si has also been demonstrated. Fine dispersions of TiB, TiC, and Ti_5Si_3 have been shown to strengthen both titanium and titanium alloys [54]. These additions also refine the grain size of RS titanium alloys [54, 58]. Fine grained equiaxed alpha microstructures are formed after consolidation of rapidly solidified Ti-6Al-4V [62, 63]. Such microstructures may lead to improvements in the fatigue properties.

Slip length refinement can improve the ductility of titanium-aluminum alloys and the intermetallic titanium aluminides by reducing slip banding or slip localization [64]. Refinement of microstructures and increased homogeneity by RS led to improvements in the low temperature ductility of some titanium aluminides [65]. RS also produced metastable structures in TiAl alloys, offering an opportunity to study the effect of composition on solidification, phase transformations and mechanical properties in heretofore inaccessible alloying regions [65, 66]. Complex shapes have been produced from niobium-modified titanium aluminides with 100% density and mechanical behavior equivalent to cast and wrought products [41, 67].

5. IMPEDIMENTS TO THE USE OF TITANIUM ALLOY PM PRODUCTS

A materials conference in Sagimore, New York in 1972 identified powder metallurgy processes as having the greatest potential for lowering cost of titanium alloy components. A number of subsequent analyses and trial applications support those early views by indicating titanium PM parts should be cost-effective and perform well. However, today there are almost no titanium alloy PM components in flight service.

Recent conversations with producers and potential users of titanium PM parts helped identify several reasons for the slow acceptance of titanium PM parts, including the following:

(a) There appears to be a reluctance in aerospace industries to replace known, dependable products and processes with ones having a relatively minor and less certain performance record unless there is a compelling need to do so. This is especially true for the more important and critical applications.

(b) Direct substitution of forged or cast parts with PM parts of the same composition, properties, and weight in established production aircraft offers no technical advantage. Nevertheless in some cases there are clear cost benefits. However, in some cases cost advantages are not obvious until buy-to-fly ratios are as great as 8:1.

(c) In the absence of extensive practical experience with titanium PM, a number of uncertainties still remain including cost trends for powders, finished PM parts and in the competing industries. Other uncertainties (which can also affect costs) include: the capacity of the present titanium PM industry to meet large-scale requirements and make timely deliveries, powder quality and part performance dependability, actual finishing requirements, and the reliability of economic estimates for producing various components. Thus it appears that more experience and knowledge would help to establish the high level of confidence desired for large-scale acceptance of titanium PM components.

(d) Timing was also cited as an important factor in the acceptance of titanium PM parts. It is important that early efforts be made to introduce new parts, technology and concepts in, for example, new prototype aircraft. This would permit remaining uncertainties about manufacturing, quality, performance, and actual costs to be resolved properly, more conveniently, and at less cost than is possible under routine production conditions. Generally, production schedules must be firm to meet delivery schedules; unforeseen delays and performance problems are costly, disruptive, and therefore, are unacceptable.

6. CONCLUSIONS AND FUTURE THOUGHTS

In the preceding sections it has been demonstrated that PM titanium alloys offer a viable low cost alternative to traditional IM for producing complex shapes.

The BE approach has been used successfully for producing components albeit with fatigue behavior below IM levels. The use of low salt sponge should increase use of this technique especially if the cost can be reduced below the $20.00/lb currently paid for this input material. An approach here would be to use low halide salt AlTi sponge [6].

For high-integrity cost-effective PA components to be accepted a number of actions must occur. The customer must insist on cost-effective products. Additionally design and performance data must be developed by industry, technical societies, and military organizations. Accompanying this, guidelines are needed for designers to use. Increased production runs would favor PM. Reduced machining costs of PM components would be maximized by only having to chemical mill components, perhaps in conjunction with shot peening to enhance fatigue behavior. Production of rare earth dispersion strengthened alloys for high temperature use can only be achieved via PM. This feature, in combination with lower costs, should stimulate use of PM. Added to this could be hybrid concepts more amenable to PM than IM fabrication. Finally PM component use will be accelerated by use in new aerospace systems rather than by replacement in current systems.

7. ACKNOWLEDGMENTS

The authors wish to thank the following for their contributions to this paper: S. Abkowitz, G. Bachvar, R. R. Boyer, G. Chanani, D. J. Czarney, E. J. Dulis, G. I. Friedman, M. F. X. Gigliotti, H. Grewe, A. Johnson, C. A. Kelto, E. F. Koch, H. Meeks, J. H. Moll, R. Peebles, I. S. Polkin, D. M. Rowell, W. Schimmel, J. A. Sutliff, E. Thellman, M. J. Trzcinski, G. E. Wasielewski, P. Winkler, R. H. Witt, and C. F. Yolton. We also want to acknowledge the able assistance of Miss Karen A. Sitzman in typing and formating the paper.

8. REFERENCES

1. Froes, F.H. et al., J. Metals, 32, No. 2, 47-54 (February 1980).

2. Froes, F.H. and Eylon, D., Titanium, Science and Technology, Vol. 1, ed. by G. Lutjering, U. Zwicker, and W. Bunk, DGM, 1985, pp. 267-286.

3. Froes, F.H. and Rowe, R.G., Rapidly Solidified Alloys and Their Mechanical and Magnetic Properties, Vol. 58, ed. by B. C. Giessen, D. E. Polk, and A. I. Taub, MRS, 1986, pp. 309-334.

4. Froes, F.H. and Rowe, R G., Titanium Rapid Solidification Technology, ed. by F. H. Froes and D. Eylon, TMS, 1986, pp. 1-19.

5. Rowe, R.G. and Froes, F.H., Processing of Structural Metals by Rapid Soldification, ed. by F. H. Froes and S. J. Savage, ASM, 1987, in press.

6. Hayes, F.H. et al., J. Metals, 36, No. 6, 70-76 (June 1984).

7. Froes, F.H., Eylon, D., and Friedman, G., ASM Metals Handbook, Vol. 7, Powder Metallurgy, Ninth Edition, ASM, 1984, pp. 748-755.

8. Abkowitz, S. and Rowell, D., J. Metals, 38, No. 8, 36-39 (August 1986).

9. Park, J. et al., Titanium Net Shape Technologies, ed. by F. H. Froes and D. Eylon, TMS, 1984, pp. 95-105.

10. Eylon, D., Froes, F.H., and Gardiner, R.W., J. Metals, 35, No. 2, 35-47 (February 1983).

11. Eylon, D. and Froes, F.H., Titanium Net Shape Technologies, ed. by F. H. Froes and D. Eylon, TMS, 1984, pp. 155-178.

12. Mahajan, Y., Eylon, D., and Froes, F.H., Powder Metallurgy of Titanium Alloys, ed. by F. H. Froes and John E. Smugeresky, TMS, 1980, pp. 189-202.

13. Welsch, G.E. et al., Met. Trans. A, 14A, No. 4, 761-769 (1983).

14. Abkowitz, S. et al., Titanium Net Shape Technologies, ed. by F. H. Froes and D. Eylon, TMS, 1984, pp. 107-120.

15. Weiss, I. et al., Met. Trans. A, 17A, No. 3, 549-559 (1986).

16. Abkowitz, S. and Rowell, D.M., Progress in Powder Metallurgy, Vol. 42, MPIF, 1986, pp. 611-624.

17. Eylon, D. and Froes, F.H., Progress in Powder Metallurgy, Vol. 42, MPIF, 1986, pp. 625-634.

18. Eylon, D. et al., Industrial Applications of Titanium and Zirconium: Third Conference, ASTM STP 830, ed. by R. T. Webster and C. S. Young, ASTM, 1984, pp. 48-65.

19. Vogt, R.G. et al., Titanium Net Shape Technologies, ed. by F. H. Froes and D. Eylon, TMS, 1984, pp. 145-154.

20. Hagiwara, M., Kaieda, Y., and Kawabe, Y., Proceedings of the 1986 International Conference on Titanium Products and Applications, TDA, 1987, pp. 850-858.

21. Brosius, E.S. et al., Proceedings of the 1986 International Conference on Titanium Products and Applications, TDA, 1987, pp. 831-839.

22. Froes, F.H. and Pickens, J.R., J. Metals, 36, No. 1, 14-28 (January 1984).

23. Savage, S.J. and Froes, F.H., J. Metals, 36, No. 4, 20-33 (April 1984).

24. Yolton, C.F. and Moll, J.H., Rapidly Solidified Materials, ed. by P. W. Lee and R. S. Carbonara, ASM, 1986, pp. 97-100.

25. Mahajan, Y.R. et al., Titanium, Science and Technology, Vol. 1, ed. by G. Lutjering, U. Zwicker, and W. Bunk, DGM, 1985, pp. 339-346.

26. Keinath, W., Mohs, R., and Bunk, W., Titanium '80, Science and Technology, Vol. 3, ed. by H. Kimura and O. Izumi, TMS, 1980, pp. 2215-2221.

27. Dulis, E.J. et al., Proceedings of the 10th National SAMPE Technical Conference, SAMPE, 1978, pp. 316-329.

28. Yolton, C.F. and Froes, F.H., US Patent 4,219,357, "Method for Producing Powder Metallurgy Articles," 26 August 1980.

29. Levin, L. et al., Titanium, Science and Technology, Vol. 4, ed. by G. Lutjering, U. Zwicker, and W. Bunk, DGM, 1985, pp. 2107-2114.

30. Smugeresky, J.E., Williams, J.C., and Froes, F.H., Powder Metallurgy of Titanium Alloys, ed. by F. H. Froes and John E. Smugeresky, TMS, 1980, pp. 303-308.

31. Schwenker, S.W. et al., Met. Trans. A, 14A, No. 7, 1524-1528 (July 1983).

32. Schwenker, S.W., Eylon, D., and Froes, F.H., Materials and Process Continuing Innovations, Vol. 28, SAMPE, 1983, pp. 436-447.

33. Schwenker, S.W., Eylon, D., and Froes, F.H., Met. Trans. A, 17A, No. 2, 271-280 (1986).

34. Singh, G.P. et al., Metal Powder Report, 39, No. 9, 504-508, 520 (September 1984).

35. Kwun, H. et al., NDT International, 17, No. 2, 73-79 (April 1984).

36. Parsons, L. et al., Metal Progress, 126, No. 4, 83-94 (September 1984).

37. Summary of Air Force/Industry Manufacturing Cost Reduction Study, US Air Force Technical Report, AFML-TM-LT- 73-1, January 1973.

38. Eylon, D. et al., Proceedings of the 12th National SAMPE Technical Conference, SAMPE, 1981, pp. 356-367.

39. Tupper, N.G., Elbaum, J.K., and Burte, H.M., J. Metals, 30, No. 9, 7-13 (September 1978).

40. Grewe, H., Potschke, J., and Weiglinin, W., Horizons of Powder Metallurgy, Part I, ed. by W. A. Kaysser and W. J. Huppmann, Verlag Schmid GmbH, 1986, pp. 75-78.

41. Grewe, H., Krupp, Essen, West Germany, private communication, August 1986.

42. Yolton, C.F. and Moll, J.H., Proceedings of the 1986 International Conference on Titanium Products and Applications, TDA, 1987, pp. 782-800.

43. Witt, R.H., Proceedings of the 1986 International Conference on Titanium Products and Applications, TDA, 1987, pp. 801-815.

44. Witt, R.H. and Ferreri, A.L., Overcoming Material Boundaries, Vol. 17, SAMPE, 1985, pp. 296-305.

45. Moll, J.H., Crucible Materials Corporation, Pittsburgh, PA, private communication, September 1986.

46. Kelto, C.A., ROC Development Center, Dow Chemical Company, Traverse City, MI, private communication, September 1986.

47. Keinath, W. and Engels, J., Titanium, Science and Technology, Vol. 2, ed. by G. Lutjering, U. Zwicker, and W. Bunk, DGM, 1985, pp. 1315-1320.

48. Winkler, P., MBB, Ottobrunn, West Germany, private communication, August 1986.

49. Jones, H., Rapid Solidification of Metals and Alloys, Monograph No. 8, The Institution of Metallurgists, 1982.

50. Vogt, R.G., Eylon, D., and Froes, F.H., Titanium Rapid Solidification Technology, ed. by F. H. Froes and D. Eylon, TMS, 1986, pp. 177-194.

51. Anderson, R.E. and Larson, J.L., Titanium Rapid Solidification Technology, ed. by F. H. Froes and D. Eylon, TMS, 1986, pp. 121-129.

52. Rowe, R.G. et al., Titanium Rapid Solidification Technology, ed. by F. H. Froes and D. Eylon, TMS, 1986, pp. 131-140.

53. Konitzer, D.G. et al., Met. Trans. B, 15B, No. 1, 149-153 (1984).

54. Sastry, S.M.L. et al., J. Metals, 35, No. 9, 21-28 (September 1983).

55. Konitzer, D.G., Muddle, B.C., and Fraser, H.L., Met. Trans. A, 14A, No. 10, 1979-1988 (1983).

56. Whang, S.H., J. Metals, 36, No. 4, 34-40 (April 1984).

57. Gigliotti, M.F.X. et al., Rapidly Solidified Alloys and Their Mechanical and Magnetic Properties, Vol. 58, ed. by B. C. Giessen, D. E. Polk, and A. I. Taub, MRS, 1986, pp. 343-352.

58. Rowe, R.G. et al., Rapidly Solidified Materials, ed. by P. W. Lee and R. S. Carbonara, ASM, 1985, pp. 107-114.

59. Sastry, S.M.L., Peng, T.C., and Beckerman, L.P., Met. Trans. A, 15A, No. 7, 1465-1484 (1984).

60. Gigliotti, M.F.X., Wasielewski, G. E., and Rowe, R. G., Titanium Rapid Solidification Technology, ed. by F. H. Froes and D. Eylon, TMS, 1986, pp. 141-152.

61. Rowe, R.G., Sutliff, J.A., and Koch, E.F., Titanium, Rapid Solidification Technology, ed. by F. H. Froes and D. Eylon, TMS, 1986, pp. 239-248.

62. Krishnamurthy, S. et al., Progress in Powder Metallurgy, Vol. 39, ed. by H. S. Nayar, S. M. Kaufman, and K. E. Meiners, MPIF, 1984, pp. 603-623.

63. Broderick, T.F. et al., Met. Trans. A, 16A, No. 11, 1951-1959 (1985).

64. Mendiratta, M.G., Sastry, S.M.L., and Smith, J.V., J. Mat. Sci., 11, No. 10, 1835-1842 (October 1976).

65. Huang, S.C., Hall, E.L., and Gigliotti, M.F.X., High Temperature Ordered Intermetallic Alloys, Vol. 81, MRS, 1987, in press.

66. Graves, J.A. et al., Scripta Met., 1987, in press.

67. Yolton, C.F. et al., Progress in Powder Metallurgy, Vol. 42, MPIF, 1986, pp. 479-488.

F. H. Froes received his Ph.D. in physical metallurgy from the University of Sheffield, England. He is currently Branch Chief, US Air Force Materials Laboratory, Structural Metals Branch, Wright-Patterson Air Force Base, Ohio. His responsibilities include the planning of research and applications related to programs on the light metals aluminum, titanium, magnesium, and beryllium. He is an adjunct professor at the University of Dayton and Wright State University and is a Fellow of ASM International.

H. B. Bomberger received his Ph.D. in metallurgical engineering from Ohio State University. He is a Visiting Scientist at the US Air Force Materials Laboratory, an adjunct professor in metallurgical engineering at Youngstown State University, and a consultant for industry. Prior to his recent retirement, he was Director of Metallurgy and Research for RMI Company. His work on titanium spans 37 years and covers all facets of the technology. He is a member of The Metallurgical Society and a Fellow of ASM International.

D. Eylon is a professor of metallurgy at the University of Dayton, Graduate Materials Engineering Department. For the past 14 years he was associated with the US Air Force Materials Laboratory doing research in the area of aerospace materials and powder metallurgy.

R. Grant Rowe is a metallurgist in the Materials Laboratory of General Electric Company, Corporate Research and Development Center, Schenectady, New York. His research is in the areas of rapid solidification of reactive metals, and the development of high temperature titanium and titanium aluminide alloys.

1st International SAMPE Metals Conference
August 18-20, 1987

ADVANCEMENTS IN ALCOA ALUMINUM P/M FOR AEROSPACE AND OTHER FUTURE APPLICATIONS

F. R. Billman and R. H. Graham
Alcoa Laboratories
Alcoa Center, Pennsylvania, U.S.A.

Abstract

Development of high performance aluminum alloy wrought powder metallurgy (P/M) at Alcoa has now resulted in the availability of two major categories of second generation materials. One alloy system is targeted for high strength applications where additional combinations of corrosion resistance, fatigue resistance and high fracture toughness are important. The second alloy category is the dispersion strengthened Al-Fe-Ce system designed for applications requiring strength up to 316°C (600°F). For both alloy categories the use of rapid solidification technology results in fine, homogeneous metallurgical structures. Performance data and projected commercial product costs are indicated. While these P/M products carry a material price premium over conventional aluminum ingot metallurgy (I/M) products, these initial material cost premiums can be offset by life cycle cost savings or cost avoidance in other areas of design or manufacture.

INTRODUCTION

Improvements in process understanding, process control and overall atomized powder cleanliness have led to new levels of strength, corrosion resistance, fracture toughness and fatigue performance in Alcoa's second generation wrought powder metallurgy (P/M) alloys. This paper will review the new alloy properties and characteristics, cover results of 1983-86 quality improvement efforts, discuss some P/M prototype components which are currently flying, highlight the ongoing AFWAL/Alcoa P/M Billet Scaleup Program, and finally indicate status and direction of Alcoa's P/M materials development program.

NEW ALLOY PROPERTIES AND CHARACTERISTICS

Table 1 lists three general application categories of P/M materials along with the respective Alcoa alloy designations. 1987 emphasis is being placed on the high strength CW67 (Al-9 Zn-2.2 Mg-1.5 Cu-0.14 Zr-0.1 Ni) and high temperature alloys (CU78 is Al-8.3 Fe-4 Ce and CZ42 is Al-7 Fe-6 Ce).

Beginning with the high strength series, Figure 1 compares strength/ toughness combinations for the venerable 7075-T73 ingot metallurgy alloy and the three Alcoa P/M alloys. The product form is extrusions of 38 mm x 114 mm (1.5 x 4.5 in.) cross-section. In effect, the new CW67 alloy material shown on the right retains the higher strength of the prior 7090 P/M alloy along with the higher fracture toughness of the earlier 7091 P/M alloy. Compared with 7075, CW67 P/M extrusions provide a 30% strength improvement along with a 40% toughness advantage.

Actual test data for P/M extruded bar are shown in Table 2. Room temperature ultimate strength values in the low 90s ksi, yields in the mid-80s with longitudinal K_{Ic} fracture toughness values near 50 ksi $\sqrt{in.}$ are noted. Short transverse toughness values are respectively in the mid to upper 20s.

Figure 2 shows similar trends for typical strengths and toughness values of forged products. Again, the strength and toughness advantages for the first generation 7090 and 7091 P/M alloys have been combined in the newest CW67 P/M alloy.

Earlier this year, several F-16 lower wing attach fittings (Figure 3) were precision forged in CW67 alloy for a General Dynamics Test and Evaluation program. Typical Alcoa lot release test data are given here in Table 3.

As demonstrated in Figure 4, the higher strength values of the CW67 P/M material can be achieved while retaining high stress corrosion cracking resistance. Short transverse samples were exposed by alternate immersion procedures for varying times at the stress levels indicated, then tested for mean breaking stress. The T6 temper of 7075 is quite sensitive to the accelerated SCC environment. In contrast to 7075-T73 I/M products, CW67-T7 P/M material can be exposed at 25% higher loads and still exhibit a significantly higher breaking load.

It's this combination of excellent strength, fracture toughness and corrosion resistance that makes CW67-T7 P/M forgings and extrusions excellent direct parts substitution candidates for existing airframe components that are now subject to very expensive maintenance and replacement sequences.

Elevated temperature P/M materials are the second material category of current emphasis. Figure 5 compares yield strength at temperature, after 100 hrs. prior exposure at the subsequent test temperature, among the Al-Fe-Ce P/M materials and three prominent I/M alloys. Currently, I/M alloy 2219 is regarded as the highest strength commercially available aluminum material for 232°C (450°F) application. At this same temperature, the Al-Fe-Ce materials provide strength capabilities about double that of the 2219 I/M. And, Figure 6 indicates that the microstructure remains stable up to 316°C (600°F) and results in essentially complete retention of the original room temperature strengths after exposures of up to 100 hrs. Additional tests have been conducted on forgings for brief exposures of 1/2 hr. up to 482°C (900°F) and 85-90% retention of prior room temperature strength was exhibited.

Recent advances in Al-Fe-Ce P/M process/structure/property understanding are demonstrated in Table 4. Through process development work at Alcoa Laboratories, K_{Ic} toughness values on die forged material have come up from the mid-teens to the high 20s with about a 10% strength decrease. In this table, Process I and Process II refer to methods of generating different strength/toughness combinations. Similar improvements in Al-Fe-Ce thin plate and sheet strength/toughness properties are reflected in the Figure 7 comparison of tear test data for CU78 P/M and two conventional I/M alloy-temper combinations.

Another major advantage of Al-Fe-Ce P/M materials is their superb corrosion resistance. Figure 8 reveals an immunity to SCC failures after 180 days alternate immersion testing at 40 ksi stress levels. General corrosion and exfoliation corrosion resistance is equally outstanding.

QUALITY IMPROVEMENTS

Alcoa's 1983-86 improvements in powder and P/M product quality are another area of competitive pride. Figure 9 shows results from use of smooth axial fatigue testing of extruded bars to measure quantitative improvements in P/M material performance as a consequence of reducing the number and size of harmful nonmetallic inclusions. Recent CW67 powder analyses indicate a <1 ppm level of foreign particles, after extensive equipment and process improvements at Alcoa's Pilot Production Atomizer near the Alcoa Technical Center. Mean fatigue performance has been increased by up to two orders of magnitude in these comparative tests of new vs. old quality CW67 P/M materials. The new P/M materials now exhibit a 10^5 cycle smooth fatigue performance at least equivalent to the best 7XXX alloy I/M materials. Additional improvements have since been directed at further reducing 10^4-10^5 cycle failure probabilities, and additional evaluation of specific prototype P/M parts and in-service loading conditions is warranted.

PROTOTYPE P/M COMPONENT TRIALS

There are several flight trials of Alcoa wrought P/M components now ongoing as listed in Table 5. Vought's leading edge flap modification on the A-7 attack aircraft includes (1) laminated 7475 I/M alloy skins with adhesive bond between skin and doubler plies, (2) 7075 doublers and (3) 7091-T7E69 P/M spar/hinge extrusions. Demonstration of this advanced flap design is part of a 56-month AFWAL/Vought program to provide structural improvements in operational aircraft. The advanced flap has shown a significant increase in life expectancies and reduction in retrofit costs at essentially equivalent weight and only 10% acquisition cost increase.

Warner Robbins Air Logistics Center is coordinating flight trials with extruded 7091-T7E69 forward longerons on the Lockheed C-141 transport as spare parts replacements for the corrosion prone 7075-T6 baseline material. Extrusions installed in 1984 now have in excess of 2,000 flight hours and are meeting the increased durability objectives.

A third part to discuss is the 7090-T7E71 P/M main landing gear door actuator forging on the Boeing 757 aircraft. As forged, these Alcoa-Vernon, California plant products weigh just over 100 lbs. each. The flying article weighs 21 lbs. and offers about a 16% weight savings over the previous 7075 I/M part designs. Approximately 100 components, assembled at two per airplane in an easily accessible location, are now being closely monitored in-service.

Other wrought P/M component demonstrations are planned by Northrop, Lockheed Georgia Company and General Dynamics as part of the ongoing AFWAL/Alcoa P/M Billet and Product Scaleup activities.

P/M BILLET AND PRODUCT SCALEUP

Table 6 outlines the general objectives and timing milestones for this important AFWAL Manufacturing Technology program. Alcoa is now about midway through the workscope, recently expanded to include both CW67 and Al-Fe-Ce P/M alloys. We are now making the initial 1200-1700 lb. P/M billets and seeking to identify other cost reduction approaches to fabricating P/M sheet and thin plate products.

OVERALL STATUS AND DIRECTION

Alcoa's Wrought P/M Program development status and direction are summarized in Table 7. We can now produce powder and billet in annual quantities of ~250,000 lbs. of CW67 or ~100,000 lbs. of Al-Fe-Ce alloys. This is adequate as a base for future customer T&E samples and prototype production studies. However, commercial capacities beyond those levels will require capitalization of new facilities. Alcoa is now actively considering joint venture possibilities with both industrial and government partners. Our current 350 lb. billet size is now being scaled-up to >1000 lbs., and we can fabricate extrusions and forgings up to the limitation of such billet sizes. To date, Alcoa has only rolled P/M plate and sheet samples on our laboratory rolling mill at the Alcoa Technical Center in widths ≤406 mm (16 in.); but we are now exploring alternate rolling facilities to generate 1219 mm (48 in.) wide sheet. We are now projecting future commercial P/M product prices in the $10-30/lb. range, pending both capitalization of new powder and consolidation facilities and improvements in overall P/M material recovery and manufacturing productivities. Alcoa's current Wrought P/M Technology Development Plan is summarized in Table 8. At this point, P/M commercialization decisions are still premature pending demonstrations of sustained improvements in P/M product performance and overall manufacturing cost reductions. Additional generations of P/M alloys and products, both monolithic and composite materials, will undoubtedly evolve at a faster rate once the generic technology has shown reliable cost/benefit advantages.

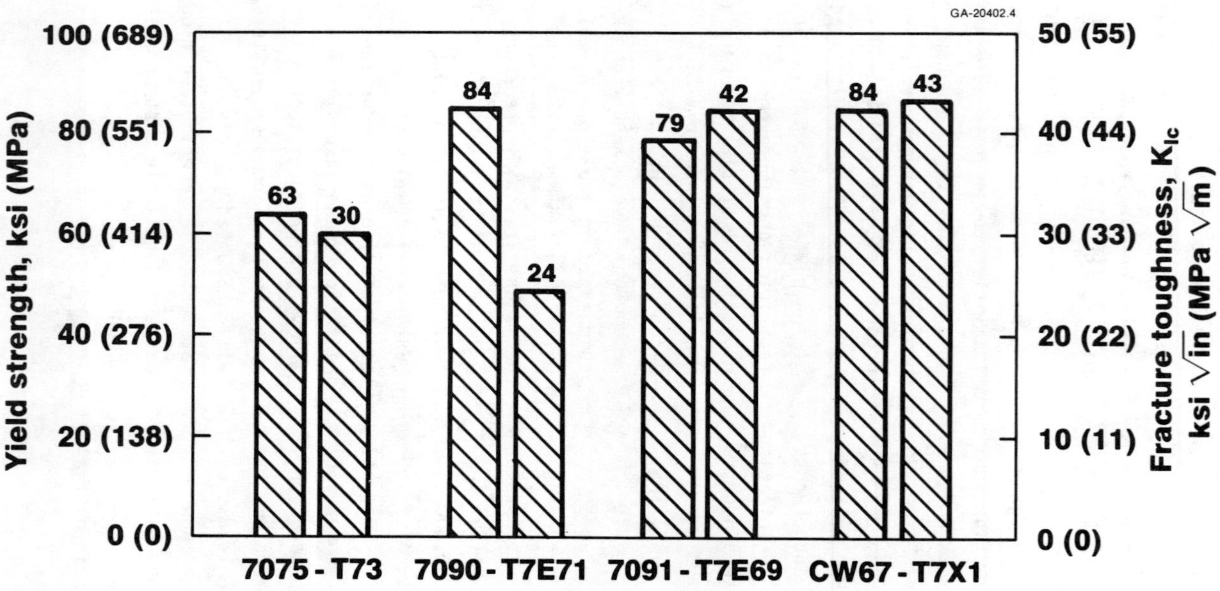

Typical Longitudinal Properties of 7XXX - T7 Aluminum Alloy Extrusions

1.5 Inch Thick X 4.5 Inch Wide (38 mm X 114 mm)

Figure 1

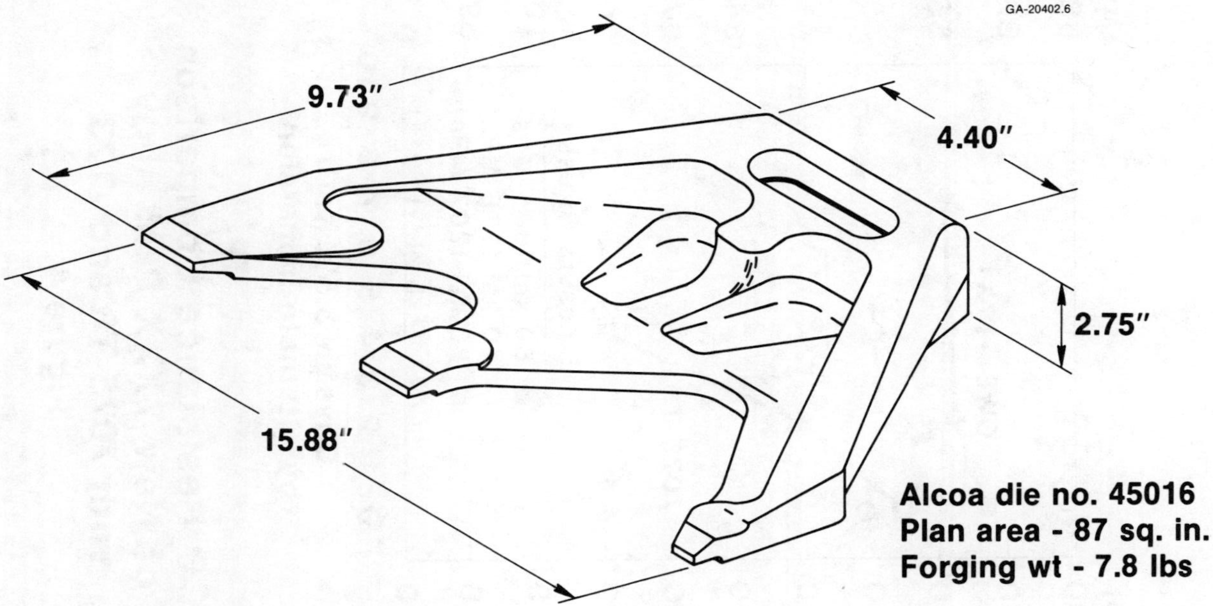

F-16 Lower Wing Attach Fitting
Figure 3

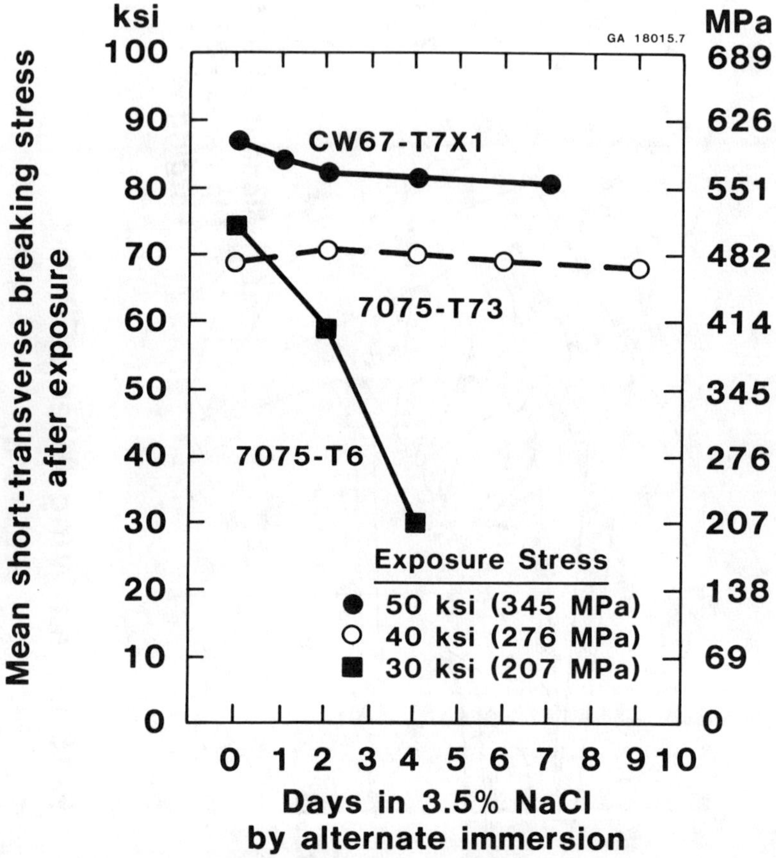

SCC Resistance Comparison of New 7XXX P/M Alloy with 7075-T6 and -T73
Figure 4

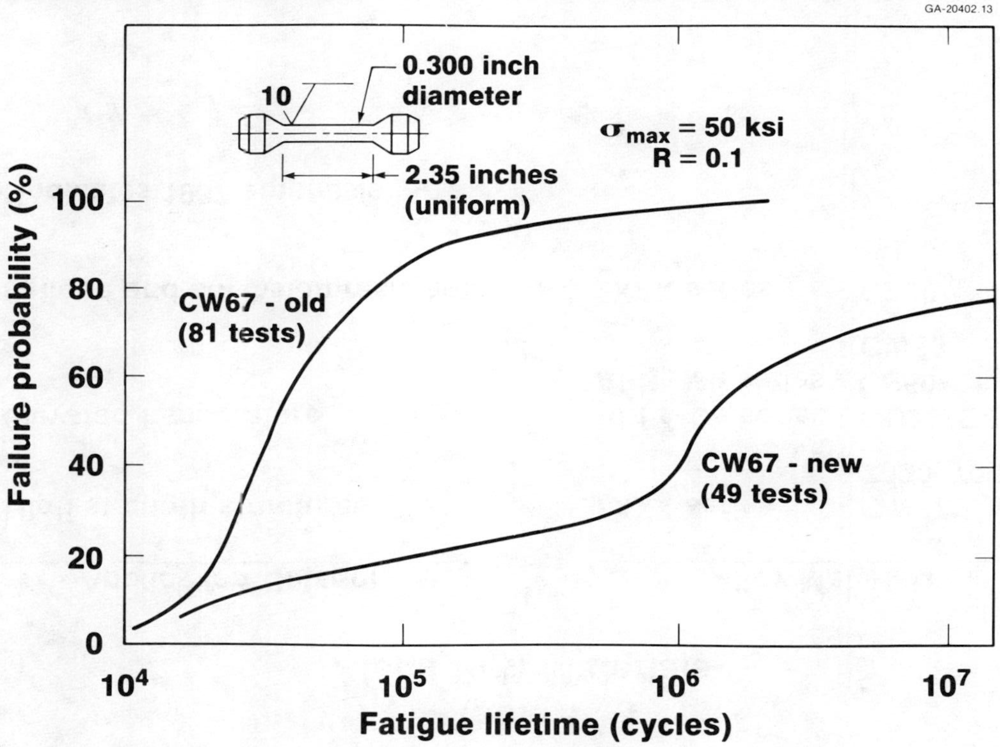

Axial Fatigue Probabilities for P/M Alloy CW67-T7X1

Figure 9

GA-20402.3

Table 1
Alcoa P/M Materials

Application Category	Alloy Materials	
High strength structures	7XXX series	- CW67* 7090, 7091
Elevated temperature	Al-Fe-Ce series	- CU78/CZ42*
	Al-Si-Mn series	- CS50 CW63
Fatigue and corrosion resistant	2XXX series	- CZ20

*Indicates 1987 emphasis

GA-20402.5

Table 2
Actual Mechanical Properties CW67-T6X/T7X Extruded Bar 1.5 Inch Thick X 4.5 Inch Wide Standard Processing

Alloy/temper	Test direction	Strength (ksi) UTS	Strength (ksi) Yield	Elong. (%)	R.A. (%)	K_{IC} (ksi $\sqrt{in.}$)
CW67-T6E192	L(L-T)	93.1	86.4	12.5	18	43.2
	LT(T-L)	86.5	79.9	11.2	—	27.6
	ST(S-L)	83.2	72.0	8.9	—	26.0
CW67-T7X1	L(L-T)	89.7	85.0	12.3	24	49.8
	LT(T-L)	84.0	79.8	10.7	26	31.0
	ST(S-L)	82.7	73.6	9.9	22	27.3
CW67-T7X1	L(L-T)	90.7	86.2	12.8	23	49.8
	LT(T-L)	85.9	81.6	10.9	23	29.8
	ST(S-L)	84.0	74.3	8.6	13	24.1
CW67-T7X8	L(L-T)	88.2	83.7	12.0	25	51.4
	LT(T-L)	83.1	78.7	10.9	26	34.0
	ST(S-L)	82.2	73.0	8.3	20	30.2

GA-20402.7

Table 3
Mechanical Property Data, F-16 Wing Attach Fitting CW67-T7E92 Forging, Die 45016

Test Direction	Strength (ksi) UTS	Strength (ksi) Yield	Elong. (%)	K_{IC} (ksi $\sqrt{in.}$)
L(L-T)	90.5	87.2	11	—
L(L-T)	95.6	93.2	13	—
LT(T-L)	83.9	78.0	10	—
LT(T-L)	90.1	85.8	11	—
ST(S-L)	83.1	78.6	11	22.7
ST(S-L)	84.0	78.4	10	—

GA 19870.11

Table 4
Mechanical Properties Al-Fe-Ce Die Forging, Die 50101

Test temp °F	Exp temp °F	Exp time hrs	Orient	Process I		Process II	
				CU78	CZ42	CU78	CZ42
Yield strength, ksi							
RT	--	--	L	54.7	53.4	62.3	59.1
RT	450	1000	L	55.3	54.0	61.7	65.7
Fracture toughness, ksi $\sqrt{in.}$							
RT	--	--	L-T	27.3	35.8[(b)]	16.6	11.9
RT	450	1000	L-T	32.2[(b)]	40.2[(b)]	13.8	12.3

(a) 1/2 Hour exposure
(b) K Q Values; all others are KIC

GA-20402 14

Table 5
Wrought P/M Flying Components

- **Extruded leading edge flap assembly**
  — Vought A-7 attack aircraft
- **Extruded forward longeron**
  — Lockheed C-141 transport
- **Forged landing gear door actuator**
  — Boeing 757 commercial airliner

GA-20402.15

Table 6
1987-88 Technical Efforts
Aluminum Alloy P/M Billet, Plate and Sheet Fabrication Contract F33615-82-C-5063

Objectives:

- Confirm baseline production capability and property data for CW67 and CU78
- Scale up P/M billet dimensions to mill size (1200-1700 lb)
- Identify cost reduction approaches to P/M plate and sheet manufacture
- Timing
  - — Pilot billet demo 3Q87
  - — Mill size 4Q87
  - — Final report 1Q89

GA-20402.16

Table 7
Wrought P/M Program Status and Direction

• Billet capacities:	CW67 - 250,000 lbs/year CU78/CZ42 - 100,000 lbs/year
• Billet size:	350 lbs current 1500 lbs 4Q87
• Product forms:	Sheet, plate to 16 inch wide (current) Extrusion, forging by billet size
• Projected commercial product costs:	CW67 - $10-$20/lb CU78 - $15-$30/lb

GA 19870.9

Table 8
Wrought P/M Development Plan

- Development for scale-up and reduced cost processing continues
- Material production support for test and evaluation continues
- P/M commercialization decisions require more cost-effectiveness demonstration:
  - Lower cost processes
  - Improved value products
  - Synergistic joint ventures
- Soliciting joint development to share facilities and technologies

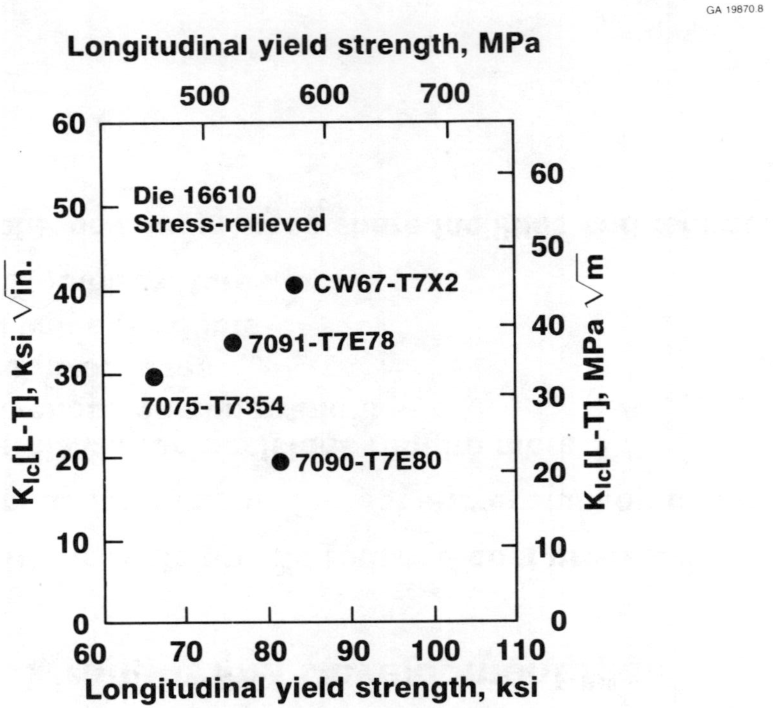

Comparison of Typical Fracture Toughness vs. Tensile Yield Strength for 7XXX P/M and 7075 I/M Forgings
Figure 2

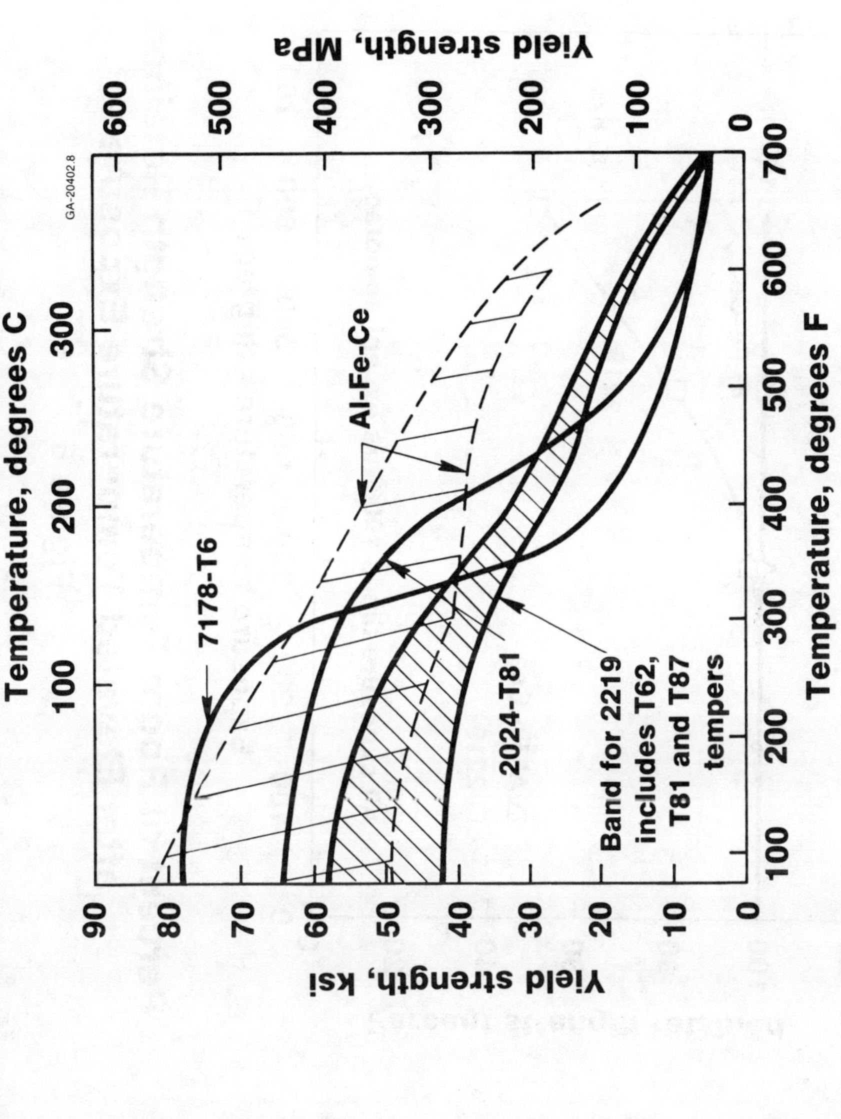

Yield Strength vs. Temperature - 100-Hour Exposure

Figure 5

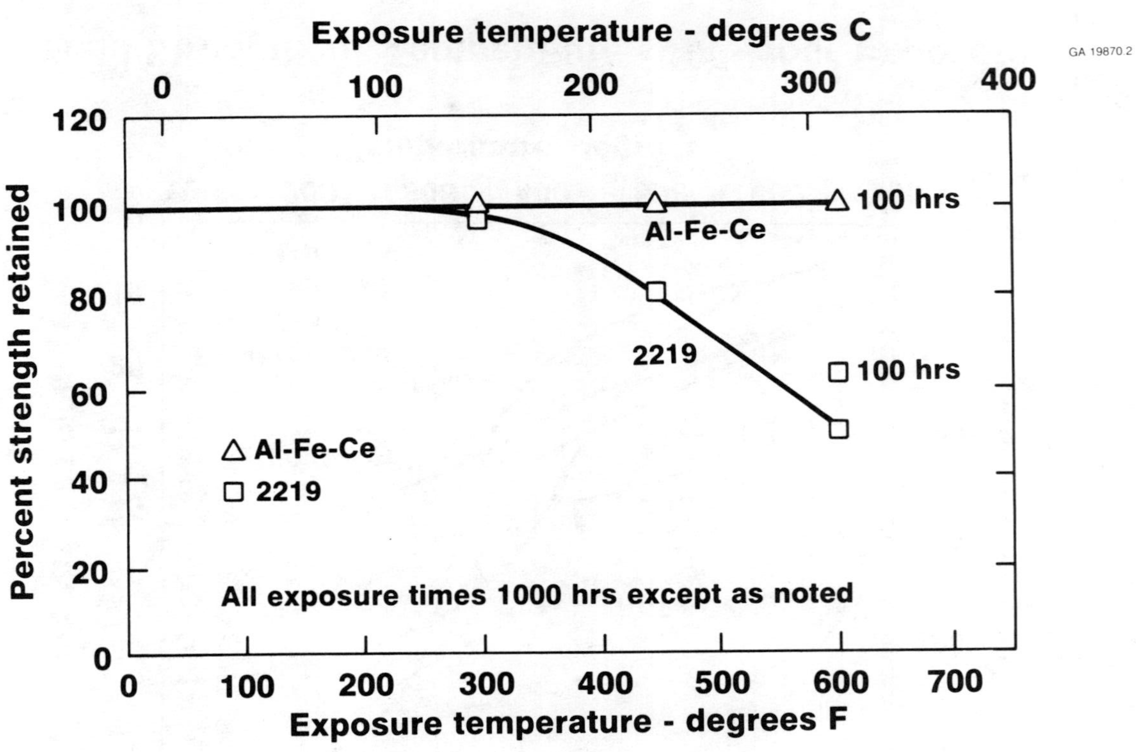

Percent of Room Temperature Strength Retained after Elevated Temperature Exposure

Figure 6

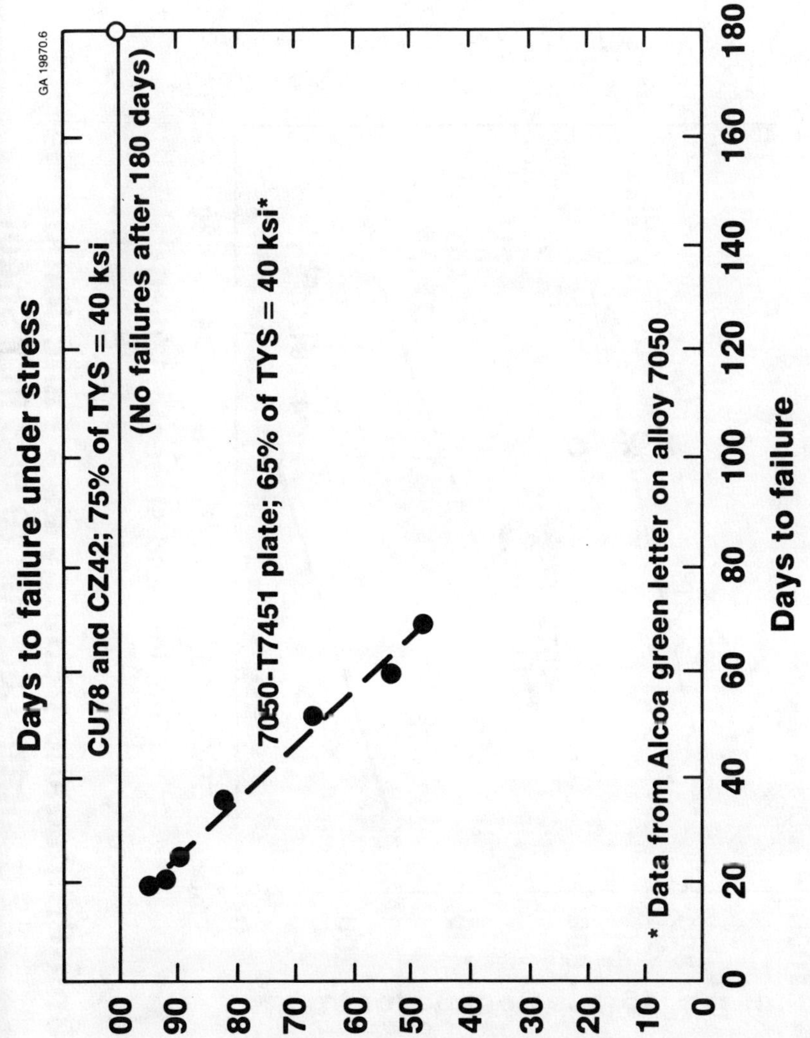

P/M Alloys CU78, CZ42 vs. I/M Alloy 7050-T7451 SCC Via Alternate Immersion

Figure 8

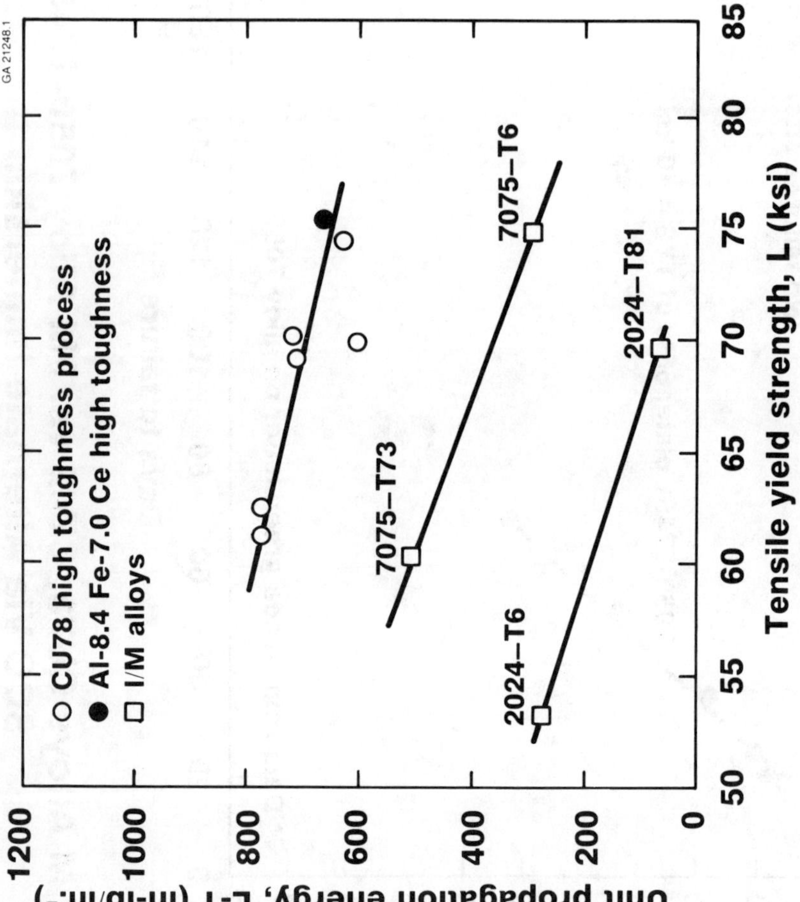

Unit Propagation Energy as a Function of Yield Strength 1.60 mm (0.063 in.) Sheet

Figure 7

1st International SAMPE Metals Conference
August 18-20, 1987

RAPID SOLIDIFICATION OF BORON-DOPED NICKEL ALUMINIDES

A. I. Taub, K-M. Chang and S. C. Huang
General Electric Corporate Research and Development
Schenectady, NY 12301, USA

Abstract

Microalloying of Ni_3Al with boron has been shown to produce high ductilities in this normally brittle material. The improved ductility is due to boron segregation to the grain boundaries where it acts to increase the cohesive strength thereby suppressing intergranular fracture. This paper reviews the unique mechanical characteristics of this ductile intermetallic system that make it useful for engineering applications. The relationship between alloy properties and the processing method are examined.

1. INTRODUCTION

The ordered intermetallic compound Ni_3Al exhibits the unusual behavior of an increasing flow stress with increasing temperature.[1,2] This property makes the alloy attractive for high temperature structural applications, but the conventionally cast material exhibits brittle intergranular fracture with zero plastic elongation at ambient temperature. Since single crystals of the compound have been shown to exhibit high ductilities in many crystal orientations even at low temperatures,[3] the low ductility of polycrystalline Ni_3Al has been attributed to the inherent weakness of the grain boundaries. Aoki and Izumi[4] found that by introducing small additions of boron to the binary intermetallic, the grain boundary cohesive strength could be improved and that consequently the ambient temperature brittleness could be overcome. The observed effect of the boron addition was quite dramatic. Doping of the binary intermetallic with only 0.5 at. pct. boron resulted in more than 30% plastic strain to failure compared to zero tensile ductility for the unmodified alloy processed under identical conditions. These results were later confirmed by Liu, et. al.,[5] who showed that through control of boron concentration and appropriate thermomechanical processing, plastic strains exceeding 50% could be obtained. The present authors[6,7] found that when processed by rapid solidification, boron-doped Ni_3Al attains high ambient temperature ductility in the as-solidified state and in addition exhibits a dramatic increase in strength as a result of boron solid solution strengthening.

In this paper, we will review two essential aspects of this ductile intermetallic. The first section deals with the mechanical properties of the boron-doped nickel aluminide. Both the boron ductilization phenomenon and the unique characteristics of mechanical behavior that make this intermetallic system useful for engineering applications will be discussed. The latter includes the positive temperature dependence of strength, high work-hardening rate and cryogenic properties. The second aspect is the relationship between alloy properties and the processing method. It will be seen that rapid solidification processing of the boron-doped nickel aluminide results in high ambient temperature ductility without the need for thermomechanical processing. This permits the fabrication of near net shape products with only thermal treatment(8) and may therefore offer a significant advantage over the conventional casting and thermomechanical processing route which in spite of successful laboratory results has encountered problems in large scale fabrication.(9) The rapid solidification techniques discussed include powder metallurgy, spray forming and plasma deposition.

2. MECHANICAL BEHAVIOR OF Ni_3Al

2.1 Boron Ductilization

Studies of the boron ductilization effect have shown that the boron segregates strongly to the grain boundaries of the intermetallic.(5,10) Grain boundary concentration enhancements of boron of about 10 times the bulk concentration have been reported. The local change in chemistry at the grain boundary suppresses the tendency for intergranular fracture permitting the matrix grains to undergo extensive deformation prior to fracture. The nature of the grain boundary strengthening mechanism is not well understood at this time and has been attributed to changes in electronic bonding,(10) surface and grain boundary energies(11) and dislocation mobility.(12)

One of the most striking aspects of the boron ductilization of nickel aluminide is the strong dependence on aluminum concentration. Figure 1 shows the plastic strain to fracture of melt spun ribbon as a function of Al and B concentration. It is seen that only aluminum-poor alloys (Ni/Al > 3/1) can be made ductile by boron addition. The ductility is also determined by the level of boron-doping. Boron concentrations of less than 0.1 at. pct. are capable of producing high ductilities, but at boron doping levels greater than about 1.5 at. pct., $M_{23}B_6$ borides precipitate at the grain boundaries and the alloy once again exhibits brittle intergranular fracture.[6] These chemistry effects on ductility are common to both the thermomechanically processed ingot product and rapidly solidified material. However, the effects on processing are much more pronounced in the ingot since segregation of aluminum and boron that occurs during solidification produces brittle regions in the castings and difficulties in post-solidification thermomechanical processing.

The increase in ductility resulting from boron doping is not limited to Ni_3Al-base alloys.[13] It has been found that Ni_3Ga behaves in a manner very similar to Ni_3Al. Ni_3Si can also be made ductile, but only if the material is properly heat treated to eliminate the multiphase structure resulting from solidification. The only other brittle, nickel-base $L1_2$ compound is Ni_3Ge. This intermetallic is not made ductile by boron-doping which is consistent with both the valence and electronegativity descriptions for grain boundary cohesive strength.[10]

2.2 Tensile Properties

The Ni_3Al intermetallic alloy system exhibits a high solubility for many elements that substitute on the Ni (eg - Co, Cu), Al (eg - Si, V, Ti, Zr) or random (eq - Fe, Cr) sites as well as for interstitial elements such as carbon and boron. This allows for extensive alloying additions to develop high strength alloys. Figure 2 compares the

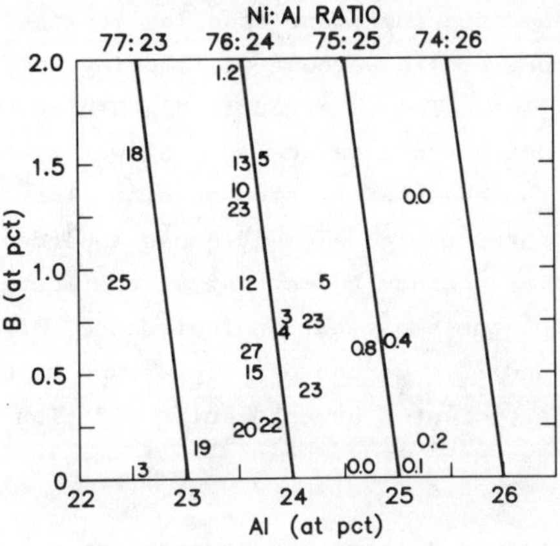

Figure 1- Plastic strain to fracture in a tensile test (see numbers) of melt spun nickel aluminide as a function of Al and B concentration. The solid lines indicate constant Ni:Al ratio. The ribbons were annealed at 1100C for 2 hours prior to tensile testing.

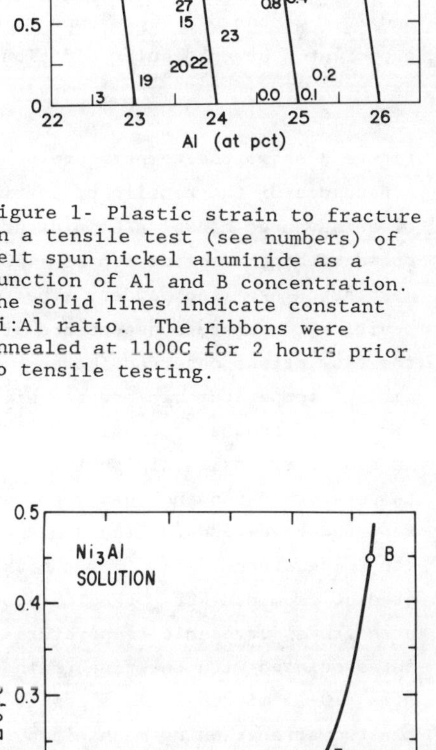

Figure 2- Correlation between strengthening potency and lattice strain for selected elements in Ni_3Al.

strengthening effect of different alloying elements defined as the average increment of alloy strength per atomic percent of alloying addition. The observed strengthening potency can be scaled nonlinearly to the observed lattice strain. The interstitial element B has a hardening potency twice that of the most potent substitutional elements, Hf and Zr. Carbon has been found to be as potent a strengthening addition as boron.[14]

Figure 3 shows the temperature dependence of the tensile properties of an advanced nickel aluminide produced by hip consolidation of -140 mesh argon atomized powder. The positive temperature dependence of the flow stress observed above ambient temperature in ordered Ni_3Al has been discussed extensively.[1,2] In single crystals this positive temperature dependence has been reported to extend to liquid nitrogen temperature (77K). For the PM product examined here, the increased strength at cryogenic temperatures is associated with the fine grain size (10-15 micron). As a result of the two strengthening mechanisms, anomalous temperature dependence of yield strength and Hall-Petch grain size strengthening, the alloy shows a yield strength minimum around room temperature. Note that the alloy retains excellent ductility down to liquid helium temperature (4K).

Boron-doped nickel aluminide exhibits a very high stage II strain hardening rate dS/de ~ 4100 MPa. This compares to about 1450 MPa for 304 Stainless (Fe-Cr-Ni), 2250 MPa for Hadfield Steel (Fe-Mn-C) and 2200 MPa for Inconel 718. Because of the high strain hardening rate, cold working results in high strengths. Powder metallurgy Ni_3Al-B can attain a yield strength above 1035 MPa with only 20% cold work and still maintain almost 20% ductility. This strengthening is important for cryogenic applications. In addition, a substantial fraction of this cold work-induced strength can be retained after a long time thermal exposure. For example, after a 30% cold reduction, a yield strength of 1725 MPa is obtained. During a hold at 500C, the strength decreases sharply in the first two hours of exposure but then stabilizes to a value of about 1200 MPa[15] which is comparable to the strength level of commercial high strength γ-γ' superalloys.

One of the barriers that limits the application of boron-doped nickel aluminide at elevated temperatures is the dynamic embrittlement problem. Figure 4 compares the plastic strain to failure of spray formed nickel aluminide in vacuum and argon contaminated with oxygen. For temperatures above 600C the ductility is seen to drop markedly in both testing atmospheres, but in the presence of oxygen the ductility loss is much more severe. In creep and fatigue testing, the embrittling effect of oxygen is observed at temperatures as low as 400C. It has recently been found that Cr additions improve the resistance to

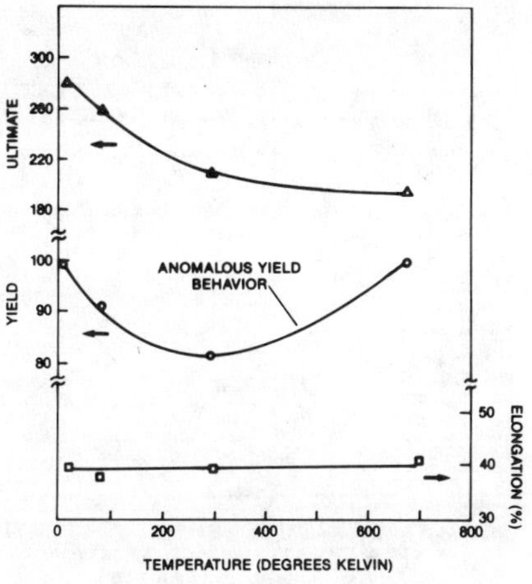

Figure 3- Temperature dependence of tensile properties for Ni-11.5Co-11.5Al- 1.75Hf-1.0Mo-0.5Nb-0.05Zr-0.05B produced by HIP consolidation of argon atomized powder.

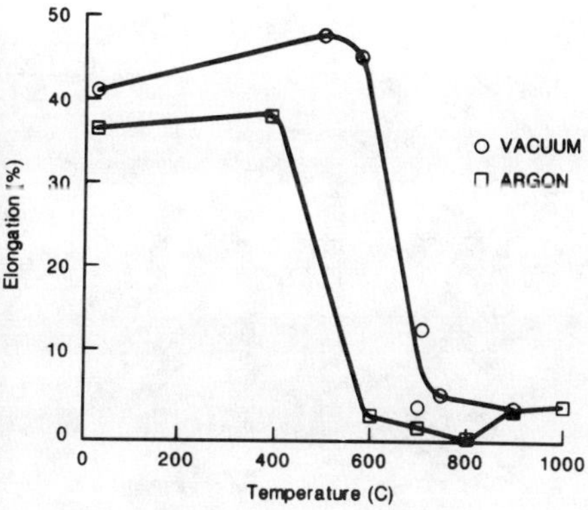

Figure 4- Plastic strain to fracture of spray formed Ni-10Co-24Al-0.25B in vacuum and oxygen contaminated argon.

oxygen embrittlement at elevated temperature.[16]

3. COMPARISON OF RAPID SOLIDIFICATION TECHNIQUES

Many rapid solidification techniques have been employed to produce boron-doped nickel aluminide that is fully ductile in the as-solidified condition. We will be concerned only with powder metallurgy, spray forming and low pressure plasma deposition processing since these techniques are capable of producing bulk specimens.

3.1 Powder Metallurgy

Inert gas atomization has been adapted commercially to produce high-strength nickel-base superalloys. Powder particles solidified from fine droplets of molten alloy cool at rates greater than 1000 C/s making this a rapid solidification process. The powder particles are subsequently consolidated by various techniques. For the nickel aluminide alloys reported here, hot isostatic pressing (HIP) at 1150C/15 ksi/2 hours has been selected as the consolidation route. The final microstructure consists of equiaxed grains with an average grain diameter of 40 μm. As seen in Table 1, the ambient temperature properties are not affected by subsequent thermomechanical processing. However, the elevated temperature ductility is improved by repeated cold working and annealing.[17]

3.2 Spray Forming

Spray forming attempts to eliminate the solid-state consolidation step of powder processing by collecting the atomized metal while it is still in the liquid state. In this process a stream of molten metal is atomized much as in conventional powder production. However, before the droplets are cooled significantly, they impact on a mandrel and form a deposit. The process as developed by Osprey Metals Ltd produces a product that is low in oxygen and which in some cases has properties superior to the conventionally cast product.(18) As seen in Table 2, the ambient temperature tensile properties of boron-doped nickel aluminide are comparable to the powder metallurgy product of Table 1. The grains are equiaxed with an average grain size of about 100 μm.

3.3 Low Pressure Plasma Deposition

A second dynamic compaction technique is the low pressure plasma deposition process. Powders are fed into the jet of a high velocity inert gas plasma where they are melted, transported and rapidly solidified at the mandrel. Low pressure during processing produces dense structures (>97% theoretical density) with oxygen contents essentially the same as that of the powder stock. With current plasma spray technology, deposits greater than 2 cm thick can be made.(19) Plasma spray is also useful for

Table 1

ROOM TEMPERATURE TENSILE PROPERTIES OF A P/M ALLOY, T56: Ni-24Al-0.75B

Processing	Yield Strength (ksi)	Tensile Strength (ksi)	Uniform Elongation (%)	Final Elongation (%)
As-HIP	66.1	192.7	41.6	44.6
Annealed	66.3	191.9	41.3	45.7
Cold Work + Annealed	67.9	192.6	37.3	41.0

Table 2

ROOM TEMPERATURE TENSILE PROPERTIES OF A SPRAY FORMED ALLOY, O58: Ni-10Co-24Al-0.75B

Processing	Yield Strength (ksi)	Tensile Strength (ksi)	Uniform Elongation (%)	Final Elongation (%)
As-Deposit	61	187	36	36
Annealed	58	182	38	38
Cold Work + Annealed	57	186	37	37

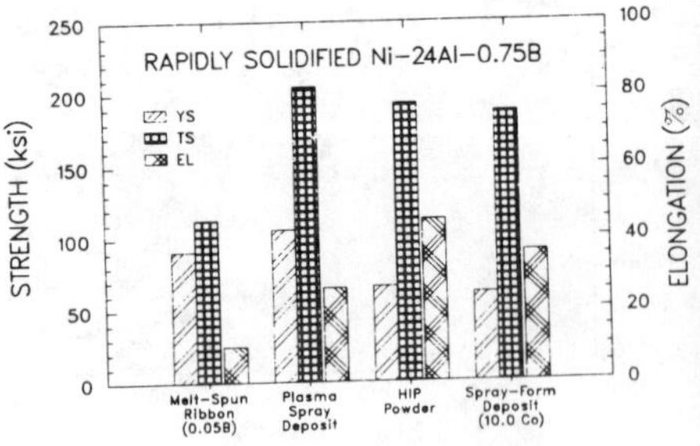

Figure 5- Tensile property comparison among boron-doped nickel aluminide materials processed by various rapid solidification techniques.

coating and fabricating thin section parts.

Plasma deposited nickel-base superalloys generally exhibit a fine grained structure. For the boron-doped nickel aluminides the average grain size was typically < 15 μm, even after annealing at 1165C. The fine grain size results in higher ambient temperature strength (see figure 5), but lower high temperature strength.

4. CONCLUSIONS

Boron additions to nickel aluminide suppress the tendency for brittle intergranular fracture of the binary intermetallic. Through additional alloying and cold working, very high strengths and ductilities can be obtained.

Rapid solidification technology has been successfully applied to the ductile Ni_3Al - B system. Unlike the conventionally cast material the rapidly solidified product is fully ductile in the as-solidified condition and subsequent thermomechanical processing is not required.

5. ACKNOWLEDGMENTS

One of the authors (AIT) acknowledges the partial support provided by the Office of Naval Research under contract N00014-86-C-0353.

6. REFERENCES

1. Thornton, P.H., Davies, R.G. and Johnston, T.L., Met. Trans. 1, 207 (1970).

2. Noguchi, O., Oya, Y. and Suzuki, T., Met. Trans. 12A, 1647 (1981).

3. Aoki, K. and Izumi, O. J., Mat. Sci. 14, 1800 (1979).

4. Aoki, K. and Izumi, O. J., Japan Inst. Met. 43, 1190 (1979).

5. Liu, C.T., White, C.L. and Horton, J.A., Acta Met. 33, 213 (1985).

6. Taub, A.I., Huang, S.C. and Chang, K.M., Met. Trans. 42A, 399 (1984).

7. Huang, S.C., Taub, A.I. and Chang, K.M., Acta Met. 32, 1703 (1984).

8. Chang, K.M., Taub, A.I. and Huang, S.C., Mat. Res. Soc. Symp. Proc. 39, 335 (1985).

9. Klarston, D.L., "Ni_3Al VIM and ESR Ingots", ECUT Workshop on Production and Fabrication of Ordered Intermetallic Alloys", Oak Ridge National Lab, TN, Sept. 25-26, 1984.

10. Taub, A.I. and Briant, C.L., Mat. Res. Soc. Symp. Proc. "High Temperature Ordered Intermetallic Alloys" (1986) in press.

11. Rice, J.R., Proceedings 1986 NATO Advanced Research Workshop, ed. R.M. Latanison and R.H. Jones, Martinus Nijhoff Publishers (1987) in press.

12. Schulson, E.M., Weihs, T.P., Baker, I., Frost, H.J. and Horton, J.A., Acta Met. 34, 1395 (1986).

13. Taub, A.I., Huang, S.C., Chang, K.M., Briant, C.L. and Jackson, M.R., Scripta Met. 20, 129 (1986).

14. Huang, S.C., Briant, C.L., Chang, K.M., Taub, A.I. and Hall, E.L., J. Mater. Res. 1, 60 (1986).

15. Huang, S.C., Chang, K.M. and Taub, A.I., "Rapidly Solidified Materials", ed. P.W. Lee and R.S. Carbonara, Amer. Soc. Metals 255 (1985).

16. Liu, C.T., Mat. Res. Soc. Symp. Proc. "High Temperature Ordered Intermetallic Alloys" (1986) in press.

17. Taub, A.I., Chang, K.M. and Huang, S.C., Scripta Met. 20, 1613 (1986).

18. Fiedler, H.C., Sawyer, T.F. and Kopp, R.W., "Spray Forming", Reports Nos. 85CRD073 and 87CRD034, General Electric Company, Schenectady NY.

19. Jackson, M.R., Rairden, J.R., Smith, J.S. and Smith, R.W., J. Metals 33, 23 (1981).

BIOGRAPHIES

Alan I. Taub - Dr. Taub received a BS in Materials Engineering in 1976 from Brown University and a MS and PhD in Applied Physics in 1979 from Harvard University. He joined General Electric Corporate Research and Development in 1979. His current research includes: development of mechanical constitutive laws for amorphous alloys; microscopic modeling of flow processes; and deformation and consolidation of rapidly solidified crystalline alloys, with emphasis on the boron modified nickel aluminide and other $L1_2$ intermetallic compounds. Dr. Taub is the author of 50 technical publications, which include three book chapters, and has been awarded 11 patents. In 1985, he was symposium co-chairman for the Materials Research Society meeting on rapidly solidified alloys.

Keh-Minh Chang - Dr. Chang received a BS in Physics from the National Tsing-Hua University, Hsin-Chu, Taiwan in 1972, a MS in Physical Metallurgy in 1977 and PhD in Physical Metallurgy in 1979 from the University of California at Berkeley. He joined General Electric Corporate Research and Development in 1980, where he works as a metallurgist in the Physical Metallurgy Branch of the Materials Laboratory. Since joining General Electric, Dr. Chang's research work has been in the area of high-strength alloys. He is a member of the American Society of Metals and the author or co-author of 40 technical publications and seven patents.

Shyh-Chin Huang - Dr. Huang received a BS in Metallurgical Engineering in 1971 from Chen-Kung University, Taiwan, a MS in Metallurgy in 1975 from the University of Connecticut, and a PhD in Materials Engineering in 1979 from Rensselaer Polytechnic Institute. He joined General Electric Corporate Research and Development in 1979. Dr. Huang has worked in a variety of areas in physical metallurgy and process metallurgy, including floating-zone melting, superconducting materials, solid-solution hardening, and solidification. Dr. Huang was a principal investigator in GE's development of high solidification rate processing for the fabrication of amorphous metal ribbons. Currently, he is investigating similar process methods for the development of high temperature crystalline alloys. Dr. Huang is a member of the American Society for Metals, AIME, the Society of Sigma Xi, Alpha Beta Gamma, and is the author or co-author of 48 technical publications and seven patents.

1st International SAMPE Metals Conference
August 18-20, 1987

INVESTMENT CAST PREMIUM QUALITY MAGNESIUM ALLOYS

A.E. Torok
P.C. Wilson
Arwood Corporation, An Interlake Company
Tilton, New Hampshire

Abstract

The process of investment casting magnesium alloys into premium quality aerospace castings has been used at Arwood for over 35 years. Fundamental considerations such as mechanical properties, alloy selection, molding and casting processes, surface protection, and safety are discussed to help the Design Engineer understand foundry capabilities and design limitations.

INTRODUCTION

For the past 35 years the Arwood Corp., Tilton Plant has been consistently producing magnesium castings in investment molds. As in the past, research and development continues to be an important aspect in casting premium quality castings. New knowledge as a result of R & D has helped to improve production methods with the advent of newer materials and machinery. Investment casting is an approach which allows the solidification of both large, thick-walled parts, (such as those generally cast in sand molds) as well as small thin-walled parts (which are usually die cast). This approach provides perhaps the best means of producing near net shapes with a high degree of control over metallurgical soundness and mechanical properties. In the following sections there will be a discussion of the breadth as well as the limitations which face a designer who is considering an investment casting.

The majority of the magnesium castings cast at Arwood are for the aircraft applications. Castings produced for non-aircraft use, enjoy the same methods used as our aerospace customers.

Alloy Selection

Investment casting has unique characteristics, and its own peculiar rating of castability of the various magnesium alloys. The best alloy in sand or in die casting may not be necessarily the best in an investment block mold. The following is a listing of the common magnesium alloys with typical properties.

AM100A - M10100

Chemistry

Al	9.3 - 10.17
Mn	0.10 - min.
Zn	0.30 max.
Si	0.30 max.
Cu	0.01 max.
Ni	0.01 max.
Others each	0.10 max.
Others total	0.30 max.
Mg	Balance

Applications- Aircraft engine parts, high strength and pressure tight castings, crankcases, housings.

Characteristics

Density	0.066 lbs/in.
Liq. Temp.	1100° F
Sol. Temp.	865° F
Latent heat of fusion	89 BTU's/lb.

Mechanical Properties T6

Minimum

UTS	YS	ELO
34,000	15,000	1.5%

Typical

	UTS	YS	ELO
$\bar{\bar{x}}$	34,200	24,600	1.9%
σ	3,400	2,200	0.9

AZ91C - M11914

Chemistry

Al	8.1 - 9.3
Zn	0.40 - 1.0
Mn	0.13 - min.
Si	0.30 max.
Cu	0.10 max.
Ni	0.01 max.
Others each	0.10 max.
Others total	0.30 max.
Mg	Balance

Applications - use in pressure tight castings where high tensile strength, moderate yield strength and good elongation is required.

Characteristics

Density	0.066 lbs/in.
Liq Temp.	1105° F
Sol Temp.	875° F
Latent heat of fusion	89 BTU/lb.

Mechanical Properties T6

Minimum

	UTS	YS	ELO
sep. cast	34,000	16,000	3.0%
cut	17,000	12,000	-

Typical

		UTS	YS	ELO
sep.	$\bar{\bar{x}}$	36,900	22,000	4.0%
cast	σ	2,500	2,400	0.68
cut	$\bar{\bar{x}}$	33,300	21,700	3.2%
	σ	1,700	2,650	0.55

AZ92A - M11920

Chemistry

Al	8.3 - 9.7
Zn	1.6 - 2.4
Mn	0.10 - min.
Si	0.30 max.
Cu	0.10 max.
Ni	0.01 max.
Others total	0.30 max.
Mg	Balance

Applications - good combinations of pressure tightness, tensile and yield strength.

Characteristics

Density	0.066 lbs/in.
Liq Temp.	1100°F
Sol Temp.	830°F
Latent heat of fusion	160 BTU's/lb.

Mechanical Properties T6

Minimum

UTS	YS	ELO
34,000	18,000	1.0%

Typical

	UTS	YS	ELO
$\bar{\bar{x}}$	40,800	24,000	4.6%
σ	2,300	2,400	0.6

EZ33A - M12330

Chemistry

Rare earth total	2.5 - 4.0
Zn	2.0 - 3.1
Zr total	0.40 - 1.0
Zr soluble	0.40 - -
Cu	0.10 max.
Ni	0.01 max.
Others each total	0.30 max.
Mg	Balance

Applications- used in the T5 temper, for service temperatures of 450°F-500°F, excellent pressure tightness, free of microporosity.

Characteristics

Density	0.066 lbs/in.
Liq. Temp.	1190°F
Sol. Temp.	1010°F
Latent heat of fusion	160 BTU's/lb.

Mechanical Properties T5

Minimum

UTS	YS	ELO
20,000	14,000	2.0%

Typical

	UTS	YS	ELO
$\bar{\bar{x}}$	22,800	15,300	4.8%
σ	1,300	420	.83%

ZE41A - M16410

Chemistry

Zn	3.5 - 5.0
Total rare earth	0.75 - 1.75
Zr total	0.40 - 1.0
Zr soluble	0.40 - -
Mn	0.15 max.
Cu	0.10 max.
Ni	0.01 max.
Others total	0.30 max.
Mg	Balance

Applications- used for pressure tightness and elevated temperature mechanical properties. Castings are used in the stress relieved T5 temper.

Characteristics

Density	0.066 lbs/in.
Liq. Temp.	1190°F
Sol. Temp.	975°F

Mechanical Properties T5

Minimum

UTS	YS	ELO
29,000	19,500	2.5%

Typical

	UTS	YS	ELO
$\bar{\bar{x}}$	32,100	21,700	5.7%
σ	1,600	1,100	1.5

QE22A - M18220

Chemistry

Ag	2.0 - 3.0
Didymium (total R.E.)	1.75 - 2.5
Zr total	0.40 - 1.0
Zr soluble	0.40 - -
Cu	0.10 max.
Ni	0.01 max.
Others each total	0.30 max.
Mg	Balance

Applications- Best combination for room temperature and elevated temperature properties (up to 550°F) alloy is pressure tight and weldable.

Characteristics

Density	0.065 lbs/in.
Liq. Temp.	1190°F
Sol. Temp.	1020°F
Latent heat of fusion	160 BTU's/lb.

Mechanical Properties T6

Minimum

UTS	YS	ELO
35,000	25,000	2.0%

Typical

	UTS	YS	ELO
$\bar{\bar{x}}$	37,000	26,300	5.3%
σ	1,000	980	.96%

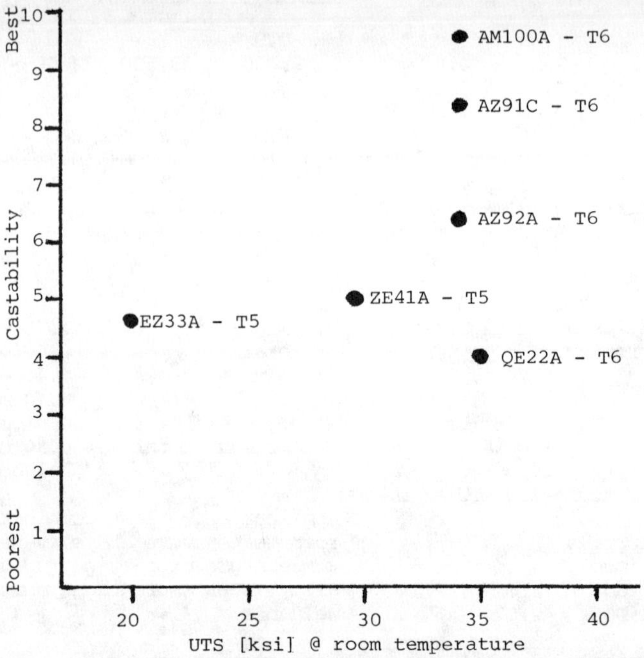

Figure 1 - Relative relationship between castability and ultimate tensile strength at room temperature.

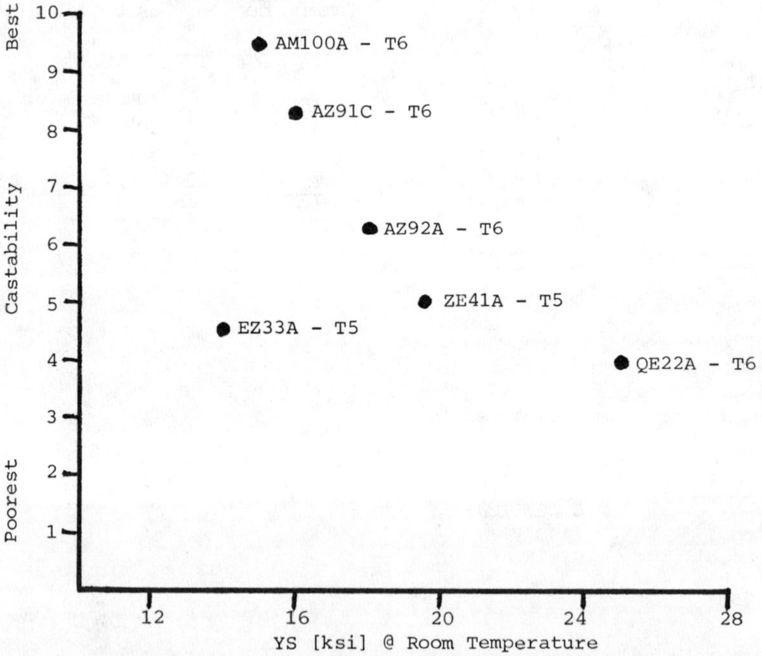

Figure 2 - Relative relationship between castability and yield strength at room temperature.

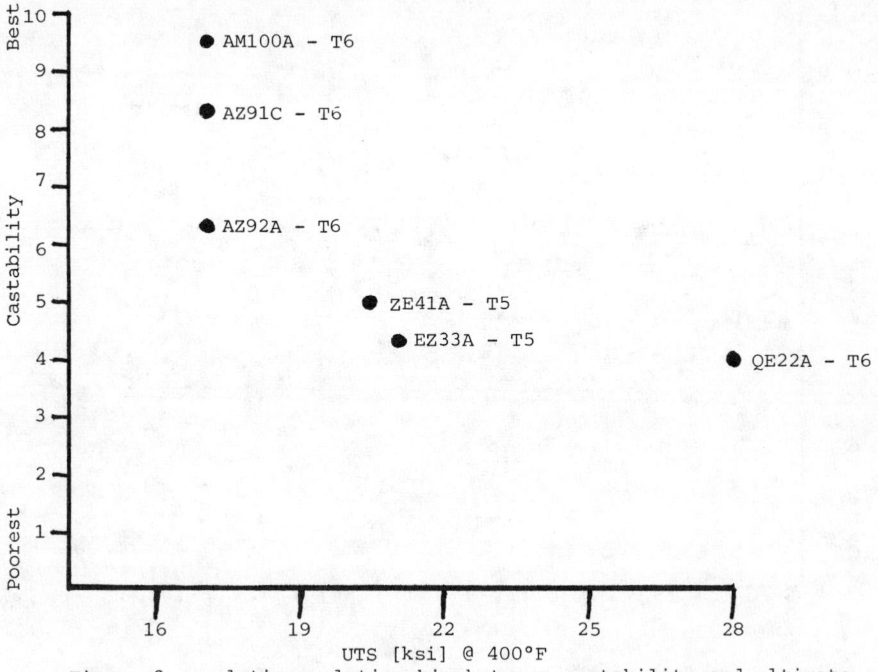

Figure 3 - relative relationship between castability and ultimate tensile strength at 400°F.

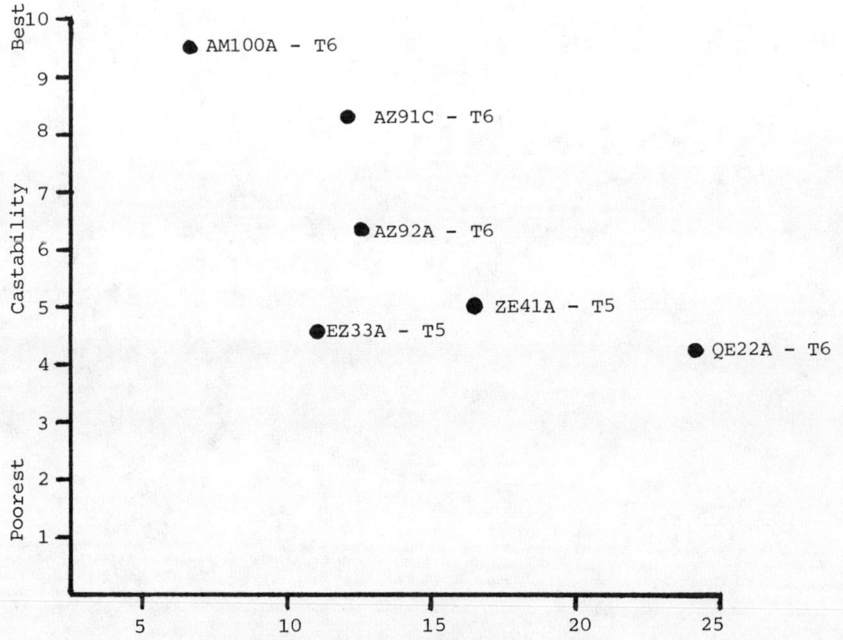

Figure 4 - Relative relationship between castability and yield strength @ 400°F.

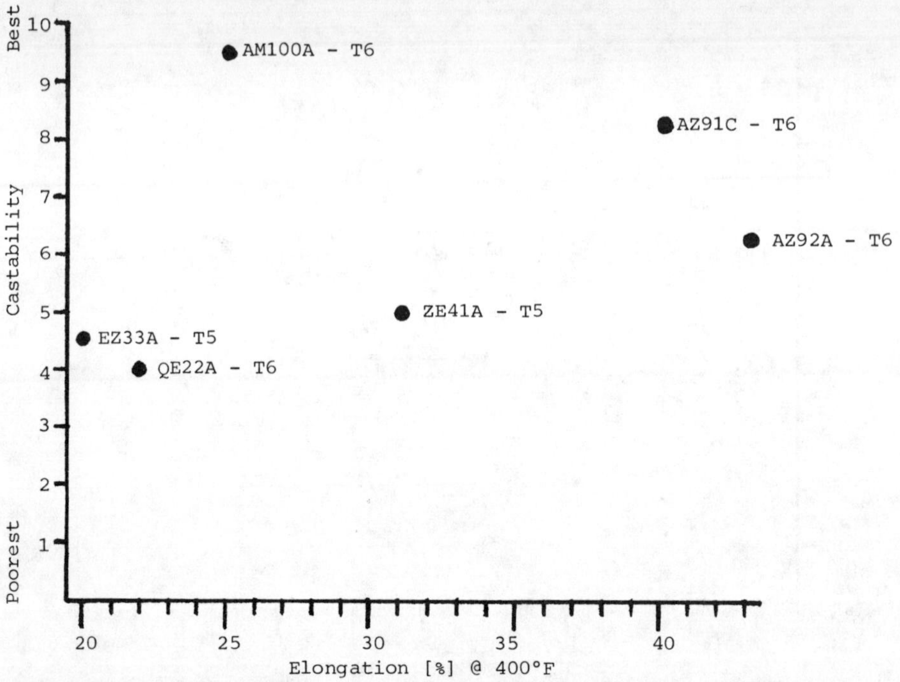

Figure 5 - Relative relationship between castability and percent elongation at 400°F.

DIMENSIONS

Complex magnesium castings can be cast by investment methods but a few design considerations must be addressed by the Design Engineer.

Linear Tolerance

0" - 1"	.005"
1" - 2"	.010"
3" - 5"	.015"
5" - 8"	.020"
8" - 12"	.025"
12" - 15"	.030"
15" - 20"	.035"
20" and up	.040"

If the tolerances stated alone are too loose for a particular part, they may be held tighter but only after a discussion of the drawing by the Design Engineer and Foundry Engineer.

Items such as straightness, flatness, concentricity, parallelism, etc., are all considerations which apply to the foundry processing methods after casting.

Typical wall sections of magnesium investment casting are 0.070 ± .015 min,. to 3/4" max. tempered with surface area.

MAGNESIUM SURFACE PROTECTION

While considering the potential design of a part in magnesium, for such considerations as weight, strength, and machinability, the designer must be cognizant of the environment in which the part is to be used. Atmosphere, alloy composition, position the part is to be used in service, and contact with a dissimilar metal are all important factors.

Magnesium is typically thought of as a metal requiring some form of protective surface coating. Yet actually it has been demonstrated that if a magensium part is clean and left to stand alone, "the susceptibility of the magnesium to corrode is far less than that of mild steel" Ref. (1). "The effect of aqueous solutions on magnesium varies considerably upon the nature of the solution. Magnesium is resistant to attack by alkalis but is attacked by most acids. The most prevalent are chloride solutions usually found in seawater. Impurities such as iron, nickel and copper contained in the magnesium alloy will also increase the rate of corrosion." Ref. 2

Corrosion is also typical when the magnesium is attacked to other metals. This corrosion is galvanic in nature and occurs when magnesium is in electrical contact with a cathodic metal in the presence of an electrolyte. Ref. (3).

Because of its electronegativity, magnesium surfaces should have some type of protective coating. These protective coatings are used as a base for paint or alone to protect the base metal. The two basic methods of protection are, a chemical treatment in the form of dipping, or an anodizing treatment.

Arwood employs a chemical treatment called chrome pickle or Dow #1 to provide some degree of protection to castings prior to machining, while in transit to the customer, and subsequent storage atmospheres. More tenacious anodic protection can be applied at outside facilities.

Specific questions on the type of protection to use for a particular part should be directed to the alloy producers such as M.E.L. or Dow who are doing a great deal of research on surface protection.

MOLD REACTION

An inherent consideration in the design of the part to the Foundry Engineer is the susceptibility to mold metal reaction. Magnesium combines with SiO_2 to form Mg_2 Si + O_2 . If we look at the binary phase diagram there appears to be a transformation phase at 1200°F (649° C). The liberated O_2 accelerates the oxidation process to such a degree that if the O_2 has not been tied up into a stable compound, extreme gas pressure and mold pressure often lead to an explosion.

Shell molds do not explode in the same manner as would a solid mold. Shell mold will break or crack long before extremely high pressure is achieved. Solid mold have very thick ceramic body, encapsulated in a metal flask. The combination ceramic mold and flask can tolerate a lot of pressure but eventually the pressure buildup overcomes the strength of the ceramic. This sudden release of high pressure as the ceramic breaks causes the explosion.

Proper mold and metal temperatures and the use of inhibitors reduce the susceptability to burning and mold reaction. Listed below are some typical inhibitors and how they rank relative to time and temperature.

Sulfur - powder - low level - used as a powder to dust melt surface to consume O_2.

SF_6 (sulfur hexafluoride) - gas - med. level - used to inhibit mold cavity and melt protection atmosphere.

SO_2 (sulfur dioxide) - gas- med. level - used to inhibit mold cavity and heat treat atmosphere.

KBF_4 (potassium fluoborate) - powder - med. level - used in molding media.

ALF_3 (aluminum fluoride) - powder - med. level - used in molding media.

BF_3 (boron trifluoride) - gas- high level - used to extinguish Mg fires only, very toxic.

MELT PROTECTION

Arwood utilizes a fluxless melting system of SF_6 and Co_2 gasses for magnesium melting. Oxidation of magnesium is a very important problem in the melting and casting process because of the extreme reactivity of magnesium with oxygen. Magnesium is protected from oxidation and subsequent burning by the use of protective atmospheres or a salt flux which liquifies at melt temperatures.

"The Dow Chemical Company started using SF_6 (sulfur hexafluoride) at a die cast facility in 1952 as a magnesium inhibitor." Ref. 4. The use of SF_6 has become more popular recently, due to in part because of environmental effects. SF_6 protection is much better than the salt fluxes, because the protective film continually renews itself, provided that the melt is not excessively agitated and sufficient presence of gas is maintained. The SF_6 protection layer is actually invisible being only a few molecules thick.

$SF_6 + Mg + O_2 \rightarrow MgO + MgF_2 + SO_2 + MgS$. Ref. 5

Sulfur hexafluoride is also preferred over the other sulfur and fluoride compound families because SF_6 gas is stable, non-corrosive and most importantly non-toxic.

The wet fluid fluxes initially form an impervious layer on the melt surface. It is this layer which protects the melt from oxidation with the atmosphere.

We all know that sometimes the cure is worse than the illness. It has been found that this is the case when melting magnesium with a fluid flux. The disadvantages of melting with a fluid flux are numerous. " Inclusions of flux can be dispersed within the casting. If found near the surface after solidfication it blooms and starts to corrode the base material." Ref. 6. Some molten salt fluxes covering a melt would often react with the oxygen in the atmosphere after long periods of time. The fluid flux would then become hard and crusty. Any disturbance to the reacted crust would cause a crack or break, exposing unprotected magnesium to the atmosphere.

The salt fluxes are composed of MgCL, KCL, BaCL, and CaF all emit fumes. "These fumes are not only toxic to foundry workers but also corrosive to the foundry equipment and building structure itself." Ref. 7. Disposal of some reacted salt flux has become a problem because it is considered a hazardous waste due to heavy metals content.

CASTING METHODS

Arwood Foundry Engineers use either of two types of casting methods. Low pressure or conventional static with vacuum assist. Each system has its own particular advantage. The conventional static pour system with vacuum assist is used when the quality requirement are not very stringent and only dimensional control is of concern. Low pressure is used for high quality requirements. The advantages of low pressure are less metal turbulance entering the mold cavity and increased feeding due to greater pressure on the transported feed metal. For example, if we take the density of magneisum alloy QE22A which is 0.065 lbs per in and apply one pound/in2 of hydraulic pressure the metal level will rise 15.38 inches in a tube. At 4 p.s.i., a typical casting pressure, there exists an effective metal head height of 61 inches. To produce the same metallostatic head in a conventional mold is by no means practical. Even at various pressures we can and do nonfill if we abuse Archimedes theory of displacement. Sometimes we are forced to use vacuum assist with low pressure on solid molds to evacuate the air inside the mold cavity. Castings with thin walls are prone to nonfill because of back pressure in the cavity and the volumetric latent heat of fusion. QE22A has a latent heat fusion of 160 BTU's/lb. and a density of 0.065 lb/in3 which converts to a latent heat of fusion value 17,971.2 BTU's/cubic foot. Aluminum alloy 201 has a latent heat of fusion of 167 BTU's/lb. and a density of 0.101 lbs/in3 which converts to a latent heat of fusion value 29,146 BTU's/cubic foot. From these simple calculations above it is easy to see that as the magnesium molten metal flows through the gating and mold cavity it has far less available energy to expend to accomplish a given temperature drop.

In general the geometry of a given part cast in magnesium can be cast in aluminum. But the inverse is not true because of the large difference in their latent heat of fusion.

The biggest disadvantage to low pressure is the low number of molds cast per unit of time. A typical cycle time for a mold using low pressure is 10 minutes. That is only 6 molds per hour, whereas with conventional static pouring the foundry is only limited to the number of molds poured by how quick the oven tender can get the molds out of the oven and to the pouring area.

SAFETY

"Nearly all metals will burn in air under the right conditions. Some metals, such as aluminum and steel are not normally thought of as being combustible, may ignite and burn when in finely divided form such as occurs during grinding and machining operations. Fine steel wool, for example, will burn if exposed to a sufficient amount of heat. Aluminum, because of its ability to burn in powder form is used in solid fuel for rockets-- in fact solid rocket fuel may contain as much as 20% aluminum powder by weight.

Magnesium is a combustible metal. Its melting point is about 1000° F and the ignition temperature is generally considered to be very close to its melting point. The ignition temperature for magnesium ribbons and shavings is about 950°F. Fine magnesium dust or powder can be ignited well below 900°F.

The case of ignition of magnesium like other metals and most solid fuels depends upon its size and shape. Ingots and castings are difficult to ignite, even with a torch. Dust particles from grinding and ribbons from machining will ignite easily from ordinary ignition sources. Friction from dull machine tooling can ignite magnesium shavings and cuttings. Friction and static electricity sparks can ignite magnesium dust.

Once ignited, magnesium burns rapidly, giving off a brilliant white light and is very difficult to extinguish. Airborne magnesium dust can explode violently. Burning dust on clothing can create a severe hazard to the person wearing it.

Magnesium shavings and dust will produce hydrogen when wet with water or water soluble coolants. Hydrogen is a highly flammable gas and as little as 4.0% in air will ignite. Wet fines will also generate heat spontaneously."Ref.8.

[Listed below are but only a few excerpts of the safety rules from Arwood Magnesium Safety Handbook.]

1. Foundry- personnel are required to wear protective apparel at all times while in the foundry area. Ordinary untreated work clothes are not permitted.

2. Finishing- proper wearing apparel is very important in safe machining and cleaning of magnesium castings. Persons that operate dust producing equipment such as grinders, buffers and wire brushers must wear flame retardant clothing.

3. Electrical equipment and wiring in the machining/cleaning area must be approved for class II, group E locations. Electrical equipment must be cleaned periodically.

Dust collectors and finishing equipment must be properly grounded to prevent the buildup of static charges.

4. Smoking, welding, open flame or use of spark producing equipment is not permitted in machining/cleaning areas.

5. Wet precipitators type dust collectors are used to collect dust from grinding operations.

6. Wet sludge not under water will evolve hydrogen and may heat spontaneously. Therefore, it is important that all sludge be removed from collectors each day. Daily inspections are necessary to insure that sludge deposits do not build up on interior surfaces of the collector or its intake and exhaust duct work.

REFERENCES

(1) "The behaviour of magnesium components in service", "Surface Treatments for Magnesium Alloys in Aerospace and Defense", Magnesium Electron Ltd., (England), pp 3-6.

(2) Ibid.

(3) Ibid.

(4) D.J. Hanawalt, "SF_6 - Protective Atmosphere for Molten Magnesium" (paper presented at the 8th SDCE International Die Casting Exposition and Congress, Detroit, Michigan, 17-20 March 1975) pg. 2.

(5) Ibid.

(6) J.W. Fruehling, J.D. Hanawalt, "Protective Atmospheres for Melting Magnesium Alloys", AFS Transactions, (date unknown) pp 159-161.

(7) Ibid.

(8) Arwood Corporation, Investment Casting Division, "Magnesium Safety Handbook" April 1984, pp 2-36.

1st International SAMPE Metals Conference
August 18-20, 1987

COST SAVINGS IN SHOT PEENING THROUGH MICRO-PROCESSOR CONTROLS

Kevin C. Tower

Metal Improvement Company, Inc.
145 Addison Road
P.O. Box 487
Windsor, Connecticut 06095

Abstract

Significant fatigue life improvement from the use of shot peening is well known. However in industries such as aerospace and automotive ever increasing demands on component performance from lower weight to strength ratios necessitate increased processing requirements. Since there is no nondestructive test currently available to assure that a metal part has been properly peened, more stringent controls on the process must be developed in order to insure the life enhancement possible from shot peening.

Computer controlled shot peening equipment has emerged to provide more precise control over the shot peening process. This new generation of computer controlled equipment allows designers to rely on the life increases measured in component testing. In addition, operational cost savings in the areas of part handling, media control and part inspection may offset increased equipment costs.

The computers are used to monitor the status of critical peening parameters and assure they stay within acceptable preselected values. Documentation of the status of the critical parameters is available in the form of printouts from the computer. Hard copy documentation provides a complete record of critical process parameters while the part was being processed. Assurance of a completely controlled and documented shot peening process is now available.

1. INTRODUCTION

Shot peening is a process for treating metal parts by controlled high-speed impact of many spherical balls (shot). Each individual ball acts as a tiny peening hammer leaving the surface of the material with a small indentation or dimple. See Figure 1 below.

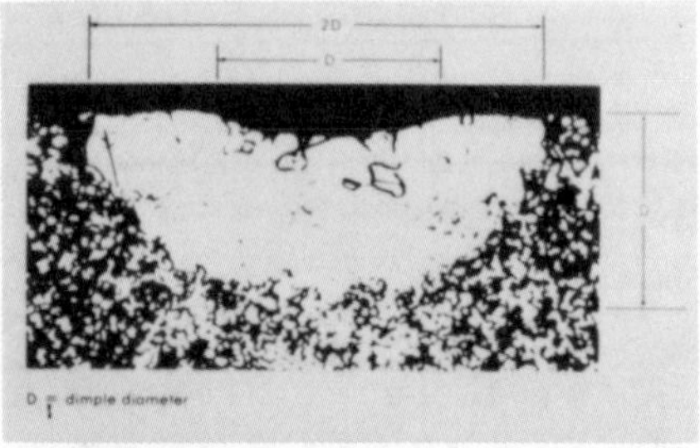

AFFECTED ZONE UNDER A SINGLE DIMPLE

Figure 1

In order for the dimple to occur the outermost surface fibers of the material must have yielded in tension. The fibers immediately below the yielded material act to restore the surface to its original shape. The action of these fibers trying to restore the shape produces substantial compressive stress in the surface. In order for the part to remain in equilibrium, offsetting residual tensile stresses are developed through the balance of the material cross section. Residual compressive stresses and offsetting residual tensile stresses may be produced by other manufacturing processes such as surface rolling, burnishing, hole expansion and surface hardening. A typical residual stress profile in a beam which has been shot peened on the upper surface is shown in Figure 2. Since the beam is in equilibrium with no external forces, the area under the stress distribution curve in the regions

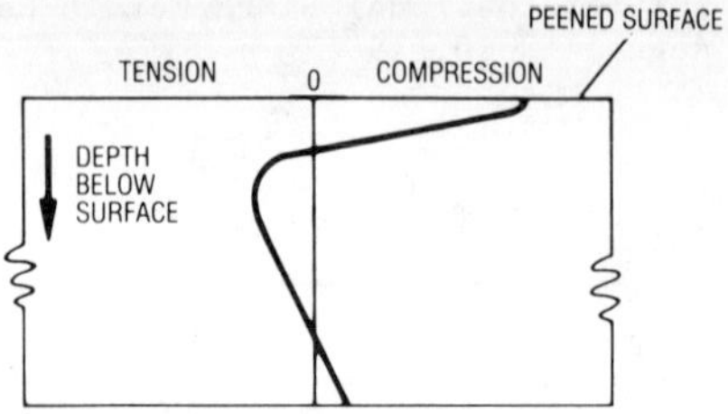

Figure 2

of compressive stress must be equal to the corresponding area under the curve in the region of tensile stress. Note that the compressive stress is relatively shallow in depth and significantly higher in magnitude than the induced tensile stress which is distributed over a broader area.

2. SHOT PEENING CONTROLS

The object of controlled shot-peening is to produce a compressively stressed surface layer in which the amount of stress, the uniformity of the stress and the depth of the layer can be held constant from piece to piece. Since there is no production method to measure the amount and uniformity of the compressive stress on a finished part, the full control of all aspects of the process becomes imperative. The four fundamental areas of shot peening controls are: 1) peening media (shot); 2) energy of the shot stream (intensity); 3) extent of peening coverage and 4) peening equipment.

2.1 Media

There are many different media which can be used for shot peening to induce a residual compressive stress. Cast steel shot is the most commonly used material with hardness varying from Rockwell C45 to Rockwell C70, however, conditioned cut wire shot (steel or stainless steel), glass beads and most recently ceramic beads are also frequently used. Whatever material is indicated, it is imperative that it be maintained round and uniformly sized at all times. As the media breaks up, the broken particles must be eliminated since the sharp corners of broken or uneven shot may produce harmful effects. Figure 3 illustrates that the use of mixed shot is uneconomical, since it takes longer to obtain the correct intensity.

EFFECT OF BAD SHOT

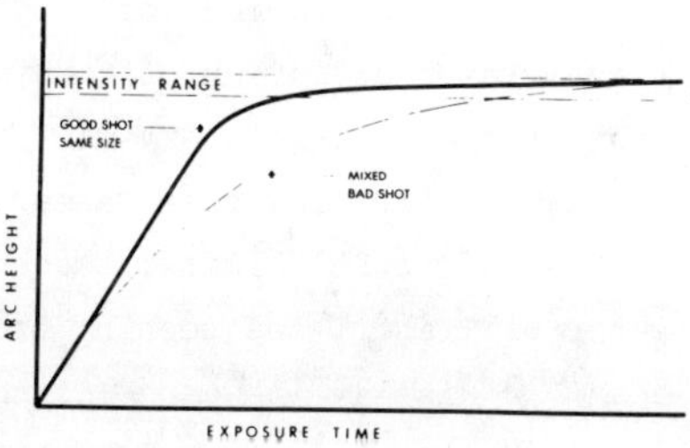

GOOD SHOT REQUIRES LESS PEENING TIME

Figure 3

2.2 Intensity

Velocity, shot size, hardness and angle of impingement combine to determine peening intensity. It is important for these parameters to be fully controlled in order to obtain consistent results from peening. The calibration procedure to achieve the required intensity of the peening is accomplished by exposing one side of a standard Almen metal test strip (SAE1070) to the same shot stream as the part will receive. Removed from the shot stream, the strip will have taken a curvature convex to the flow of the media. Figure 4 shows a typical curvature of an Almen test strip.

ALMEN INTENSITY

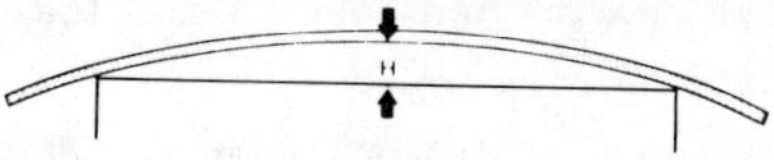

H ARC HEIGHT ALMEN INTENSITY

Figure 4

The arc height of the strip measured in a special depth gauge serves as a measurement of the peening intensity and also a relative indication of kinetic energy. The curvature is proportional to the bending moment of the compressive stresses generated by the peening. Specification standards provide for test strips of three different thicknesses, suitable for a range of intensity levels.

2.3 Coverage

Shot peening coverage is defined as the extent (percentage) of obliteration or dimpling of the original surface by the impact from individual shot particles. It is necessary to achieve 100% coverage on parts peened for fatigue enhancement. Otherwise there may be areas of the components that are not protected by the compressive stress.

All of the parameters that affect peening intensity can also affect peening coverage. Other parameters such as run time, shot flow, nozzle traverse rate and part rotation rate affect peening coverage as well. These parameters also must be closely monitored.

3. COMPUTER CONTROLLED EQUIPMENT

Once the correct peening parameters have been identified for a particular application, control of the entire peening process can be exercised by the use of computer controlled peening equipment. This new generation equipment incorporates important peening practices such as continuous shot separation, hard fixturing, machine controls, monitoring systems, automatic shut down and documentation via computer printout of the critical machine parameters throughout the peening process.

3.1 Continuous Shot Separation

This equipment includes both screen and spiral systems which separate and reject oversize, undersize and unacceptable shape shot particles to specified quality levels. Since these systems operate continuously to maintain shot quality parameters, operating savings result from the elimination of machine down time and from efficient shot usage.

3.2 Hard fixturing

Traditional shot peening machinery utilizes scrap parts to which Almen fixtures are secured in areas which require intensity verification.

The advent of computer controlled peening equipment locks in part fixturing and nozzle fixtures to ensure that each part is positioned to receive the proper angle of shot flow throughout the production run. Additionally, operating time savings can be gained from reductions in machine set-up time and intensity verification requirements.

3.3 Machine calibration

After determining process variables and arriving at a successful shot peening process, computer control allows the user to precisely repeat and constantly monitor the procedure. Variables that can be monitored include shot flow, air pressure, turntable or roller speed and nozzle oscillation. Nozzles can be programmed for varying travel rates or for on-off operation at prescribed intervals.

In addition if variables wander outside of preset limits, the computer will abort the peening process and display the cause of shut down. Once corrections have been made processing is resumed at the point of shut down.

3.4 Documentation

At the conclusion of each peening cycle a final print-out lists all process variables and verifies that the process has been completed successfully. Input from the process monitoring systems is continuously fed to the computer and evaluated against maximum and minimum limits. This information is available for either hard copy documentation or archival storage. All process interrupt information is also available in print-out form. This could include time, position and cause of abort, status of process parameters at time of abort and position of restart.

4. DISCUSSION OF BENEFITS

The primary benefits from microprocessor controlled shot peening may be documented in a number of general categories. First, this class of peening can now allow a primary vendor to incorporate shot peening benefits at the design level rather than as an insurance treatment. For example, an aerospace engine manufacturer was recently able to qualify one of its military engines under a more rigorous standard for the larger commercial market due in part from microprocessor controlled shot peening of cirtical parts. Equally important, it is now possible to extend component warranties of existing parts thereby improving market position in the competitive aerospace industry.

Secondly, micro-processor controlled shot peening can significantly reduce maintenance costs of life-limited parts. For example, aircraft engine overhaul service requirements can be extended thereby reducing engine downtime and maintenance service costs.

Thirdly, this type of advanced peening technology can in many cases greatly reduce shot peening operating costs thus decreasing the price to peen certain complex parts. This is particularly true for parts with long peening cycle times having difficult areas to inspect. As an example, ID inspection requirements for a 25 foot long helicopter spar depicted in Figure 5 were greatly reduced with the employment of micro processor controls.

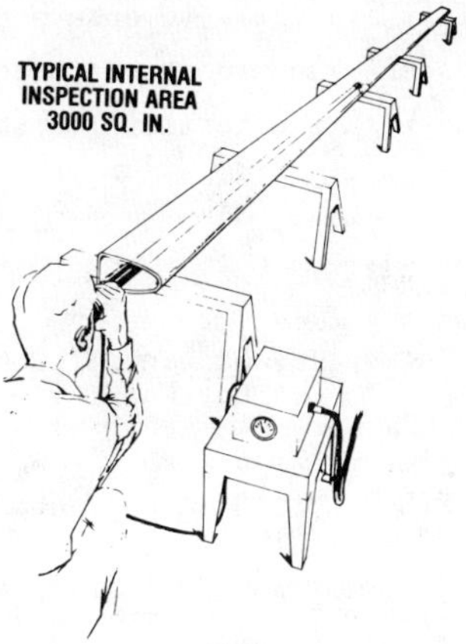

Figure 5

Also manual survey of shot peening operating parameters during the 50 minute peening cycle were eliminated. The price to peen this part was reduced from \$200.00 to \$135.00.

5. CONCLUSION

Computer controlled shot peening equipment now enables designers to utilize the full benefit of shot peening in the design of fatigue critical components. Production costs may be reduced because less expensive materials and lower surface finishes may be acceptable. Additional cost savings can be realized from approval of standard components for higher stress environments.

New generation computer controlled equipment is designed to provide an accurate technical approach to the shot peening process. The new computer equipment will not permit critical processing parameters to deviate from preapproved preprogrammed limits. Documentation for each part is available to ensure the component was processed in accordance with the applicable process specifications.

REFERENCES

1. Fuchs, H., Mechanical Engineers Handbook, Shot Peening. 1986

2. Shot Peening of Metal Parts, MIL-S-13165B Amendment 2, 25 June, 1979.

3. Almen, J.O. and P.H. Black, Residual Stresses & Fatigue In Metals 1983.

4. Shot Peening Applications, Metal Improvement Company, Sixth Edition, 1980.

BIOGRAPHY

Kevin Tower received his Bachelor of Arts degree in Chemistry from the University of Massachusetts in 1972 and an MBA from Western New England College in Springfield Massachusetts in 1978. Mr. Tower has extensive business and science related experience through his ten years as an environmental scientist and five years in metallurgical and shot peening process applications.

1st International SAMPE Metals Conference
August 18-20, 1987

QUALIFICATION OF HIGH TEMPERATURE VACUUM CARBURIZING FOR AN AIRCRAFT GEAR STEEL

Roy J. Cunningham
Supervisor Metallurgy
Boeing Vertol Company
Philadelphia, Pennsylvania

Abstract

Carburizing in a vacuum is a relatively new and unique method of case hardening gears for helicopter transmissions. It has been found that vacuum carburizing:

- Reduces the time required to heat treat a component by as much as 50 percent
- Improves production scheduling
- Requires less energy
- Produces less part distortion
- Eliminates preheat

The successful development and qualification of the high temperature vacuum carburizing process for 9310 steel, along with the method for implementing this process into production, is detailed. Some of the problems encountered and solutions developed are presented.

1. INTRODUCTION

Carburizing in a vacuum is a relatively new process for case hardening helicopter transmission gears. It has been said that any heat treating procedure done in an atmosphere furnace can be accomplished in a vacuum furnace. However, in comparison with conventional carburizing, significant benefits in both time and cost can be obtained by carburizing in a vacuum.

The purpose of carburizing is to provide a hard, wear resistant case, (surface) while maintaining a soft, tough core. This increases both the load carrying capability and the resistance to scuffing and scoring of the gear tooth surface. For aerospace components, there are three primary requirements to be met:

1. Specified effective case depths at R/C 50 must be maintained around the tooth profile
2. Uniform and repeatable carburization must be obtained
3. Surface hardness must meet an R/C 60 minimum.

The primary advantage of vacuum carburizing over conventional processing is the reduction in cycle time as shown in Figure 1, which depicts the variation in carburizing time for both vacuum and atmospheric carburizing.

Examination of Figure 1 shows that to achieve an effective case depth of 0.050 inch, carburizing takes approximately 7.5 hours conventionally in an enriched endothermic atmosphere furnace as compared to only four hours in a vacuum furnace. This amounts to a 210 minute savings in carburizing time. The obvious advantages are as indicated below:

1. Savings in operating costs – If heat treating costs are equal to

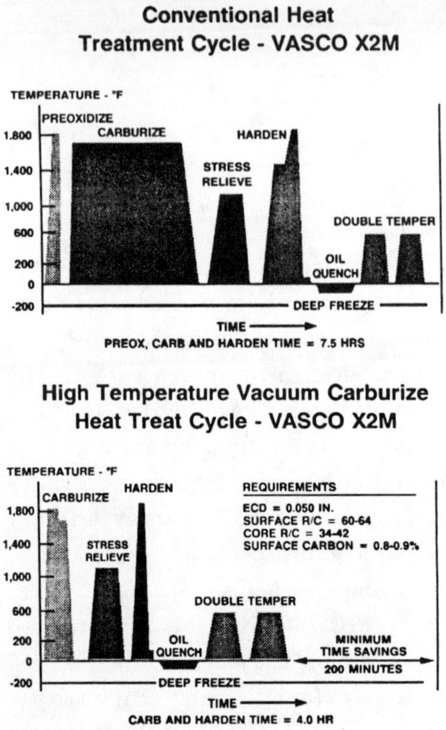

Figure 1. Typical Conventional and High-Temperature Vacuum-Carburizing Cycles

$60 per hour, then $210 per carburizing run can be saved. If 10 carburizing runs are made in a week's time, this amounts to $2,100 per week or approximately $105,000 per year.

2. Savings in production scheduling – Conventionally, only one carburizing run per eight-hour shift can be done. With vacuum carburizing, two runs can be achieved with time to spare. Thus, two lots of production parts can be processed in vacuum where only one would normally be done.
3. Savings in energy usage – Since a vacuum carburizing cycle takes only four hours to complete, from the example a 3.5 hour savings of natural gas and electricity results. As some vacuum furnaces can be brought to temperature in less time than a conventional furnace of the same capacity, an additional savings occurs.
4. Less distortion – The majority of distortion from heat treatment occurs during the carburizing cycle (as compared to the hardening or austenitizing cycle). Naturally, the longer the time spent at the carburizing temperature, the greater will be the distortion. It therefore follows that since vacuum carburizing requires significantly less time at temperature than does conventional carburizing, less distortion of the components will result. This could have an advantageous domino effect on both grinding stock (less grinding stock being required) and grinding burn susceptibility (less distortion equals less chance of burning).
5. No pretreatment necessary – Certain alloys require pretreatments for carburization uniformity. Vacuum carburizing eliminates the need by vaporizing any detrimental passive oxide on the surface of the component. This is also illustrated in Figure 1.

2. PROCESS DEVELOPMENT

Given the stated advantages of vacuum

carburizing the question arose as to how to develop and adapt the process for aerospace gearing components. It is not considered enough just to develop the proocess; the process must be proven to be effective. This can only be accomplished by testing pertinent material characteristics to assure that they are equivalent to or better than that obtained from conventional carburizing.

Under Government sponsorship*, a three-phase program entitled, "Aircraft Quality High Temperature Vacuum Carburizing," was initiated. It was directed at developing a high temperature vacuum carburizing technique for 9310 and Vasco X2M steels. The chemical analysis of these materials is as shown in table I.

Table 1. Chemical Analysis of 9310 and Vasco X2M Steels. Steel are double vacuum melted, VIM-VAR processed.

	NOMINAL %	
ELEMENT	9310	VASCO X2M
Carbon	0.10	0.15
Silicon	0.27	0.90
Manganese	0.55	0.30
Sulphur	0.010 Max	0.010 Max
Phosphorous	0.010 Max	0.015 Max
Chromium	1.20	5.00
Molybdenum	0.19	1.40
Nickel	3.25	–
Tungsten	—	1.35
Vanadium	—	0.45

The program was organized as follows:

1. A development phase (Phase I) whereby the process would be established and would be considered acceptable only after successfully testing for surface contact fatigue, bending fatigue, and scoring as detailed in Figure 2.
2. A test gear phase (Phase II) where both spur gear and spiral bevel gear components were evaluated for gear tooth scoring characteristics.
3. A transmission test phase (Phase III) in which actual full size production components are manufactured and tested in a production transmission environment. During this test, sufficient test cycles at 100 percent of load must be obtained and documented such that qualification of the process for production is obtained.

*Army Contract DAAG46-82-C-0034-AMMRC COTR

This procedure is considered a minimum standard for production qualification of new aircraft quality materials and/or processes. Each phase is necessary as was demonstrated early in the program.

2.1 Development Problems

The first problem encountered developed during Phase I of the program for the Vasco X2M material. Testing was nearing completion. The process had been sufficiently developed, surface contact fatigue data had been generated, and the data available appeared significantly better than conventional carburizing data. Scoring tests gave similar results, and at this point not only did the program look

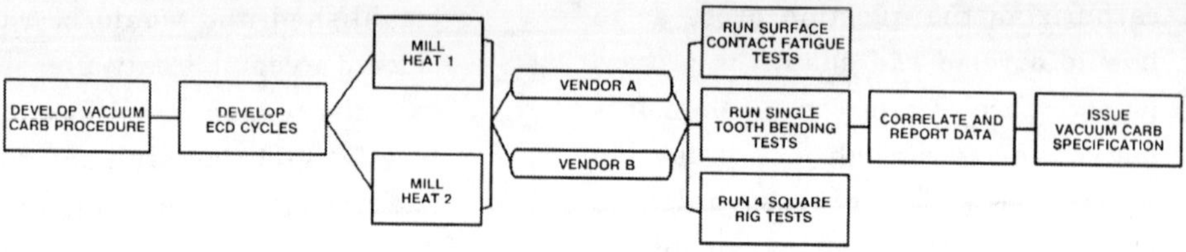

Figure 2. High Temperature Vacuum Carburizing-Phase 1

successful, but a substantial improvement to the previously indicated properties was observed. However, when evaluating the data from the single tooth bending fatigue test gears, the bottom dropped out. This data showed a lower than expected bending fatigue life for the vacuum carburized components. Metallurgical evaluation of the test gears revealed an unacceptable case microstructure displaying an interconnected carbide network. The testing was stopped at this point and redevelopment of the process was initiated in order to improve the case microstructure. This work is on-going.

It should be noted that had we not evaluated all three types of test specimens, this problem would not have been revealed, causing possible catastrophic results later down the line. Thus it is important that each segment of the program be successful, not just one or two.

Process development for 9310 started with a problem. Since vacuum carburizing equipment operates at much higher temperatures than conventional furnaces, it was decided to develop the 9310 steel vacuum carburizing procedure at 1,900F (instead of the conventional 1,700F temperature). Briefly, two major problems resulted. After developing the process from a case/core hardness and effective case depth standpoint, surface contact fatigue testing revealed extremely low lives. Conventional practice provided fatigue lives of 770 hours. With vacuum carburizing at 1,900F, the life was reduced to 5-10 hours. Metallurgical examination of the carburized case microstructure of several test specimens revealed a very large grain size (greater that ASTM no. 5). The detrimental effect in fatigue life was attributed to the large grain size which had occurred as a result of the increased carburizing temperature.

Additionally, during carburizing of the scoring/bending fatigue test gears, an inadvertent furnace overheat condition occurred which resulted in temperatures in the 2,000F range. This would not have been a problem (except for that mentioned above) had the parts not been copper plated for carburizing stop-off. As a result, the copper melted and dripped down over the gear teeth and prevented carburizing of certain gear teeth. Needless to say, the gears in this cycle had to be scrapped.

2.2 Process Refinement

Because of the foregoing problems, the vacuum carburizing temperature was lowered to 1,800F. Redevelopment of the process was then accomplished. A brief synopsis of the optimized cycle is shown in Table 2:

Table 2.
9310 Optimized Vacuum Carburizing Cycle

- Evacuate Chamber to 500u
- Soak at 1,800F – 35 minutes
- Carburize at 1,800F – 45 minutes, CH-250 Torr
- Diffuse at 1,800F – 80 minutes
- Gas Quench
- Stress Relief – 1,275F – 2 hours, vacuum
- Harden per D210–12023–1 requirements

A typical microhardness survey showing Rockwell C hardness versus the depth below the surface for one of the test specimens is shown in Figure 3. This sample was carburized to the requirements detailed above. Other test samples were uniformly carburized to effective case depths at R/C 50 of 0.082 inch. It is noted that the total vacuum carburizing time for this particular group of specimens was 425

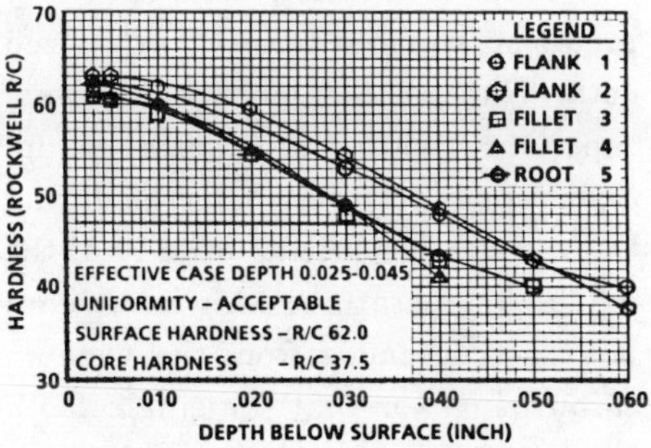

Figure 3. Hardness vs Depth Curve for 9310 Vacuum Carburized Steel

minutes or approximately 7 hours total time. Typically, to conventionally carburize a sample to this case depth requires 24-30 hours. However, the total vacuum carburizing cycle including carburizing, hardening and tempering per the Boeing Vertol process documents was only 21 hours. This is a significant time savings. Where conventional carburizing alone took a minimum of three work shifts, with vacuum carburizing, the total cycle including carburizing, hardening, deep freezing, and tempering can be accomplished in three shifts. It is easy to see that a significant improvement in production scheduling can be obtained utilizing this new carburizing technique.

3. RESULTS

Figure 4 illustrates the excellent uniformity that can be obtained through vacuum carburizing. Illustrated are the results of micro-

Single Tooth Bending Fatigue Test - Spur Gear Configuration

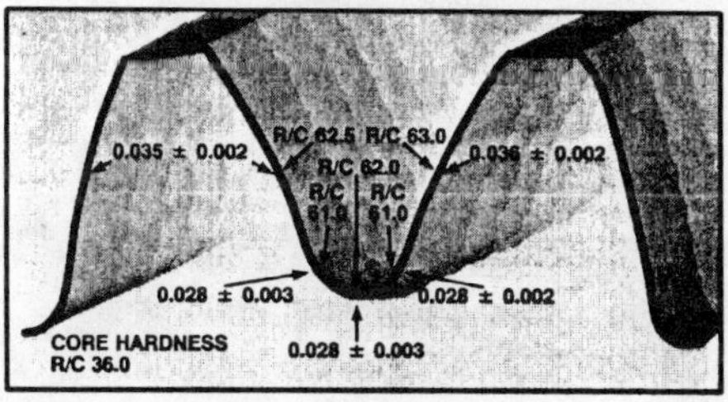

Figure 4. ECD Results Showing Uniformity of Carburized Layer

hardness surveys at various positions around the tooth profile of two single tooth bending fatigue test gears. The data shows that the effective case depths at different locations varied only two to three thousands maximum between test gears. It can also be observed that the flank surface hardness was in the range of R/C 63. Conventional carburizing of 9310 steel (with shallow cases such as these) usually results in hardness values at or below R/C 60. Of the many samples that were produced during this program, none of the samples had surface hardness below R/C 60 even with the shallow case depths detailed herein.

From a metallurgical standpoint, a chief concern of any new process would be the metallurgical characteristics of the case and core microstructures. Typical vacuum carburized test sample microstructures are shown in Figure 5 and 6. The case microstructure consists

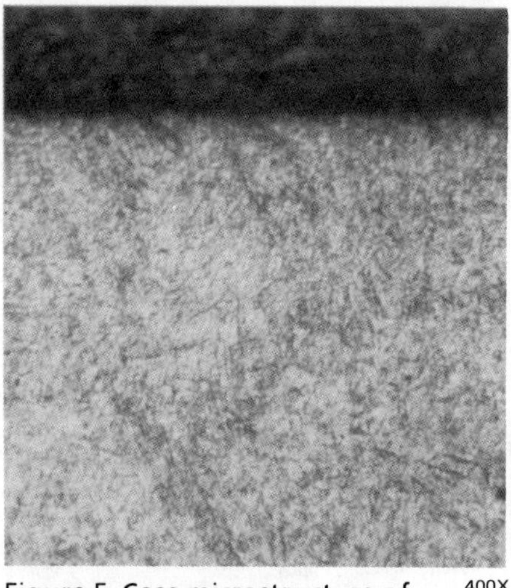

400X

Figure 5. Case microstructure of vacuum carburized single-tooth bending fatigue specimen showing typical fine grainbo martensitic structure.

of a typical hardened and tempered martensite matrix with some evidence of small, round carbides. The core structure was primarily tempered martensite. No metallurgical

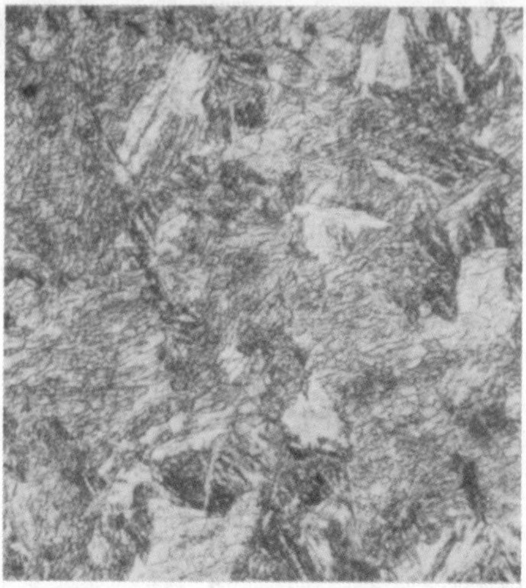

400X

Figure 6. Vacuum carburized core microstructure consisting primarily of tempered martensite.

anomalies were observed. In addition, electron beam microprobe analysis for carbon content in the carburized layer indicated that the surface carbon was in the desired range of 0.8 to 0.95 percent. A typical carbon diffusion curve similar to that obtained from conventional carburizing was observed.

It is significant to note that all 9310 testing required in Phase I and II of the program has been successfully completed. In each case the vacuum carburizing data that was generated was equal to or better than similarly produced conventionally carburized data. None of the problems that were experienced with the 1,900F vacuum carburizing cycle were observed during the vacuum carburizing of 9310 at the 1,800F temperature. An additional unexpected benefit from the vacuum carburizing was that the single tooth bending fatigue test data was found to be greater than that for the conventional data. All of the data mentioned herein will be forthcoming in a future report when Phase III is completed.

Phase III is now in progress. Full size production components are being vacuum carburized and will be tested in a helicopter production transmission test. The gears are of the spiral bevel type and are very highly loaded. While carburizing of these gears conventionally takes approximately two days total time, the gears will be processed by vacuum carburizing in less than four hours. It is interesting to note that these gears were removed from a production run of gears and are being shipped to a vacuum carburizing vendor approximately 1,000 miles away. There the parts will be vacuum carburized and then returned to the original vendor for final hardening and finish machining. This will all be done within the original production scheduling of the components at the machining vendor and without any interruption. This again illustrates the improvement to production scheduling which can be obtained by vacuum carburizing.

4. CONCLUSIONS

The benefits of vacuum carburizing to both cost and scheduling in the production of aerospace components paves the way for the production use of this process. With the cost of energy skyrocketing, this technique is a must for the future.

Someone once said, "Vacuum is really nothing." Well, vacuum carburizing is something else!

1st International SAMPE Metals Conference
August 18-20, 1987

THE EFFECT OF THERMOMECHANICAL TREATMENTS ON THE MECHANICAL PROPERTIES OF AL-LI-CU-MG-ZR ALLOYS

W.S. Miller, J. White and M.A. Reynolds
Alcan International Limited
Southam Road
Banbury, Oxon.
OX16 7SP. England.

Abstract

This study explores the effect of solution heat treatment, deformation prior to ageing and ageing temperature on the grain boundary microstructure of Al-Li alloys. Emphasis is placed on 8090 but comparison is made with other alloy systems where appropriate. It is concluded that all three variables affect grain boundary precipitation and that the latter two can be used to refine the grain boundary microstructure. These observations have led to the development of final thermomechanical treatments (FTMT's) for 8090 in various product forms which produce an improved strength/toughness relationship. The properties obtained on commercial materials by use of FTMT will be shown.

INTRODUCTION

The objective of this study is to examine the influence of final thermomechanical treatment (FTMT) on the grain boundary microstructure of complex aluminium alloys containing lithium.[1] Emphasis will be placed on 8090 but comparisons will be made with other alloy systems where appropriate. The fracture behaviour of aluminium lithium based alloys has been shown to be dominanted by grain boundary failure (1,2). This has been attributed to both deformation behaviour within the matrix and to grain boundary structure. In the former the presence of the coherent ordered δ' (Al_3Li) phase within the matrix leads to intense planar slip resulting in stress concentrations at high angle grain boundaries. In the latter case, large quasi-equilibrium phases are precipitated at high angle grain boundaries whilst associated δ' precipitate free zones (PFZ's) are formed. All of these factors would be expected to reduce grain boundary strength thus leading to a decrease in fracture resistance in this region.

In order to improve fracture resistance there are several courses of action; modify either the deformation characteristics of the alloy, the grain structure or the grain boundary microstructure. These are not alternative courses of action and may be combined to produce the optimum combination of properties in a final product (3).

The main attempt to alter the deformation chracteristics of Al-Li alloys has been to incorporate additional precipitation. The prime example of this is S' (Al_2CuMg) precipitation in Al-Li-Cu-Mg-Zr alloys (4). This phase is precipitated heterogeneously on sub-grain boundaries and dislocations and hence there is considerable scope for modification of the distribution of this phase using mechanical working treatments prior to ageing.

The effect of grain structure on toughness has been discussed particularly in relation to sheet products (3). It has been shown that under plane stress conditions toughness is controlled by the amount of deformation occurring in the plastic

[1] Note. The thermomechanical treatments discussed in this paper are identical to those previously applied to 2000 seies alloys which are also strengthened by the S′ phase,

zone. Under these conditions of "strain control" a fine recrystallised grain size is required to generate maximum strains to failure in the plastic zone. The situation is different for plate products. Under the extra constraint of plain strain, failure is initiated ahead of the induced fatigue crack at the nearest weak feature. This will be the nearest grain boundary particle or precipitate. Thus the ideal structure for thick products will be a coarse unrecrystallised structure. In thick product forms therefore, grain boundary microstructure is of primary importance.

Precipitates on grain boundaries are of two main types; constituent phases (primarily iron containing intermetallics) and precipitates arising during the solution treatment and ageing practice. The effect of constituent phases on mechanical properties has been characterised by Owen (5). Decreasing the volume fraction of constituent phases is achieved by minimising the iron and silicon content. In commercial alloys the practical limit for these elements is typically 0.1 wt%. The most practical method of modifying the grain boundary microstructure, therefore, is to control solution heat treatment practice, deformation prior to ageing and ageing temperature. This is the subject of the present paper. The observed changes in grain boundary microstructure are related to mechanical properties and a final section of the paper will demonstrate how the results of this study have been applied to commercial material.

EXPERIMENTAL

All of the 8090 alloy used within the present study has been laboratory produced with compositions within the published specification ranges (6) (see Table 1). The product forms and pertinent fabrication details are described with the associated experimental results when this is appropriate. TEM foils were prepared using a commercial twin jet polishing unit (Struers Tenupol). The electrolyte was 25% nitric acid in methanol and the specimens were thoroughly washed in methanol following perforation. The specimens were examined in a JEOL 2000FX transmission electron microscope fully equipped with microanalytical facilities. PFZ widths were measured with the boundary vertical with respect to the electron beam. Mechanical properties were determined using round bar tensile specimens, Charpy impact samples and compact tension fracture toughness samples. All tests were carried out in accordance with the relevant ASTM standard.

RESULTS

1. Effect of Cooling Rate from Solution Heat Treatment

The effect of cooling rate from solution heat treatment (SHT) on the T-L toughness of 13 mm thick extruded 8090 has been investigated using Charpy impact samples. Figure 1, shows the effect of cooling rate on impact energy for material aged 48 h at 150°C. The two extremes of cooling rate are typical of still air cooling and water quenching. Decreasing the cooling rate, as in an inefficient quench, produces a decrease in impact resistance. Typical grain boundary microstructures in slow cooling rate and high cooling rate samples are shown in Figure 2. The grain boundary precipitation is clearly affected by the change in cooling rate. Slow cooling has resulted in both an increase in apparent volume fraction and a change in morphology of the precipitates at the grain boundary. Examination of similar samples in the as-quenched condition clearly showed that this additional precipitation arises during the quenching (or rather air cooling) phase of the heat treatment cycle. The presence of these precipitates in both the quenched and the aged conditions indicates that they do not dissolve during ageing.

Two types of grain boundary precipitates are observed. The more common of these (Figure 3) exhibits five-fold symmetry in several zone axis diffraction patterns and is probably the icosahedral phase that has been the subject of much recent attention (see for example Ball et al (7)). The other phase, which has been seen only rarely, exhibits zone axis diffraction patterns consistent with a cubic structure. This may be the C-phase (8). Both phases have similar morphology and extend along the grain boundary often nucleating at triple points. It would thus be expected that they would have a particularly deleterious effect on both impact energy and fracture toughness (2). EDX analysis in the TEM shows the presence of aluminium and copper: an associated δ' PFZ suggests the presence of lithium in the precipitates.

The grain boundary microstructure in solution treated material has been examined in a variety of product forms in order to assess the effect of section thickness on the quench sensitivity of 8090. In relatively thin section material, that is sheet or thin extrusion (typically less than 6 mm), the conventional cold water quenching treatment following solution treatment entirely suppresses grain boundary precipitation. In thicker section up to 50 mm (typical of plate, forgings and extrusions) cold water quenching suppress grain boundary precipitation to the extent that isolated precipitates are observed only infrequently in the centre of the thickest sections. They thus are not expected to have a significant effect on fracture toughness.

The amount of grain boundary precipitation induced during the quench becomes significant at section thicknesses of > 100 mm (Figure 4). At these section thicknesses precipitation during quenching cannot be suppressed by the use of a more efficient quenching medium (e.g. iced brine) as the rate of heat extraction is governed by the heat flow through the material rather than transfer to the quenching medium. The extent of precipitation is, however, much less than in the air cooled samples from the initial study and in the fully aged condition the grain boundary precipitation is dominated by the ageing treatment rather than the quench.

2. Effect of Mechanical Work Prior to Ageing

The effect of a 2% stretch prior to ageing on the level of grain boundary precipitation in 8090 is illustrated in Figure 5 which shows 8090 plate in the underaged condition. Qualitatively there appears to be a decrease in the volume fraction of precipitates at high angle grain boundaries in the stretched compared with the unstretched material. In an attempt to quantify this difference the number of precpitates per unit length of grain boundary was measured in foils taken from stretched and unstretched plate aged for different times at 190°C (Figure 6). The differences in the level of grain boundary precipitation are not as marked as initial qualitative observation would suggest and clearly a more systematic survey is required to improve the statistics (this would involve accurate measurement of foil thickness which has not been made in this instance). There is, however, a statistically significant decrease in grain boundary precipitations with stretching in the underaged temper. This difference is not so marked in the peak and overaged tempers. The level of grain boundary precipitation also saturates in the peak aged condition.

Figure 7 shows that fracture toughness is higher in stretched material at short ageing times at 190°C in accord with the data in Figure 6. The fracture toughness values in Figure 7 were determined using 100 mm centre cracked panels cut from 1.6 mm rolled sheet. Stretching prior to ageing also increases strength as a result of additional S' precipitation (4). The overall result is a significant improvement in the strength/toughness relationship (Figure 8).

3. The Effect of Ageing Temperature

The effect of ageing temperature on grain boundary microstructure has been assessed using 25 mm 8090 plate solution heat treated, cold water quenched and given a 2% stretch prior to ageing. Three ageing temperatures were used: 150°C, 170°C and 190°C. Qualitative observations, using bright field TEM, confirmed the expected increase in grain boundary precipitation as temperature or time is increased. In order to quantify these observations, the width of the δ' PFZ has been used as a measure of the variation in the grain boundary structure. This is because of the difficulty in obtaining statistically valid measurements of the number of grain boundary precipitates (as was demonstrated in the previous section).

In agreement with previous workers (9,10,11,12) the half PFZ width has been plotted against the square root of time and the results are shown in Figure 9. The results appear to fit the parabolic relationship seen in other Al-Li alloys within the range of experimentation. Much longer ageing times, however, would be required to confirm that the PFZ width varies with the square root rather than the cube root of time. Such ageing times are unlikely to be used in a commercial heat treatment. For the range of temperature and time investigated, temperature has a far more pronounced effect than time on grain boundary microstructure. Within the limits of experimental error the data can be extrapolated to conclude that no PFZ is present following solution treatment. This, combined with the continuous growth of the PFZ's during ageing, indicates that they are formed by a solute rather than a vacancy depletion mechanism. In order to confirm this rate constants were determined for each temperature and used to calculate the activation energy for PFZ growth in 8090. This was found to be 1.11 eV (107 kJ/mol). This agrees with the range of published data for the activation energy for PFZ growth in Al-Li binary alloys (9,10) and Al-Cu-Li alloys (11). This suggests that the rate determining step in all three types of alloy is the same; the diffusion of lithium in aluminium (activation energy 1.43 eV) (13).

Figure 10 shows the influence of ageing temperature on the T-L orientation strength and fracture toughness of the 25 mm 8090 plate used in the TEM studies. The strength/toughness relationship is improved as the ageing temperature is decreased from 190°C to 135°C with the expected reduction in PFZ width. The microstructural and property benefits associated with low temperature ageing, however, must be balanced by practical and commercial considerations. Low ageing temperatures require long ageing times to reach the required strength levels (e.g. 100+ hours at 150°C), such times require increased investment in ageing ovens with a resultant cost penalty on the finished product. The use of lower ageing temperatures thus has a commercial limit and further improvements must be sought by other means. One method is to increase the level of cold work beyond 2% prior to ageing.

The effect of increasing the level of stretch on the grain boundary microstructure of 8090 has been investigated in a preliminary study. Qualitatively increasing the stretch from 2% to 6% has no observable effect on the level of precipitation at high angle boundaries. PFZ widths are also unaffected (Figure 11). These observations differ from those of Ashton et al (14) who show a decrease in grain boundary precipitation with increasing stretch in 8091 and 2090. This difference may be attributable to the higher copper contents of 2090 and 8091.

Figure 12 shows the effect of stretching up to 8% and ageing at 170°C on the T-L orientation strength/toughness relationship of the 25 mm 8090 plate considered earlier. As the stretch is increased the fracture toughness increases for a constant value of 0.2% proof strength. This improvement is not due to any significant modification of the grain boundary microstructure resulting from the increased stretch. Instead it is largely attributable to the fact that to achieve the same level of 0.2% proof strength a shorter ageing time can be used (Figure 13). This results in a more underaged temper and hence a higher fracture toughness. It is also possible that the increased stretch affects the precipitation of matrix S' and also introduces dislocation tangles into the matrix. Both of these will reduce planar slip and thus the likelyhood of stress concentration build up at grain boundaries.

The next obvious step is to combine lower ageing temperatures with an increase in the level of stretch prior to ageing. The result of this is shown in Figure 14 in which the strength and fracture toughness of plate stretched 6% and aged for 24 h at 170°C is compared with that of the same plate stretched 2% and aged for 16 h at 190°C. The material aged using the modified FTMT shows both improved strength and fracture toughness.

4. Application of FTMT's to Commercial Material

This final section deals with the application of optimised FTMT's to wrought products manufactured through the commercial processing factories of British Alcan.

Plate Product

Tables 2 and 3 show the properties obtained on 8090 plate given a 7 ± 1% stretch and aged to either a damage tolerant or peak aged condition. Comparison is made with 2024-T351 and 7070-T7651 respectively.

In the case of the damage tolerant temper the 8090 plate meets all the mechanical properties required to replace the existing damage tolerant alloy used in airframe construction - 2024-T351. The one important property of damage tolerant materials not given in Table 2 is fatigue crack growth rate. Figure 15 shows that this property is significantly improved in 8090 compared with 2024-T351. This is attributed to bifurcation of the developing fatigue crack due to planar slip (15).

Table 3 shows that peak aged 8090 meets the minimum requirements for 7050-T7651, with the exception of a slightly lower 0.2% proof stress, in the short transverse orientation. However, the advantages of 8090 are more clearly demonstrated when its density is taken into account.

Forgings

Table 4 details results from heavy section 8090 hand forgings where cold compression rather than stretching is used. Comparison is made with 2014-T652 as this alloy is typically used in medium strength aerospace forgings. An important feature is that the results are from section thicknesses > 100 mm. As discussed earlier material > 100 mm thick is susceptible to grain boundary precipitation during quenching which can deleteriously effect fracture toughness. The data in Table 4 clearly shows that by application of a FTMT an acceptable combination of properties can be achieved.

Extrusion

Stretching > 2-3% is not necessary in extruded product due to the inherently high strengths generated from their fibrous structure. Nevertheless significant improvements in the strength toughness relationship can be obtained by modification of the grain boundary microstructure by use of low ageing temperatures as illustrated in Figure 16.

Sheet

In general sheet products are not amenable to FTMT treatment since they are often resolution treated by the manufactures to improve formability. However, grain structure modification can be used to give a significant improvement in fracture toughness (3).

DISCUSSION

The observations relating to the precipitation of grain boundary phases during quenching following solution treatment demonstrate that 8090 is insensitive to this phenomenon in products with section thicknesses of less than 50 mm. The recent report of Bretz (16) of quench sensitivity in 2090 in section thicknesses of greater than 25 mm is in broad agreement with this conclusion. The difference in critical section thickness may be indicative of an effect of alloy content (probably copper as the main precipitating phase is copper containing). In section thicknesses of greater than 50 mm a quench sensitivity in 8090 would be anticipated on the basis of the microstructural observations and is observed in practice. It has been demonstrated, however, that satisfactory toughness can be achieved in 100 mm plate (17) and forgings (18) by suitable modification of processing and heat treatment variables.

The decrease in grain boundary precipitation arising from stretching prior to ageing can be related to improvements in the proof strength/fracture toughness relationship as observed by other workers (2,3,14). A further correlation with properties has been observed by Gray (19) who has observed an improvement in stress corrosion cracking susceptibility in stretched 8090. In a preliminary examination of 2090 by the authors a more marked decrease in grain boundary precipitation with stretching was found. This probably arises because of the different heterogeneously nucleated matrix precipitates. $T_1(Al_2CuLi)$ precipitation in 2090 will affect the diffusion of copper and lithium to the grain boundary whereas S' precipitation in 8090 will affect the diffusion of copper and magnesium. As the main grain boundary precipitates in both alloys systems contain predominately copper and/or lithium (e.g. δ (AlLi) and I-$(Al_6Cu(Li,Mg)_3)$ (7,8,11) matrix precipitation of the copper and lithium containing T_1 would be expected to have a greater influence than the copper and magnesium containing S'. This will make 2090 more sensitive to stretching than 8090.

A logical extension to the measurement of an activation energy for PFZ growth in 8090 is the development of an effective model for diffusion to the grain boundary in the alloy involving calculation of the underlying diffusion coefficients. This requires accurate measurement of the volume fraction of matrix precipitates present and is currently in progress. The use of PFZ width as a measure of grain boundary precipitation is perhaps questionable. The number of grain boundary precipitates per unit length of grain boundary, however, is directly related to PFZ width nd can be shown to be related to time through a parabolic relationship. To a first approximation, therefore, the assumption is valid. The results demonstrate that within the range of commercially viable ageing practices temperature has a more significant effect on grain boundary microstructure than time. The implication of this is that in order to obtain the optimum combination of mechanical properties, 8090 should be aged at as low a temperature as possible consistent with achieving the desired strength. It is this philosophy which has been used to develop the FTMT's incorporating high levels of stretch (to achieve strength) and low ageing temperatures (to achieve high toughness).

CONCLUSIONS

1. The cooling rate from solution treatment, stretching prior to ageing and ageing temperature and time significantly affect the grain boundary microstructure of 8090. Apart from cooling rate these variables can be used to control the grain boundary microstructure of 8090.

2. The growth of PFZ's in 8090 is controlled by a solute depletion mechanism the activation energy of which is 1.11 eV and the rate determining step is the diffusion of lithium in aluminium.

3. Lowering the ageing temperature and increasing the level of stretch from 2% to 6-8% either singularly or in combination improve the strength/toughness relationship in 8090.

6. ACKNOWLEDGEMENTS

This paper is published with the permission of Alcan International.

7. REFERENCES

1. W.S. Miller, M.P. Thomas, D.J. Lloyd and D. Creber. Proc. 3rd Int. Al-Li Conference - Oxford U.K. 1985. Ed. C. Baker, P.J. Gregson, S.J. Harris, C.J. Peel. Inst. of Metals, London 1986, p.584.

2. A.K. Vasudevan and R.H. Doherty, Acta Met. 35, 1987 p.1193.

3. W.S. Miller, J. White, M.A. Reynolds, D.S. McDarmaid and G.M. Starr. Proc. 4th Int. Al-Li Conf. Paris 1987. In press.

4. C.J. Peel, B. Evans, C. Baker, D.A. Bennett, P.J. Gregson and H.M. Flower. Proc. 2nd. Int. Al-Li Conf. Monteray CA. 1983. Ed. T.H. Sanders and E.A. Starke, TMS-AIME p.363.

5. N.J. Owen, D.J. Field and E.P. Butler - in Ref 1 p.576.

6. M.A. Reynolds, A. Gray, E. Creed, R.M. Jordan and A.P. Titchener. In Ref 1 p.57.

7. M. Ball and H. Lagace, In Ref 1 p.555.

8. A. Loiseau and G. Lapasset. J. de Physique. 47, 1986 p.C3-331.

9. T.H. Sanders, E.A. Ludwiczak, and R.R. Sawtell. Mat. Sci. Eng. 43, 1980, p.247.

10. O. Jensrud and N. Ryum. Mat. Sci. Eng. 64, 1984 p.229.

11. J.C. Huang and A.J. Ardell, In Ref 1 p.455.

12. S.C. Jha, T.H. Sanders and M.A. Dagananda. Acta Met. 35, 1987 p.473.

13. L.P. Costas (referred to in Ref 12).

14. R.F. Ashton, D.S. Thompson, E.A. Starke and F.S. Lin. In Ref 1 p.66.

15. S. Suresh and A.K. Vasudevan in Ref 3.

16. P. Bretz in Ref 3.

17. R. Grimes, T. Davis, J. Fearon and H.J. Saxty in Ref 3.

18. R.E. Lewis et al in Ref 3.

19. A. Gray in Ref 3.

TABLE 2 - Typical Mechanical Properties of 8090 Plate Produced Via FTMT Processing Compared with Minimum Requirement for 2024-T351 ~ 50 mm Thick (Mil Handbook V)

Property	Direction	8090	2024-T351
Fty MPa	L	425	322
	T	368	288
	ST	330	N.Q.
Ftu MPa	L	470	425
	T	459	425
	ST	448	N.Q.
Fcy MPa	L	356	268
	T	386	302
	ST	357	N.Q.
El %	L	6.0	6
	T	7.0	N.Q.
	ST	4.0	N.Q.
K_{1c}*	L-T	44.0	35
	T-L	37.0	27
	ST-L	28.0	24

N.Q. - Not quoted in Mil Handbook V

* Typical values for K_{1c}

TABLE 1 - 8090 Alloy Chemistry

	Li	Cu	Mg	Zr	Fe	Si
8090 Range	2.2-2.7	1.0-1.6	0.6-1.3	0.04-0.16	≤ .30	≤ .20
Alloy Used in Study	2.52	1.18	0.78	0.13	0.07	0.04

TABLE 3 - Typical Mechanical Properties of 8090 Plate Produced Via FTMT Processing with Compared Minimum Property Requirement for 7050-T7651 (Mil Handbook V)

Property	Direction	8090	7050-T7651
Fty (MPa)	L	495	453
	T	470	453
	ST	390	411
Ftu (MPa)	L	524	521
	T	519	521
	ST	486	487
Fcy (MPa)	L	457	446
	T	472	472
	ST	439	466
El (%)	L	6.2	8.0
	T	6.7	6.0
	ST	3.3	2.5
K_{1c} (MPa$\sqrt{m}$)	L-T	34	25.0
	T-L	28.5	24.0
	ST-L	19.5	18.0

TABLE 4 - Typical Mechanical Properties of 8090 125 mm Forgings Produced Via FTMT Processing Compared with Minimum Property Requirements of 2014-T652 (Mil Handbook V)

Property	Direction	8090	2014-T652
Fty (MPa)	L	400	370
	T	390	370
	ST	345	343
Ftu (MPa)	L	470	425
	T	465	425
	ST	450	412
El (%)	L	7	7
	T	4	2
	ST	3	1
K_q (MPa$\sqrt{m}$)	L-T	37	32
	T-L	27	24
	ST-L	20	20

* Data for 2014-T652 K_{1c} are typicals

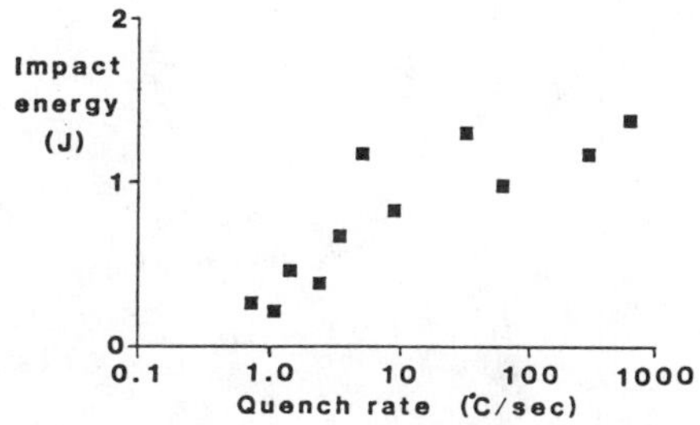

Figure 1. Effect of quench rate from the solution treatment temperature on the T-L orientation Charpy impact energy: 8090 Aged 48h at 150 °C.

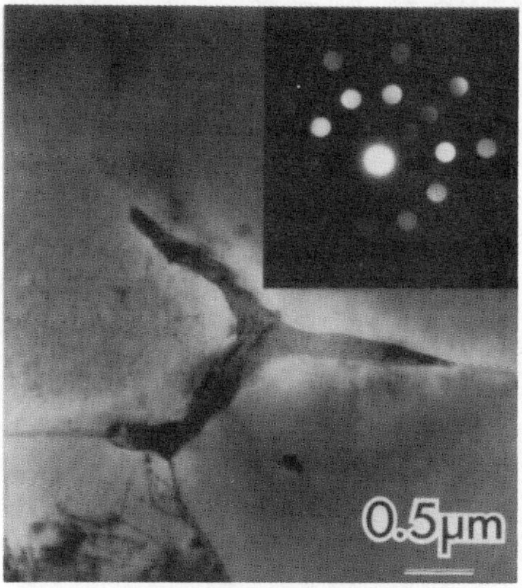

Figure 3. High angle grain boundary precipitate nucleated at a triple point during air cooling (1 C/sec) from the solution treatment temperature. 8090: aged 48h at 150 °C.

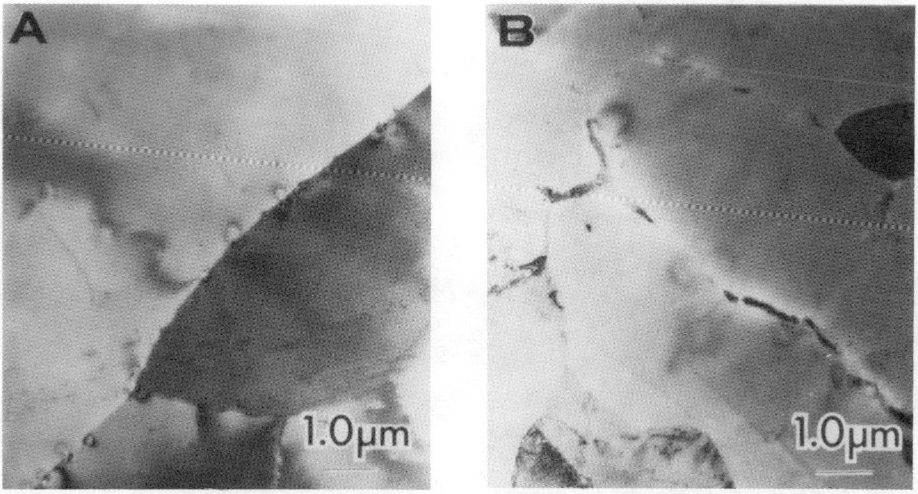

Figure 2. The effect of cooling rate on grain boundary microstructure of 8090.
a. water quenched (800 °C/sec.) aged 48 hours at 150 °C.
b. air cooled (1 °C/sec) aged 48 hours at 150 °C.

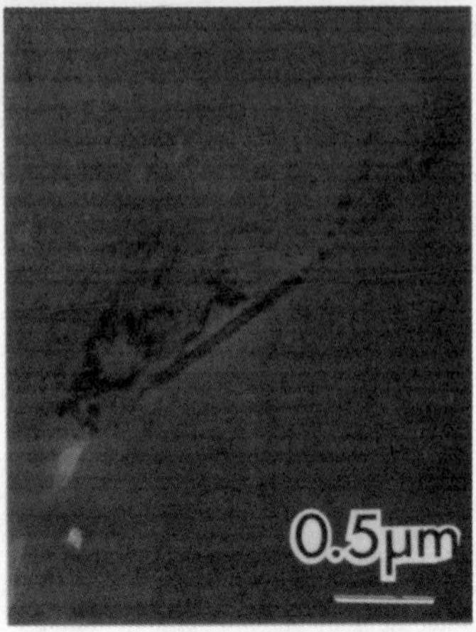

Figure 4. High angle grain boundary precipitation at the core of 100mm cold water quenched 8090 plate in the as-quenched condition.

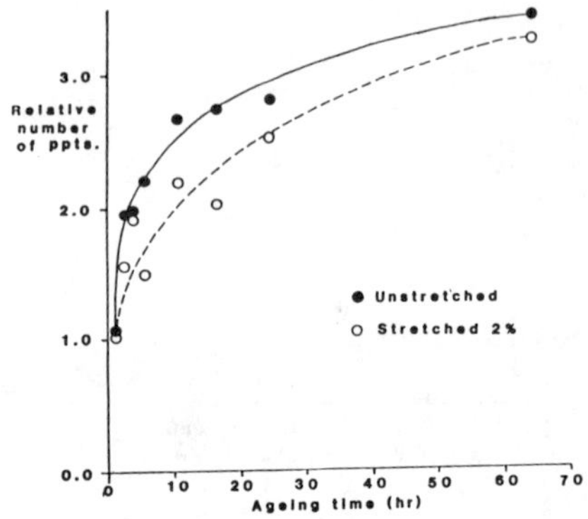

Figure 6. Comparison of the relative number of grain boundary precipitates between stretched and unstretched 25mm 8090 plate aged at 190°C.

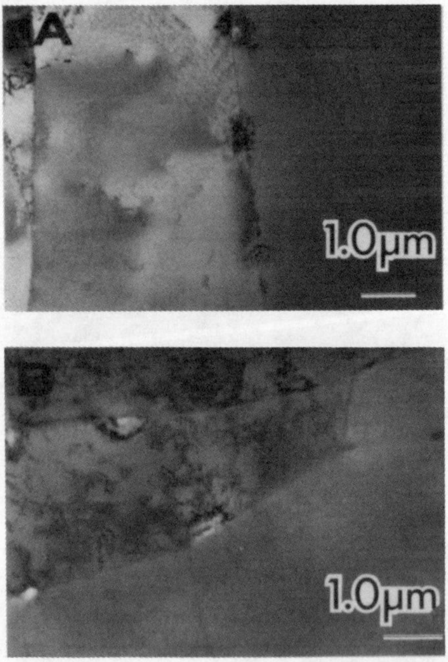

Figure 5. High angle grain boundary precipitation in 25mm 8090 plate aged 5h at 190°C. a. unstretched prior to ageing b. stretched 2% prior to ageing.

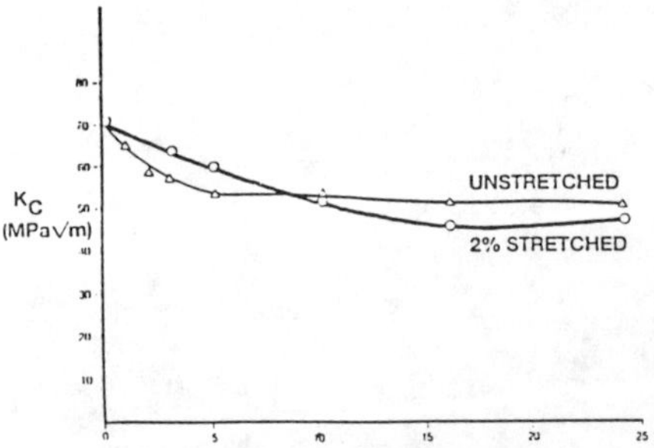

Figure 7. The influence of stretching on the variation in fracture toughness (Kc) for 1.6mm 8090 sheet with ageing time at 190°C.

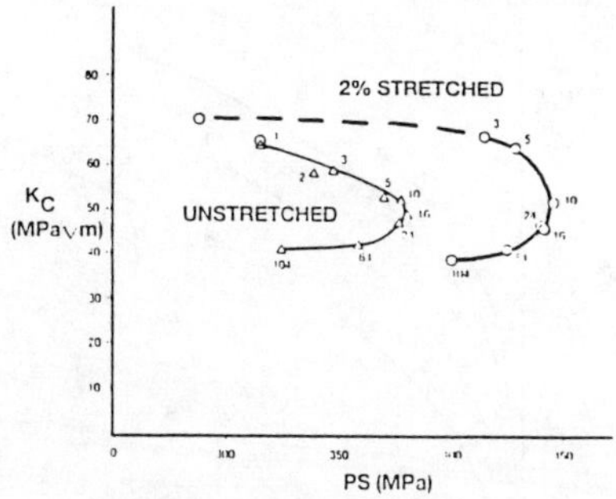

Figure 8. The strength/toughness (0.2%PS v Kc) relationship of unstretched and stretched 1.6 mm 8090 sheet aged at 190 °C.

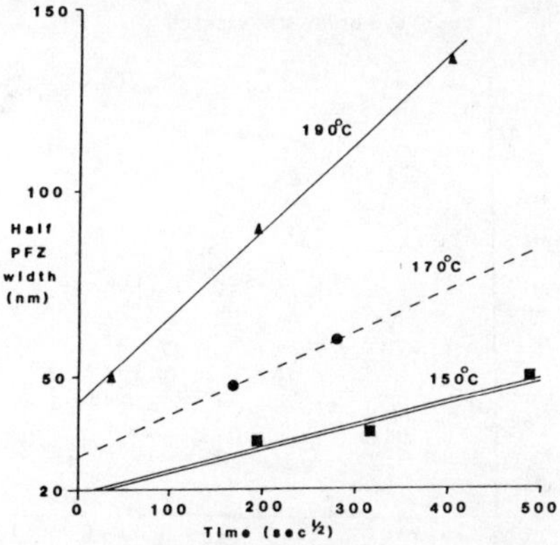

Figure 9. PFZ width plotted against $\sqrt{t}$ for 2% stretched 25mm 8090 plate.

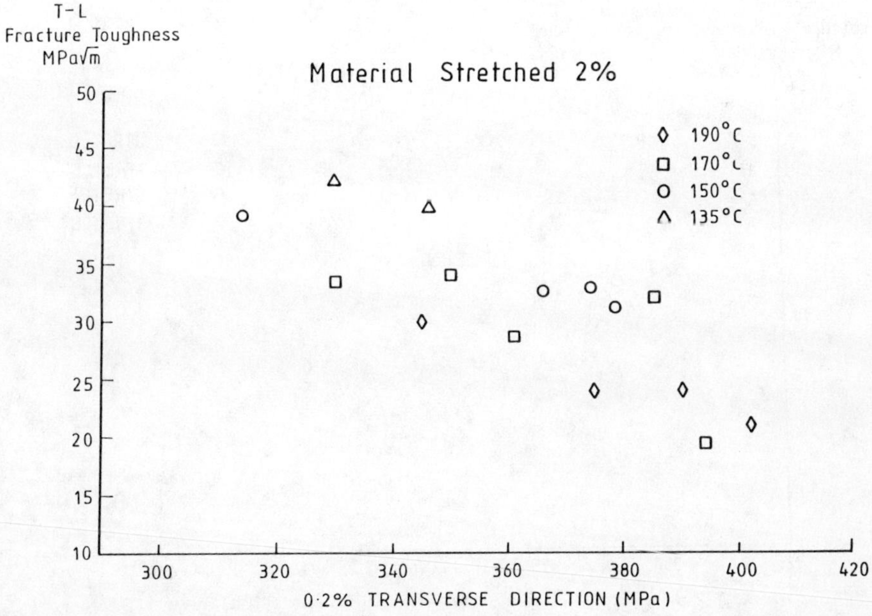

Figure 10. The effect of ageing temperature on the T-L orientation strength/toughness relationship for 25mm 8090 plate stretched 2% prior to ageing.

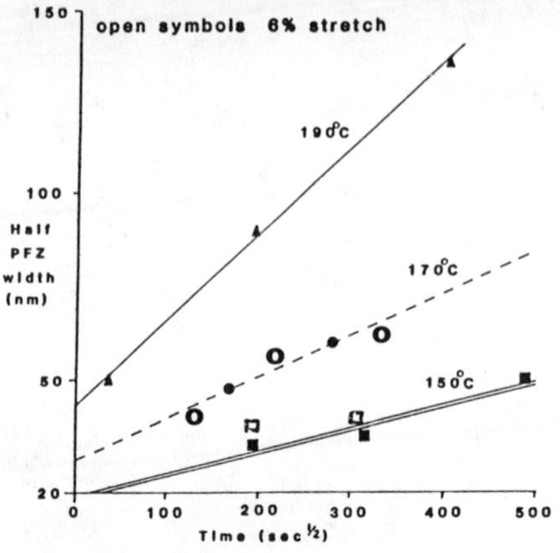

Figure 11. The effect of increasing the degree of stretching from 2% to 6% on the PFZ width.

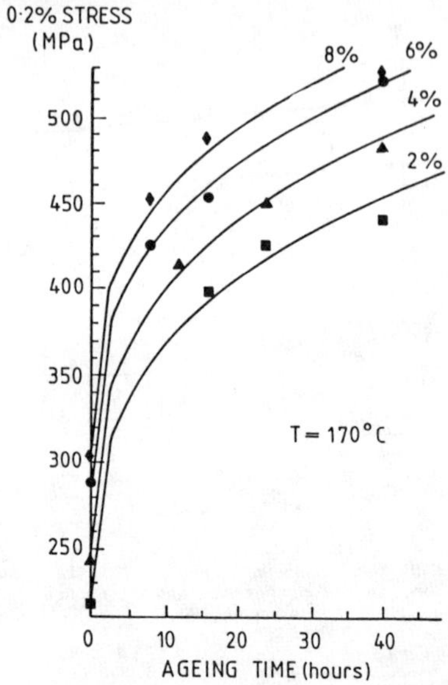

Figure 13. The artificial ageing response at 170°C of 8090 25mm plate with the degree of prior stretch. Longitudinal 0.2% proof stress.

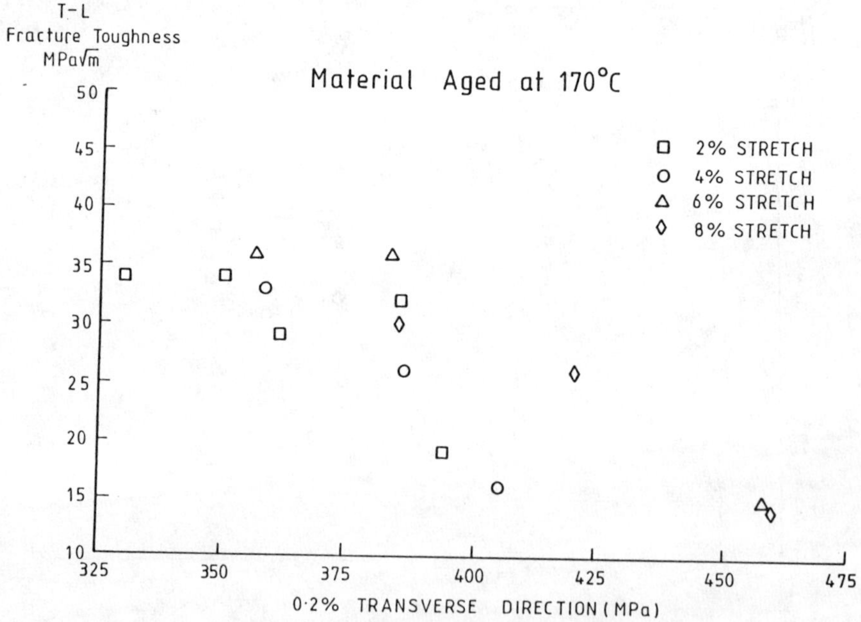

Figure 12. The effect of prior stretching on the T-L orientation strength/toughness relationship for 25mm 8090 plate aged at 170°C.

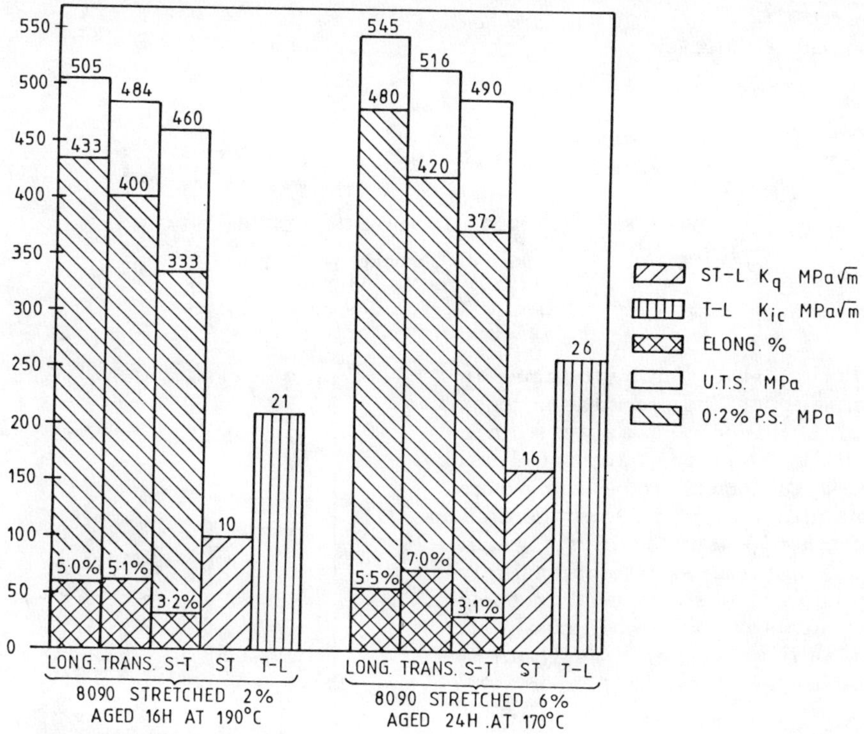

Figure 14. Comparison of the mechanical properties of 25mm plate given an optimised FTMT (stretched 6% age 24 hours at 170°C) with the properties of the same plate stretched 2% and aged 16 hours at 190°C.

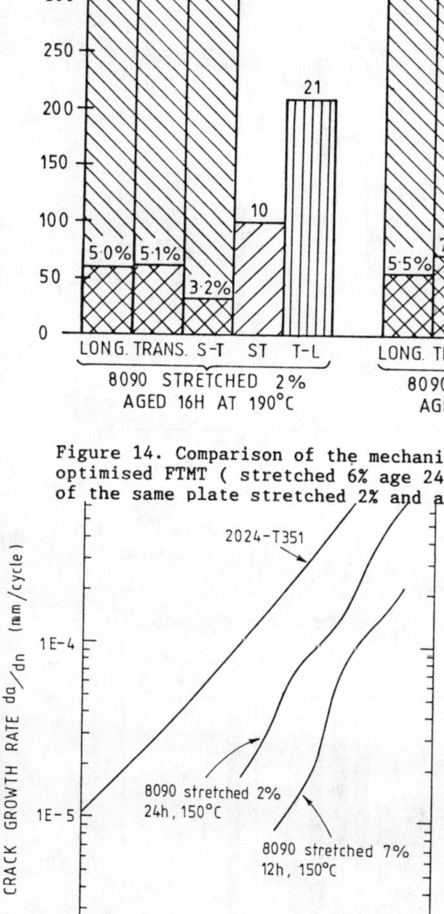

Figure 15. The fatigue crack propagation rates of 8090 50mm plate stretched 2% and 7% prior to ageing to the damage tolerant condition in comparison with 2024-T351 plate.

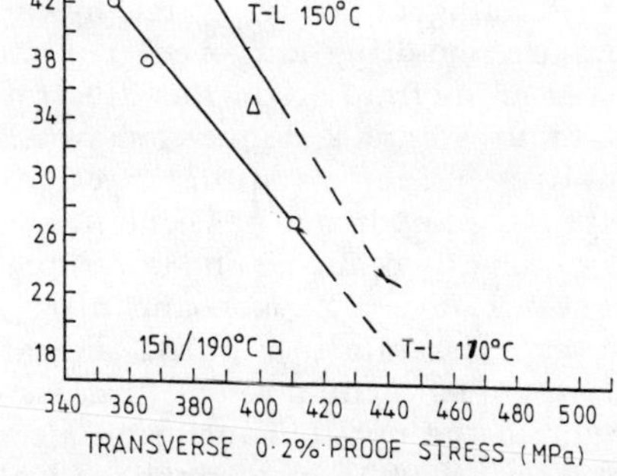

Figure 16. The effect of ageing temperature on the strength/toughness relationship of 100mm x 25mm section 8090 extrusion given a 2% stress relieving stretch.

PROCESS DEVELOPMENTS IN ADVANCED GEAR MATERIALS

Roy J. Cunningham
Supervisor Metallurgy
Boeing Vertol Company
Philadelphia, Pennsylvania

Perhaps the dominant theme in steel selection and specification today is pinpointed by a quote from a specialty steel company's annual report, "Customers are becoming more demanding in their requirements." (Reference 1.) Manufacturers of weight critical structures are increasingly requiring that steels used in their components exhibit much higher levels of cleanliness than was thought possible a few years ago.

A major step in improving cleanliness occurred when melting practices began incorporating vacuum atmospheres in conjunction with the standard air-melt processing techniques. The following detrimental features were considerably reduced:

- Entrapped gas content
- Degree of alloy segregation
- Inclusion frequency/severity (number/size) ratios.

Refining, or purifying steel by vacuum-arc remelting improved characteristics of the final product as illustrated in Figures 1 and 2. Figure 1 shows that melting under vacuum practically eliminated hydrogen gas content while significantly reducing and lowering the concentration of oxygen and nitrogen gases. The data in Figure 2 indicated that single vacuum-remelt processing reduced the nominal inclusion size to less than 0.0005 of an inch and also that vacuum-arc remelting (VAR) produced slightly cleaner material than did vacuum-induction melting (VIM).

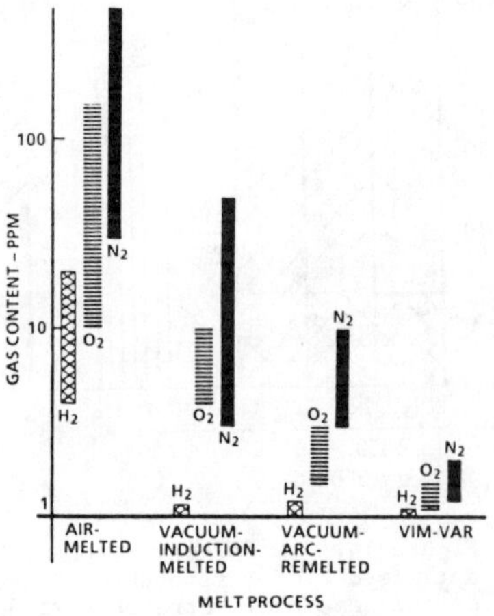

Figure 1. Comparison of Gas Content with Various Melt Processing

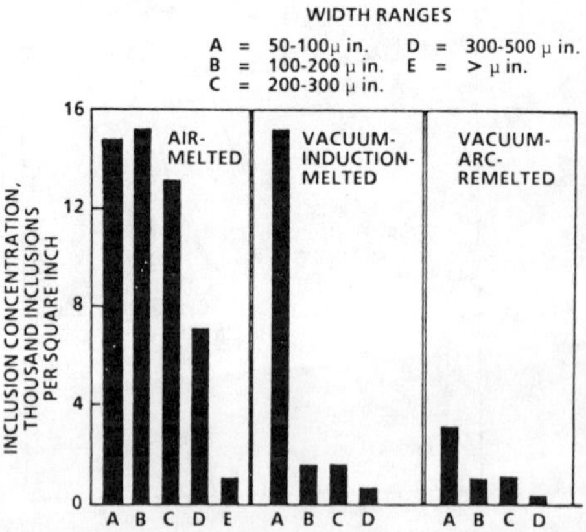

Figure 2. A Comparison of the Inclusion Concentrations in Various Width Ranges of Air-Melted, Vacuum-Induction-Melted and Vacuum-Arc-Remelted 4340 Steel

The VIM process, illustrated in Figure 3, is carried out by placing a ceramic crucible containing the charge inside a coil within a vacuum chamber, and charging the coil with an AC current of the proper magnitude and frequency. (Reference 2.) The very low gas pressures and magnetic stirring result in extremely low gas contents. However, since the ceramic lining erodes during melting, slag and refractory particles can be trapped in the VIM melt and, unless subsequently removed, will be found in the final product.

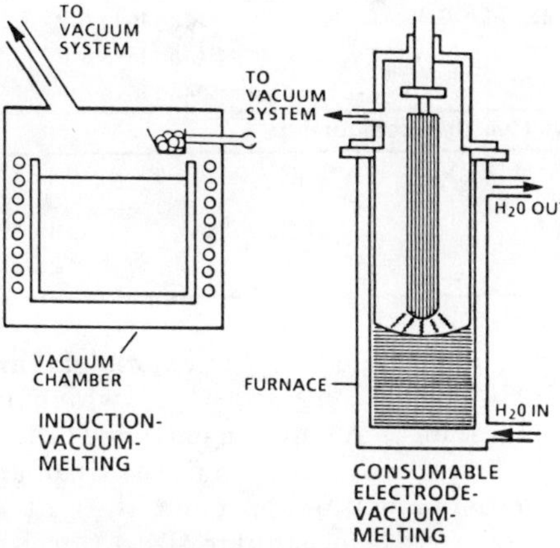

Figure 3. Schematic of Melting Processes - Vacuum-Induction and Vacuum-Arc-Remelted

The consumable VAR process is started by striking an arc between a charge electrode and a molten pool of metal in the melting crucible and maintaining the arc until the charge electrode has been completely melted. In consumable arc-vacuum melting, also illustrated in Figure 3, a water cooled copper crucible is used as the ingot mold thus increasing the rate of solidification and reducing the chance of slag/refractory type inclusions from being present in the final product. The combination of very high arc temperatures and low pressure results in a VAR product with very low gas and inclusion content.

One primary benefit of single-vacuum melting (SVM) over air melting (AM) for helicopter transmission applications, is an increase in endurance fatigue life as shown in Figure 4. (References 3 and 4.) These curves compare the endurance lives of planetary pinion bearings made of carburized and hardened 8620 steel. The SVM fatigue life is essentially double that of AM. This fatigue life improvement is directly related to the vacuum melting of the steel.

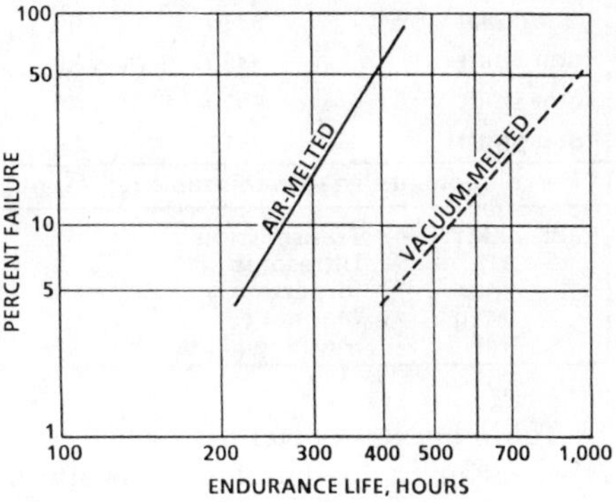

Figure 4. Comparison of Vacuum-Melted vs Air-Melted Carburized AMS 8620 Endurance Fatigue Lives

Even though improving the quality of steel by vacuum melting results in improved fatigue properties, more homogeneous structures, and uniform mechanical properties in the longitudinal, transverse and short transverse directions, large inclusions still inadvertently find their way into production components. Rejectable slag and refractory type defects which originated due to melting practice, have been found during:

- Ultrasonic inspection of raw material
- Magnetic particle inspection of forgings
- Magnetic particle inspection of finished machined parts.

TABLE 1. BRIEF HISTORY OF COMPONENTS REJECTED* DUE TO LARGE SLAG/REFRACTORY TYPE INCLUSIONS

Part Name	Matl	Melt Processing	Inclusion Size			Method of Finding
			Length	Width	Depth	
Spiral Bevel Ring Gear	9310	VAR	–	–	–	XMSN Failure
Spiral Bevel Ring Gear	9310	VAR	0.72	0.15	0.042	UT Insp of Gear
Billet	9310	VAR	0.7	0.004	0.021	UT Insp
Quill Shaft	9310	VAR	0.4	0.004	0.02	Mag Insp MP
Planet Gear	9310	VAR	0.096	0.022	0.115	Mag Insp MP
Quill Shaft	9310	VAR	0.8	0.03	–	–
Planet Gear	9310	VAR	1.1	0.08	0.01	Visual Insp MP
Rotor Shaft	9310	VAR	0.5	–	–	Mag Insp MP
Rotor Shaft	9310	VAR	0.5	0.2	–	Machining
Billet	4340	VAR	2.5	–	–	UT Insp
Planet Gear	9310	VAR	0.12	–	–	Mag Insp MP
Rotor Shaft	9310	VAR	0.25	0.06	–	Mag Insp MP
Rotor Shaft	9310	VAR	0.75	–	–	Mag Insp MP
Rotor Shaft	9310	VAR	0.75	–	–	Mag Insp MP

*The components listed in this table represent $125,000 worth of components.

KEY:
- XMSN – Transmission
- UT – Ultrasonic
- Insp – Inspection
- Mag – Magnetic
- MP – Machined Part

Table I gives a brief history of parts rejected for inclusions found in single vacuum-arc remelted 9310 and 4340 steels. In the majority of these cases the material was single vacuum melt VAR 9310 steel. The maximum size inclusion found was 1.1 by 0.08 by 0.01 inches. However, in a heat of single vacuum-arc remelted 4340 steel a defect 2.5 inches long was found.

The data presented in Table I gives an indication as to the type and number of parts rejected due to large slag/refractory type inclusions. In the majority of examples, the defects were found after finish machining had been accomplished. This is one of the most costly methods of finding these defects. It is noteworthy that the parts detailed in Table I came from mill heats from seven various steel mills which indicated that lack of cleanliness may be a common problem.

Inclusions are of great concern to aircraft manufacturers because these defects can act as stress raisers and shorten fatigue life. For example, if an inclusion is long and thin, which is typical for that found in extruded billet, the stress concentration at the tip of an inclusion could approach that of a fatigue crack of similar size. Certain inclusions are more detrimental than others in this regard due to thermal expansion characteristics of the inclusion and the surrounding matrix material. Tensile stresses can result at the inclusion tip as a result of thermal processing. Determinations using high temperature X-ray diffraction and dilatometry by Brooksband and Andrews (Reference 5), show that slag refractory exogenous type inclusions have greater tensile stress raising potential than do indigenous type (i.e., those inherent in the melt, such as manganese sulphide inclusions in 4340 steel). This is illustrated graphically in Figure 5.

Figure 5 reveals the stress raising/void forming potentials of various types of inclusions in 1 percent C – 1.5 percent Cr bearing steels. In the figure, the mean coefficient of thermal expansion for the matrix material, α_1, is plotted with that for various types of inclusions, α_2. From the data given, inclusions with $\alpha_1 > \alpha_2$ can be considered to be deleterious to fatigue properties since they provide tensile stresses in the matrix during cooling of the material from 800°C. Inclusions with $\alpha_1 > \alpha_2$ should not be detrimental since tensile stresses will not develop around the particle. Instead, the stress-raising potential represented by the difference (α_2-α_1) for the former detrimental types becomes replaced by the void-forming potential for the latter cases (particularly the sulphides).

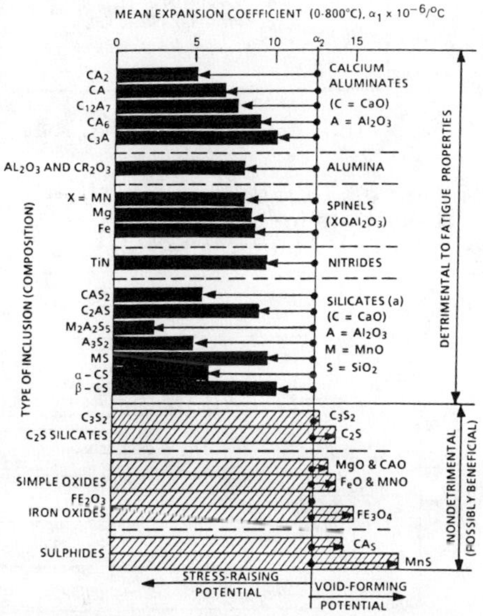

Figure 5. Stress Raising Properties of Inclusions in 1% C-Cr Bearing Steels

The following is a linear mechanistic approach to explain the generating of tensile stresses around an elongated inclusion/particle due to thermal expansion characteristics. When heating a material which contains such a particle, both the steel matrix and the particle expand at rates defined by their individual thermal expansion coefficients. If the thermal expansion coefficient of the particle is less than that for the matrix, the particle cannot expand at the same rate as the matrix. Since the particle and matrix are integrally tied together, the matrix being the stronger of the two, strains the particle beyond its elastic limit, and causes plastic deformation to occur. On cool down, the particle cannot return to its original size but it is now some delta size larger due to the amount of plastic deformation experienced. The matrix, however, attempts to return to its original size but cannot do so in the vicinity of the particle. The force generated by the inability of the matrix to return to its original shape results in squeezing the particle and straining the matrix.; If the particle is elongated in shape, this strain in the matrix will be concentrated at the tip of the particle and tensile stresses will result round the tip of the particle. The mechanism would be exactly opposite in the case where the thermal expansion coefficient of the matrix is less than that of the particle. In this case, a void is produced at the tip of the elongated particle.

The results from several experimental studies (References 6, 7, 8 and 9) have indicated that exogenous type inclusions, i.e., calcium aluminates, alumina, spinel, silicates and titanium nitride inclusions (in that order) are the most detrimental to fatigue life. The stress-raising potentials (Figure 5) for these compounds are clearly quite consistent with such observed relative effects. The most pronounced effect in the opposite direction is given by manganese sulphide which correlates well with the harmless effect of this type of indigenous inclusion. Interestingly, sulphides, because of their high coefficient of expansion and void forming potential, have reduced the sensitivity of welded C-Mn steel forgings to hydrogen cracking since the voids act as hydrogen sinks. (Reference 10.)

Because of the inclusion type defects found in single vacuum-melt steels, Boeing Vertol now requires a second vacuum-melt processing cycle during the manufacture of the steel used in all critical transmission components. Metallurgical evaluation of steel produced in this manner, i.e., by double vacuum-arc remelting VAR-VAR and VIM-VAR, gave very encouraging results. Significant improvements were observed particularly in fracture mechanics properties.

Fracture toughness results were nearly doubled as shown in Figure 6. (References 11 and 12). Figure 6 illustrates that for BMS 7-223 steel, VASCO X2M, the fracture toughness was increased from values in the 50 Ksi$\sqrt{}$in. range for single vacuum melt to values in the 80 Ksi$\sqrt{}$in. range for double vacuum melt. More important, however, was a 5:1 improvement observed in threshold stress intensity level properties. (Reference 13). This improvement can be directly related to the reliability of the product. This is illustrated in Figures 7 and 8.

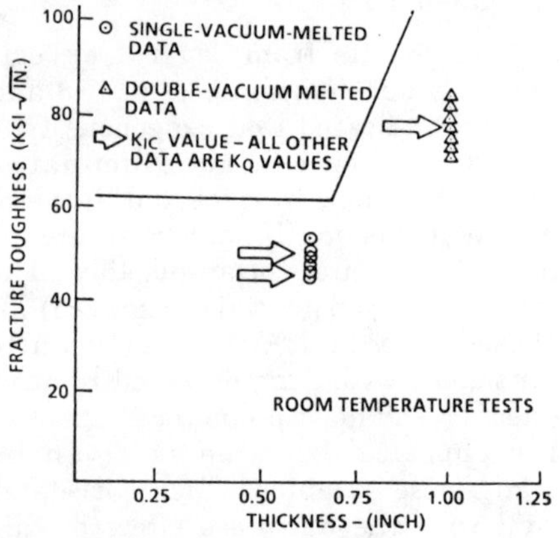

Figure 6. Fracture Toughness Test Results, Vasco X2M Steel

Figure 7 shows the effect of double-vacuum melting over single-vacuum melting on threshold stress level intensity values. The data obtained for VAR VASCO X2M basically shows a threshold stress intensity value of 1.7 Ksi$\sqrt{}$in. Note the significant increase in threshold value from single vacuum-melt material – 1.7 to double vacuum melt – 8.2 Ksi$\sqrt{}$in.

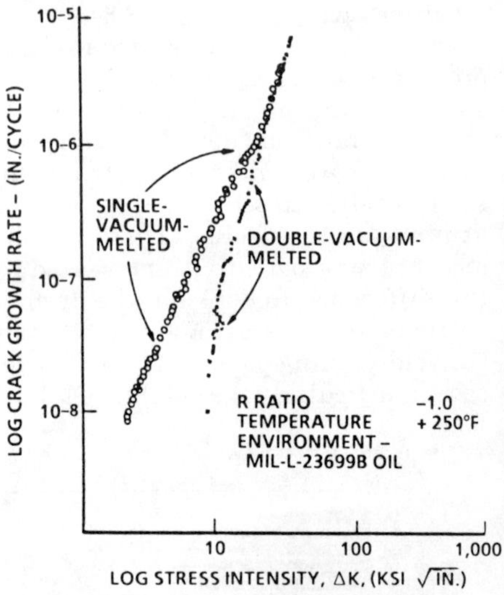

Figure 7. Crack Threshold Stress Level Data for Vasco X2M – Single vs Double Vacuum-Melted Material

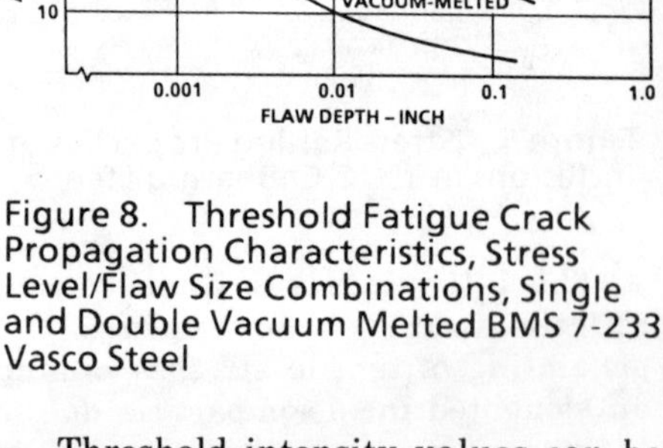

Figure 8. Threshold Fatigue Crack Propagation Characteristics, Stress Level/Flaw Size Combinations, Single and Double Vacuum Melted BMS 7-233 Vasco Steel

Threshold intensity values can be used to define the maximum or limiting crack size permitted in a component which will not propagate when the component is loaded to a particular stress level. Knowing this limiting crack size the component can then be sized to assure that defects smaller than this size will not propagate during the service life of the component.

Naturally, a factor of safety would be added to this critical size to further insure nonpropagation of the defect. This is illustrated graphically in Figure 8. These curves, for single and double vacuum melt steel have been developed from the information contained in Figure 7, and a flaw size ratio, with a 5:1 surface length to depth. This ratio is typical of fatigue cracks found in gear shaft and web failures. In the figure there are two curves, one representative of the critical threshold stress level for single vacuum-melt steel and the other representative of that for double vacuum-melt material. At any point to the left of each curve, propagation will not occur, while to the right, the defect stressed at any level will propagate. Recent work has shown that defects purposely placed in actual test specimens react very closely to the data in Figure 8 when tested above and below the critical stress threshold level. When the test specimen was stressed above its threshold level, based on the size of its known defect, the defect propagated or grew. When the specimen was tested with a stress below its calculated threshold stress level, the crack did not grow. Figure 8 also reveals a major benefit of using double, rather than single vacuum-melt steel. If a defect 0.15 of an inch long by 0.03 of an inch deep is present in double vacuum-melt steel, it will not propagate at an alternating stress less than 28 Ksi. However, if this same defect is present in single vacuum-melt steel, it will propagate, thus reducing the service life of the component. It is important that gears be designed such that any allowable defects that cannot be detected by nondestructive techniques will be below the critical size necessary from a stress point so that propagation during the service life of the part will not occur.

With this information, the designer can:

- Design to single vacuum-melt threshold criteria and thus build into the component a significant factor of safety for defect tolerance while increasing the reliability of the component
- Design to double vacuum-melt threshold properties and thereby decrease the size and weight of the component
- Design to single vacuum-melt critical flaw size properties and use the delta difference in allowable alternating stress between the single and the double vacuum-melt threshold values for future growth considerations.

This defect tolerance analysis, however, does not preclude slag inclusion type defects from getting into the melt and subsequently costing the producer and consumer many times the price of the initial part. As an example of this, Table II indicates that double vacuum-melt processing by itself does not insure that large detrimental inclusions will be eliminated by processing alone. As a result, specified controls have now been placed on the mills processing our steel. These controls are detailed in material specifications for not only the BMS 7-223 material but also the traditional 9310 and 4340 steels. In addition to the standard requirements for chemistry, workmanship, grain size, macrocleanliness, etc., the following controls have been added:

TABLE II. SLAG TYPE INCLUSION FOUND IN VAR-VAR VASCO X2M STEEL

	Inclusion of Size			
Part Name	Length	Width	Depth	Method of Finding
Rotor Shaft	2.5	0.13	0.13	Machining Forging
Spiral Bevel	1.0	0.02	–	Mag Part
Sun Gear	0.05	0.013	–	Visual During Machining
Planet Gear	0.13	0.0135	0.004	Visual During Machining

- The material shall be multiple melted by vacuum induction melting followed by the consumable electrode vacuum arc-remelt process (VIM-VAR).
- The machined round mill products shall be immersion ultrasonic inspected in accordance with the Company approved procedure and shall have a specified amount of material removed from the surface after the ultrasonic inspection.
- The steel producer shall have a written procedure for melting the material which is agreed upon by the producer and the purchaser, and which details the requirements for each operation and the extent to which each operation can vary.
- The producer shall record any significant variation which occurs during production of the heat. A significant variation is any variation from the standard procedure which has or may have an effect on the quality and/or the life of the finished part.
- Examples of significant variations to be reported are as follows:
  - Primary ingot construction for vacuum-arc remelting.
  - Consumable arc interruption.
  - Abnormal or excessive cropping of the vacuum-arc ingot for reasons of quality.
  - Change in heat treatment of finished mill product.
  - Nonspecified additions to the composition.

An interesting sidelight of this approach to utilize additional controls to insure cleanliness is that all billet material is ultrasonically inspected using a technique which insures that any defects remaining in the material will be less than one fifth the size required to cause crack propagation in service. Observing Figure 8 and using a maximum-design-allowable alternating stress of 25 Ksi, the allowable defect size would be 0.04 of an inch deep by 0.2 of an inch long. Noting that the ultrasonic inspection technique utilizes a number 2 flat bottom hole for acceptance/rejection criteria (number 2 FBH = 0.031 inch diameter), a factor of 6:1 between allowable defect size vs propagation of that defect is obtained.

For the past several years BMS 7-223 steel has been produced by vacuum-induction melted followed by vacuum-arc remelted, i.e., VIM-VAR. (References 14 and 15). Shown in Figure 1 is the comparison of the gas content of VIM-VAR with that from other types of processes. Note that the gas contents are much reduced.

Since the addition of mill processing controls to procurement specifications, problems due to lack of cleanliness have not been experienced with parts. Approximately 60 mill heats of steel from at least two different mills have been produced. These heats represent approximately 0.5 million pounds of the Vasco X2M material.

The benefits obtained from double vacuum-melt processing and the controls required can be directly related to the reliability and performance of structural components. These benefits strongly justify the additional cost of the steel incurred because of the increased care in processing. Improving the cleanliness of the steel used will most assuredly have both short range (decreased scrap rates) and long range (improved reliability) benefits, both to the manufacturer and the customer.

REFERENCES

1. Chandler, H.E., WHAT USERS SHOULD KNOW ABOUT CLEAN STEEL TECHNOLOGY, Metal Progress, October 1981, PP52-59.
2. Henry, R.J. and Cary, R.A., EVALUATION OF VACUUM PROCESSED STEELS, Paper Given at Golden Gate Metals Conference, September 1968.
3. A SUMMARY OF COMPARATIVE PROPERTIES OF AIR MELTED AND VACUUM MELTED STEELS AND SUPERALLOYS, DMIC Report No., 1960.
4. Statter, R., DOUBLE VACUUM MELTING OF HIGH PERFORMANCE BEARING STEELS, Industrial Heating, September 1974, PP40-55.
5. Brooksbank, D. and Andrews, K.W., STRESS FIELDS AROUND INCLUSIONS AND THEIR RELATIONSHIP TO MECHANICAL PROPERTIES, The Iron and Steel Institute, Volume 4, April 1972, P246.
6. Frith, P.H., FATIGUE TESTS ON ROLLED ALLOY STEELS MADE IN ELECTRIC AND OPEN HEARTH FURNACES, The Iron and Steel Institute, London, 1954.
7. Murray, J.D. and Johnson, R.F., CLEAN STEEL, The Iron and Steel Institute, Volume 110, London, 1963.
8. Johnson, R.F. and Sewell, J.F., JISI, Volume 196, 1960, P414.
9. Howard, J.I. and Cummings, R., LOW ALLOY STEELS, The Iron and Steel Institute, Volume 105, London, 1968.
10. Hewitt, J. and Murray, J.D., British Welding Journal, July 15, 1968, P151.
11. Cunningham, R.J., TOUGHNESS EVALUATION OF DOUBLE VACUUM MELTED BMS7-223 STEEL, Boeing Vertol Company Document D210-12033-1, Boeing Vertol Company, Philadelphia, Pennsylvania, 1979.
12. Cunningham, R.J., PARAMETERS OF AEROSPACE GEAR HEAT TREATING, Heat-Treating Magazine, July 1979, PP21-23, August 1979, PP30-32.

13 Cunningham, R.J., THRESHOLD STRESS LEVEL EVALUATION OF VASCO X2 STEEL (BMS7-223) AND AISI 9310 STEEL (BMS7-249), Boeing Vertol Company Document D210-12034-1, Boeing Vertol Company, Philadelphia, Pennsylvania, 1979.

14 Cunningham, R.J., THRESHOLD STRESS LEVEL EVALUATION OF VASCO X2 STEEL (BMS7-223) AND AISI 9310 STEEL (BMS7-249), Boeing Vertol Company Document D210-12034-1, Boeing Vertol Company, Philadelphia, Pennsylvania, 1979.

15 Cunningham, R.J., METALLURGICAL EVALUATION OF PRODUCTION MILL HEATS OF VIM-VAR BMS7-223 STEEL, Boeing Vertol Company Document D210-12043-1, Boeing Vertol Company, Philadelphia, Pennsylvania, April 1982.

1st International SAMPE Metals Conference
August 18-20, 1987

FLO-FORGE -- NEAR NET SHAPE

Joseph J. Mele
J.M. Enterprise, Inc.
188 Warner Road
Huntington, N.Y. 11743

Abstract

FLO-FORGE is a new development in shaping metal to very near net shape. It can also shape parts to sizes not possible with existing equipment. The process is unique in that a part is shaped progressively in a die configuration described herein. The part size is not limited as in the conventional forging of low volume, high cost parts.
The progressive application of force over a small area at a time delivers high localized uniform pressure that forms a part to very near net shape. The pressure is supplied by a hydraulic cylinder that pulls a patented die between platens, the upper of which has an inclined surface. The inclined surface magnifies the force provided by the cylinder. This concept allows the FLO-FORGE system, with only one hydraulic cylinder to accomplish what existing forging equipment can using up to nine hydraulic cylinders.

1. INTRODUCTION

Forging to net shape and near net shape represents a big savings in material and subsequent machining after forging. Interest in this area of shaping metal has attracted a great deal of attention by the Forging Industry since the early 1970's. Due to the limitations of current forging technology for shaping high cost, low volume items (pressing a one piece male portion of a die into the female portion), the anticipated growth of net shape applications, especially in larger parts, has not come close to expectations. A recently published report from a leading forging company discusses their development work making near net shapes up to 600 square inches.

FLO-FORGE a recent development in forging has the potential to forge to very near net shapes exceeding 600 square inches by a factor of ten. The results of experimental work accomplished and a description of this patented process will be described in this report.

2. THE FLO-FORGE PROCESS

FLO-FORGE is a process that differs from conventional isothermal forging in predominately two ways:

1) The male portion of a matched die set is segmented perpendicular to the longitudinal axis of the die (as in a loaf of sliced bread).
2) The power that forces the male portion of the die into the female portion is provided by a hydraulic cylinder that pulls the die assembly past an inclined surface.

Unlike conventional forging (where the platens move towards each other during the shaping operation), the FLO-FORGE process requires that the spacing between platens be established and positioned. The piston of a hydraulic cylinder pulls the die assembly between the platens causing the male segments of the die to come in contact with the inclined surface and positively and gradually force the male segments into the material to be formed. {see figure 1}.

The spacing between the upper {inclined surface} and the lower platen controls the amount of penetration that the male segments make into the material to be formed. Therefore, controlling the spacing between the platens to close tolerances results in forged parts controlled to similar tolerances. With the proper spacing between platens obtained, one pull of the die assembly past the inclined surface forms the subject part to very near net shape.

3. EXPERIMENTAL RESULTS

The experimental equipment shown in figure 2 was used to make parts of different shapes and configurations to evaluate the FLO-FORGE theory. The first part formed that is shown in figure 3 was selected to

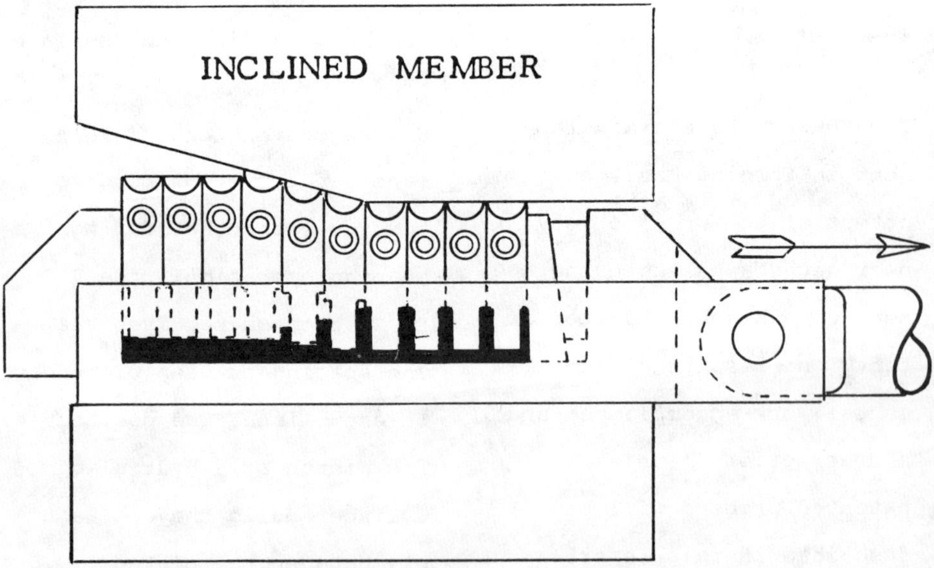

Fig. 1 Segmented FLO-FORGE Die Being Pulled Past Inclined Surface

Fig. 2 Experimental FLO-FORGE Equipment

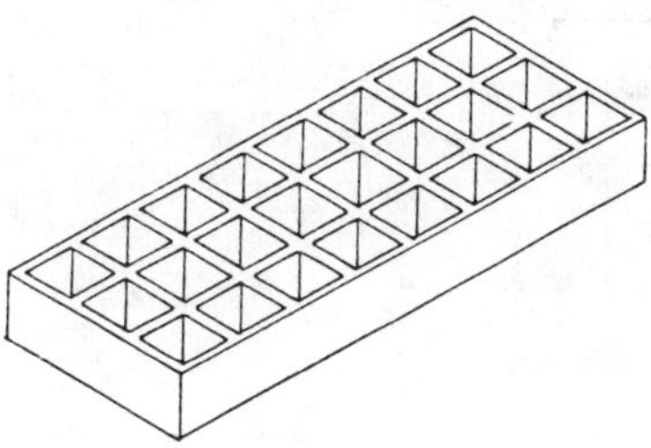

Fig. 3 First Part FLO-FORGED

show forging feasibility and determine flange height to thickness ratios. It was also discovered that it was possible to reproduce the finest of detail as shown in figure 4. Due to the high localized incremental pressures, H/D (height to thickness) ratios of 14 were realized. A summary of physical properties of specimen cut from the FLO-FORGE part shown in figure 3 is given in table 1.

Other configurations were selected to simulate actual

Fig. 4 Photograph Shows Fine Detail on FLO-FORGED Part

Table 1 Physical Properties of Specimen Cut from Part in Fig. 3

TENSILE TEST REPORT — REPORT NUMBER: — MATERIAL: 2024-T6 ALUM ALLOY — DATE: 2/22/85

TEST MACHINE: RIEHLE — TEST ENGINEER: W.K. — PLANT/GROUP: — — EXTENSION:

EXTENSOMETER: DN-20 — REQUESTER: — PLANT/GROUP: — — EXTENSION:

1. SPECIMEN NUMBER	A	B	C	R2	R3	R4	AVER.
2. GRAIN DIRECTION	LONG.	LONG.	LONG.	LONG.	LONG.	LONG.	LONG.
3. TEST SECTION (INCHES)	.238×.324	.246×.380	.240×.330	.121×.411	.115×.430	.140×.380	
4. INITIAL GAGE LENGTH (INCHES)	2.00	2.00	2.00	2.00	2.00	2.00	
5. HARDNESS READING ROCKWELL B	79.3	79.5	79.3	78.5	78.3	78.1	
6. TEST TEMPERATURE (°F)	71	71	71	71	71	71	
7. SOAK PERIOD (MIN.)	N/A	→	→	→	→	→	
8. STRAIN RATE IN/IN /MINUTE	.005	.005	.005	.005	.005	.005	
9. ULTIMATE LOAD (LBS.)	5140	6180	5260	3330	3270	3540	
10. YIELD LOAD (LB.) 0.2% OFF-SET	4320	5150	4420	2750	2720	2910	
11. GAGE LENGTH AFTER FAILURE (INCHES)	2.175	2.130	2.160	2.150	2.165	2.10	
12. DIA. AT FRACTURE (INCHES)	(.0678) .224×.303	(.0814) .230×.354	(.0679) .224×.303	(.044) .111×.396	(.0424) .109×.389	.0461 .120×.384	
13. INITIAL SPECIMEN AREA (IN.²)	.077	.093	.079	.050	.049	.053	
14. ULTIMATE STRESS (PSI)	66,753	66,452	66,582	66,600	66,735	66,792	66,652
15. YIELD STRESS (PSI)	56,103	55,376	55,949	55,000	55,510	54,906	55,474
16. % ELONGATION IN INCH(ES)	8.30	6.50	8.55	7.30	7.05	7.8	7.58%
17. % REDUCTION OF AREA	11.9	12.5	14.0	12.0	13.5	13.0	12.82%
18. E (MODULUS OF ELASTICITY) (PSI X 10⁶)	10.85	10.92	10.96	10.90	10.79	10.80	10.87

aircraft structural parts and are shown in figures 5,6 and 7. Figure 5 represents a scaled down version of a 757 landing gear support beam. Although the actual landing gear had flanges on one side only, the sample part was designed with flanges on both sides. Numerous tests with this configuration showed very conclusively that FLO-FORGED parts could be shaped with consistent repeatability. The fine detail achieved required no machining except for the removal of flash as shown in figure 5a.

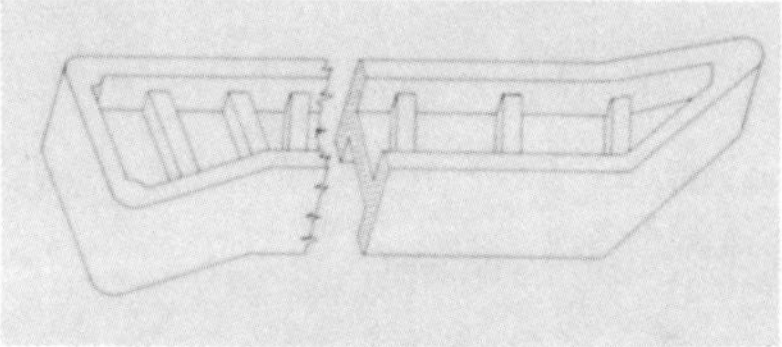

Fig. 5 Scaled-Down Version of 757 Landing Gear Support Beam

Fig. 5A FLO-FORGED Part with Less Than 3% Flash

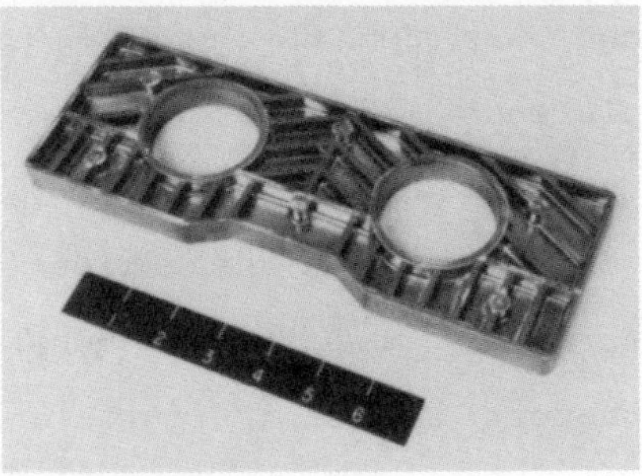

Fig. 6 Model of Typical Landing Gear Bulkhead

Fig. 7 Model of Typical Long Structural Aircraft Wing Component

With the selection of the proper blank, the amount of flash removed was less than 3 %. By comparison, the actual 757 landing gear support beam weighed 4000 pounds as conventionly forged and approximately 1000 pounds after machining, representing 75 % scrap.

The part made as shown in figure 6 was forged net except for the removal of 3 % flash. The two cutouts represent engine cutouts found in a typical landing gear bulkheads of military aircraft. It is anticipated that most if not all of the CNC machining usually associated with the manufacturing of aircraft primary structural components will not be required.

An effort was made to select still another application where the FLO-FORGE operation could save material and reduce machining. The part is shown in figure 7 and is typical of long structural members found in wing structures. These parts are usually machined from "H" extrusions and the "buy to fly" weight ratio averages 7:1. Figure 8 shows the "H" extrusion from which the final shape is machined. Figure 9 shows the blank used to FLO-FORGE the same part where the "buy to fly" ratio will be approximately 1.1:1 when performed on a full scale production basis.

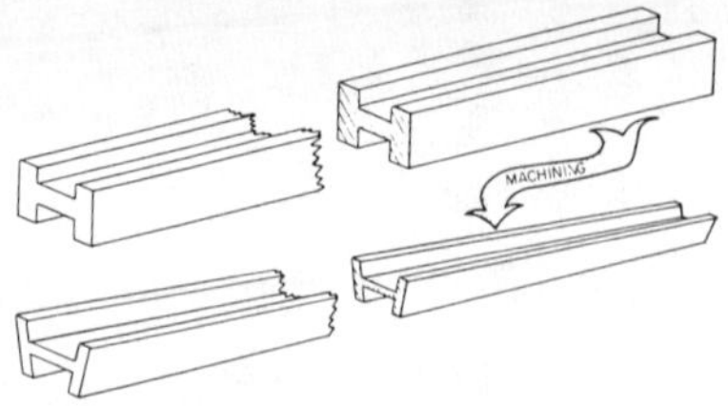

Fig. 8 Machining of Wing Component from "H" Section

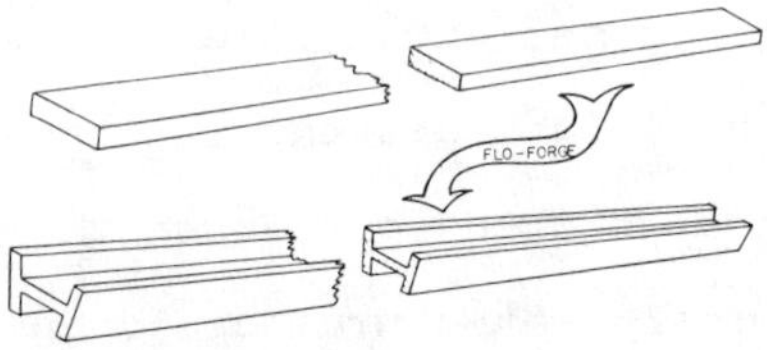

Fig. 9 Wing Component Made by FLO-FORGE Process

4. FLO-FORGE DIE CONCEPT

The segmented die design is an integral part of the FLO-FORGE system. The method of applying force to each successive male segment as it contacts the inclined surface provides high localized pressure that results in the net shape configurations shown in figures 5,6 and 7.

Furthermore, any increase in the length of the die does not require any additional force to form the subject part. Figure 10 shows that dies of any length can pass through the FLO-FORGE platens and near net shape can be obtained throughout. By comparison with conventional forging, (power requirements are proportional to the surface area of the part) FLO-FORGE power requirements are greatly reduced. There is no need for a pumping station and accumulating tanks to supply hydraulic fluid to the numerous cylinders on large conventional forging presses. Figure 11 shows the comparison of power requirements for forging large structural components.

A typical male segment is shown in figure 12. It has a concave surface on top that supports a rocker element that can swivel to maintain contact with varying angles of the inclined surface.

It also has a roller at each end that is used to facilitate the exiting of the male segments from the female after the part has been FLO-FORGED. This can be seen in step 4 of figure 13.

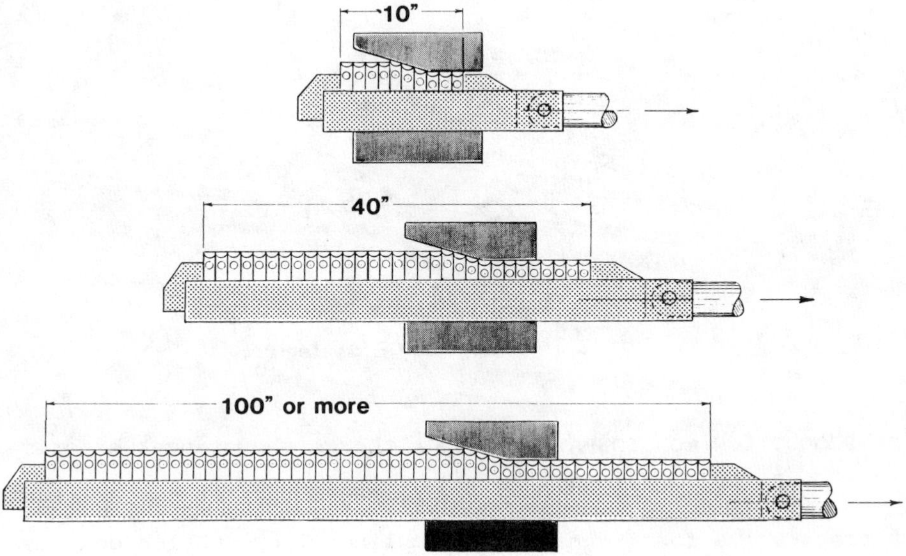

Fig. 10 Near-Net Shape of Any Length Can Be Processed by FLO-FORGING

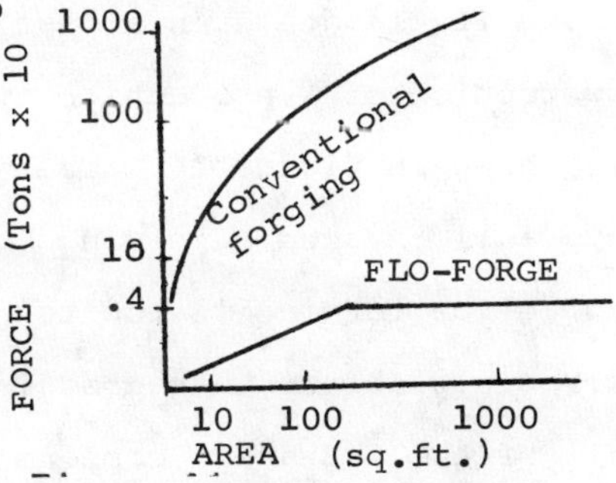

Fig. 11 Power Requirements -- Conventional Forging versus FLO-FORGING

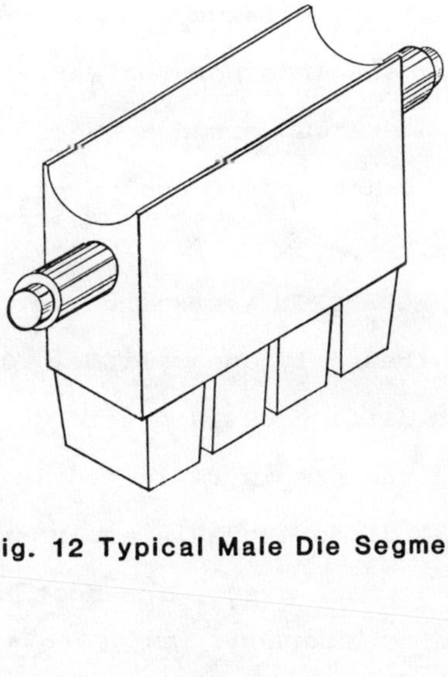

Fig. 12 Typical Male Die Segment

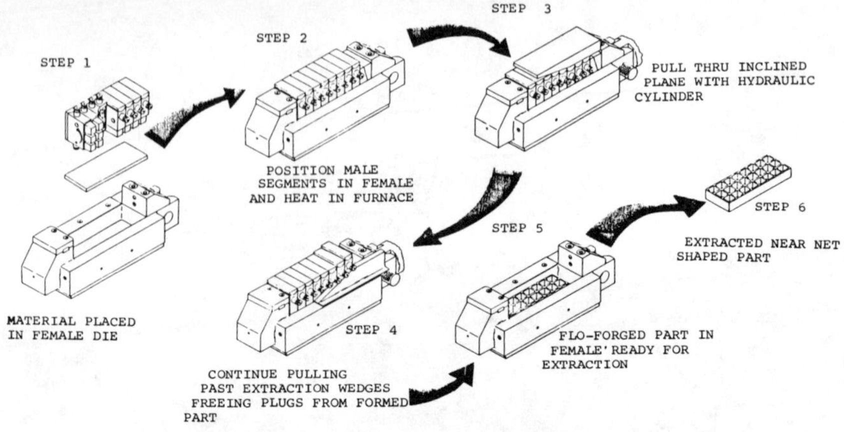

Fig. 13 The FLO-FORGE System

5. PRODUCTION CONSIDERATIONS

The progressive forming nature of the FLO-FORGE system permit the forging of very large parts not possible with current forging capability. Large conventional equipment incorporates many cylinders supported by accumulating tanks and a pumping station.

The FLO-FORGE system however, requires only one cylinder. This vast difference in power requirements is reflected in large differences in equipment procurement costs, in addition to lower maintenance and operating costs.

Although the present experimenta FLO-FORGE system has a fixed cylinder, production equipment will have a cylinder that can be indexed. Thusly a six foot cylinder can forge parts many times its length. An artist conception of a production syste is shown in figure 14. The hydraulic system supplying the force for forming is located on movable platform that contains all of the hydraulic components for the total system. Indexing holes in the side plates permit the cylinder to reposition itsel after each retraction of the piston as it pulls the die past the inclined surface.

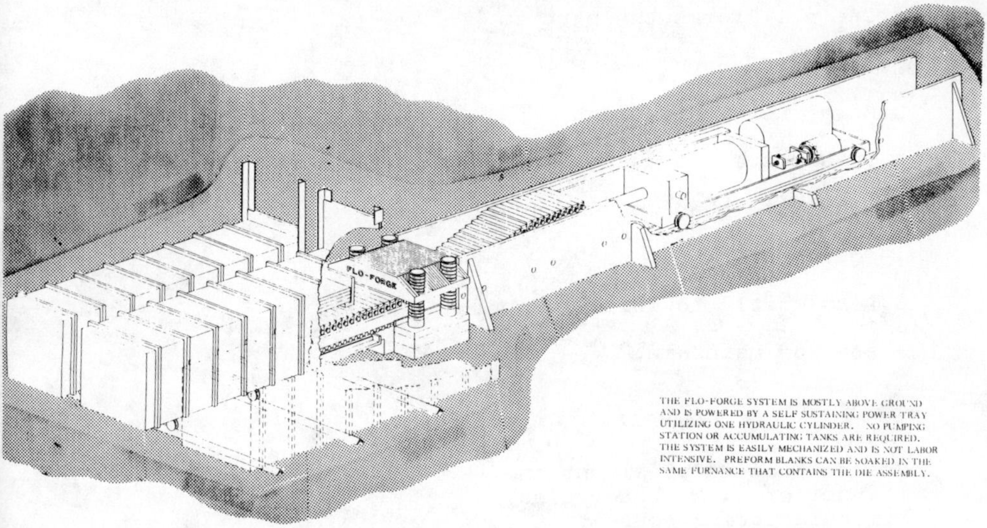

Fig. 14 Artist's Conception of FLO-FORGE Production System

Should it be necessary to make parts longer than the existing capacity of the system, all that is required is the lengthening of the side plates and adding indexing holes. The inclined surface and the hydraulic power package do not change. The furnace which is of modular design is readily expanded by adding the required sections.

6. SUMMARY

The FLO-FORGE system has demonstrated feasibility and potential for making low volume, high cost structural components. The cost to design, fabricate and test a production prototype system is estimated to cost several million dollars. J M Enterprise is seeking partners to achieve this goal. The skills of a machine tool builder, a forging company and one or more end users will contribute to the continued development program that hopefully can accomplish this goal within a two year period.

The significant advantages of the FLO-FORGE system are:

1. Material savings {reduce or eliminate machining}.

2. One pull forms the part to near net shape.
3. More energy efficient.
4. Less costly to design and fabricate than conventional equipment.
5. Less costly to install, operate and maintain.

7. BIOGRAPHY

Joseph Mele is president and founder of J. M. Enterprise, a manufacturing consultant company that developed the FLO-FORGE process. Mr. Mele is an Aeronautical Engineer with master's degrees in Mechanical Engineering and Applied Mechanics. He is a Certified Manufacturing Engiineer and the holder of six patents, some of which have been filed abroad. With 40 years diversified experience in manufacturing, design, materials and research and development, he has held positions at American Airlines, Sikorsky, Piasecki and Grumman. Published articles authored by Mr. Mele include developments in aircraft monocoque structures, plastics, ceramics, tool design, special bushings and CAD/CAM applications. He has been the recipient of cost savings awards, certificates of outstanding service to the Society of Manufacturing Engineers and an honorary member of Sigma Xi. Concurrent with his industrial experience, he lectured on the technology of plastics and resins for a period of eight years. For a two year period starting in 1980 Mr. Mele left industry to become a full time professor of Mechanical Engineering.

1st International SAMPE Metals Conference
August 18-20, 1987

THE ENVIRONMENTAL CRACKING BEHAVIOUR OF ALUMINIUM-LITHIUM BASED ALLOYS

A. Gray, N.J.H. Holroyd, W.S. Miller,
Alcan International Limited,
Southam Road,
BANBURY,
Oxon. OX16 7SP.

Abstract

The objective of this paper is to highlight the current understanding of the environmental cracking behaviour of Al-Li base alloys. Particular emphasis is placed on the Al-Li-Cu-Mg-Zr alloy system (alloys 8090 and 8091). The effect of alloy chemistry, alloy temper and test environment on both crack initiation and propagation is discussed. It will be shown that these parameters significantly influence crack initiation behaviour but have a reduced effect on crack growth rate. The initiation behaviour of the alloys can be explained largely by invoking an anodic dissolution mechanism whereas propagation behaviour appears dependent upon a mechanical effect as well as on crack tip chemistry and anodic dissolution. The stress corrosion behaviour of the alloys is compared with that of conventional alloys used in airframe construction.

1. INTRODUCTION

The potential benefit of aluminium alloys containing lithium for aerospace applications is now well established (1,2) and this has stimulated extensive research into a number of alloy systems. Whilst the environmental behaviour of these alloys is a critical property, particularly in aerospace applications, the subject has received only limited attention in the literature compared to extensive publications on microstructure and mechanical properties (3-9). A number of general trends have emerged to date; Al-Li, Al-Li-Zr, Al-Mg-Li-Zr alloy systems exhibit a high resistance to stress corrosion crack initiation (3-4) whereas copper containing alloys such as Al-Cu-Li-Zr and Al-Li-Cu-Mg-Zr have been shown to exhibit a greater susceptibility to crack initiation (3,5-9). Within the copper containing alloys, under ageing decreases the susceptibility to crack initiation for Al-Cu-Li-Zr alloys, however, for copper and magnesium containing Al-Li alloys the reverse trend is observed (overageing decreasing susceptibility) (3,9).

The aim of this paper is to examine the factors influencing both crack initiation and propagation with particular reference to the Al-Li-Cu-Mg-Zr alloys, 8090 and 8091, developed by the Royal Aircraft Establishment and Alcan International Limited (10). The stress corrosion performance of these alloys will be compared with established alloys used in airframe construction.

2. EXPERIMENTAL METHOD

2.1 Materials

The environmental cracking behaviour of a range of aluminium lithium alloys was studied. The alloys were chosen to cover most of the basic aluminium systems at appropriate alloying levels of lithium, copper and magnesium to represent the majority of the potential alloy systems (Table 1).

The alloys were D.C. cast as 300 Kg ingots and fabricated to 25 mm plate on commercial mills. In the case of the 8090 alloy investigated samples were also examined from 3000 Kg D.C. cast commercial ingots fabricated into plate up to 100 mm in thickness. All plate was solution treated, cold water quenched, and stretched 2.5% and artificially aged. The actual ageing conditions varied from alloy to alloy and are detailed in the text.

2.2 Stress Corrosion Testing; Crack Initiation

Standard specimen designs used were either C-rings (ASTM G38) or tension bars (ASTM G49). All specimens were machined so that the susceptibility to stress corrosion cracking was assessed in the most susceptible direction i.e. the short-transverse orientation. Strict machining practices were employed to minimise any induced surface stresses and heating effects.

Alternative immersion testing was carried out according to ASTM G44 using a 10 min wet cycle followed by a 50 min dry cycle. Air temperature (27 ± 1°C) and relative humidity (45 ± 5%) were closely monitored. Total immersion tests were conducted at the free-corrosion potential in standardised environments e.g. 3.5% NaCl (ASTM G44), 2% NaCl/0.3% Na_2CrO_4 pH = 3 using HCl and additionally in 3.5% NaCl with the pH adjusted to 3 and 9 using HCl and NaOH.

Time to failure of the C-ring specimens was defined as the time of detection of the first observable crack using a low-power optical microscope (magnification X20). Using this criteria good agreement was obtained between the C-ring and tension bar data. This involved frequent inspection of the samples during the initial period of testing and although many of the cracks subsequently blunted to broad intergranular fissures and thus did not propagate the specimens were diagnosed as having failed. Catastrophic failure of C-ring specimens was rarely observed, particularly for samples in the peak aged temper.

2.3 Stress Corrosion Propagation

Double cantilever beam (DCB) specimens (½ width = 12.7 mm notch width = 3.2 mm) were loaded to mechanical pop-in and exposed to an environment maintained at 55 ± 5% relative humidity and 25 ± 2°C. A 3.5% NaCl solution was added dropwise three times a day. Crack growth was monitored using an optical microscope and measurements taken on both sides of the specimen. Other tests were carried out in a fully automated multi-specimen facility where crack length was monitored using a potential drop technique (11). These test used specimens of the same design as Smith and Piper (12) and were exposed to either a 95% relative humidity (RH) environment or to total immersion in 3.5% NaCl.

2.4 Pre-exposure Embrittlement

Prior to testing in laboratory air (27 ± 1°C, 45 ± 5% relative humidity) or vacuum tension bars were continuously immersed in 3.5% NaCl or synthetic sea water for periods of up to 50 days. After pre-exposure specimens were cleaned using practices ranging from a rinse and ultrasonic wash in demineralised water followed by warm air drying to mechanically polishing to remove surface attack and films developed during pre-exposure. Additionally slow strain rate tests were carried out on selected samples following immersion in both continuous and alternate immersion environments.

2.5 Surface Attack Studies

Metallographically polished sections in the short-transverse plane were exposed to 3.5% NaCl under condition of total or alternate immersion and the resultant attack morphologies examined using optical metallography. Surface imperfections, e.g. hardness indentations were used to produce local stress concentrations, were introduced onto the as-polished surface prior to exposure.

3. RESULTS

3.1 Crack Initiation

Crack initiation data taken from both C-ring and tension bar tests is given in Figure 1 for a number of aluminium alloys containing approximately 2.5 wt% lithium (see Table 1), when subjected to alternate immersion stress corrosion testing. Immediately apparent from the data is the importance of the alloy's copper content as is more clearly shown in Figure 2. Crack initiation in peak aged alloys is moderately rapid for all alloys containing above about 1 wt% copper (and without zinc). Cracking occurs within a few days for stress levels greater than approximately 200 MPa. However, cracking susceptibility may be significantly reduced by overageing when copper levels are below about 2.0 wt%. It should be noted that this behaviour, observed in Al-Li-Cu-Mg-Zr alloys contrasts with that observed in Al-Cu-Li alloys with copper levels of about 2.7 wt% (9). The addition of zinc to copper containing Al-Li-Cu-Mg-Zr alloys results in a significant reduction in susceptibility to stress corrosion crack initiation. Crack initiation was not observed in Al-Li-Zr, Al-Mg-Li-Zr or Al-Mg-Li-Mn alloys after 50 days exposure to the alternate immersion environment. This indicates that, for the compositions studied, lithium additions either singularly or in combination with magnesium do not inherently promote susceptibility to stress corrosion cracking. Figure 3 considers the effect of copper and magnesium concentration within the Al-Li-Cu-Mg-Zr alloy system. The data shows the residual strength of longitudinal samples following 400 days exposure to alternated immersion. As the copper/magnesium ratio increases the susceptibility to SCC initiation increases. The loss in strength appears to reach a plateau once the copper to magnesium ratio exceeds 3.0. An additional point to note is that no failures occurred in this orientation for any of the alloy compositions examined (stress level 240 MPa).

The influence of temper on the stress corrosion susceptibility of the 8090 (nominally Al-2.5%Li-1.2%Cu-0.7%Mg -0.12%Zr) alloy is shown in Figure 4. In the under aged temper the alloy exhibits the greatest susceptibility to stress corrosion cracking. It should be noted that in this temper the alloy is aimed as a replacement for 2024-T351 (the conventional damage tolerant alloy used in airframe construction). Table 2 shows that even though the 8090's threshold stress for a 20 day life under alternate immersion testing conditions is 60 MPa, this is comparable with that of 2024-T351 at identical strength and fracture toughness levels (50 MPa). In the peak aged temper the threshold stress for a 20 day life is increased to 120 MPa. In the peak aged temper the alloy has similar strength and toughness to conventional medium strength alloys such as 2014-T651 as shown in Table 2. However, it has a greater inherent resistance to stress corrosion initiation compared with that of 2014-T651 (Table 2). In the isothermally overaged condition the stress corrosion threshold is increased to approximately 230 MPa which is comparable with 7075-T73, albeit at lower strength and toughness values.

As with other established aluminium alloys final thermomechanical treatments can be applied to improve the balance between strength and fracture toughness (13,14). The effect of factors such as cold working prior to ageing and the ageing temperature have been used to modify the microstructure at grain boundaries in 8090 and 8091 thereby improving the alloys fracture toughness for a given strength level (15). Table 3 shows that cold working prior to ageing improves the stress corrosion susceptibility and Figure 5 indicates that changes in the isothermal ageing temperature (from 190°C to 170°C) have little effect on stress corrosion initiation for a given ageing time. However, for a given strength level some improvement is observed by ageing at the lower ageing temperature. Recent reports suggest that this trend continues as ageing temperature is decreased further to 150°C (16). It is also noteworthy that material in the naturally aged condition is at least as susceptible as lightly aged material.

The processing route used to achieve high resistance to stress corrosion cracking in isothermally overaged tempers can lead to low fracture toughness (see Table 2). This is attributed to the large area fraction of grain boundary precipitates formed during ageing (15). However, duplex ageing treatments are under development to improve stress corrosion susceptibility whilst maintaining a satisfactory level of fracture toughness (16). Figure 6 compares the stress corrosion characteristics of 8090 50 mm plate in the damage tolerant temper with that of duplex aged material. Both samples having identical strength and fracture toughness in the short transverse direction (0.2% proof stress = 290 MPa; ST-L K_{1c} = 19 MPa $\sqrt{m}$). The duplex aged material exhibits a fourfold improvement in the threshold stress for a 20 day exposure period. Ongoing work is aimed at application of such treatments to alloys with strength and toughness values at or near to those obtained in peak aged material.

Tests on 8090 peak aged plate under continuous rather than alternate immersion in 3.5% NaCl, resulted in no initiation in tension bars after 40 days exposure, which contrasts with rapid failure in alternate immersion (see Table 6). This difference in behaviour is more marked than that observed in conventional aluminium alloys although, as expected, alternate immersion is the more severe test.

The influence of bulk solution pH on crack initiation has not yet been fully established. Occasional failures occur in non-buffered saline solutions with pH's adjusted to 3 or 9 (Table 6), but the majority of specimens survived a test duration of 30+ days. Non-buffered solutions with pH's of 3 and 9 induced both blunt fissures and general corrosion attack. Buffered pH 3 solutions (2% NaCl/0.5% Na_2CrO_4) promoted reproducible failure with fine cracks and little evidence of crack blunting.

3.2 Crack Propagation

Figures 7 and 8 show the crack extension and the change intensity with time respectively for 50 mm 8090 plate in the under aged and peak aged tempers. Tests were carried out using DCB's drip fed with 3.5% NaCl and crack lengths were measured optically. The crack velocity against stress intensity (V-K) curves from this data is shown in Figure 9. The under aged samples exhibits a slightly lower crack growth rate and a higher K_{Iscc} compared with peak aged material. Also shown in Figure 9 is previously published data (3) for peak aged 8090 in a 95% RH environment at 40°C. A significant shift in apparent K_{Iscc} to higher values was observed in the non-saline environment.

Figure 10 shows the crack extension (measured using potential drop) for 8090 25 mm plate in the under, peak and overaged conditions in the high humidity atmosphere and the effect of changing the test environment to 35 g/l NaCl continuous immersion on crack extension. The crack growth rate (plateau velocity) in the non-saline environment decreases from 6 x 10^{-9} m/sec to 1.4 x 10-9 m/sec as the alloy temper changes from under to over aged. Introduction of a saline environment increased plateau velocity for all alloy tempers. In the under and peak aged tempers the crack velocities (8 x 10^{-9} m/sec) are similar to those shown in Figure 9 (this fact also indicates that optical and potential drop measurement techniques are self consistent). In the overaged condition the plateau velocity is increased to 4 x 10^{-9} m/sec.

Figure 11 shows that for peak aged 8090 there is little difference in either crack velocity or K_{Iscc} for continuous immersion, constant immersion or drip fed saline environments.

Table 5 shows the plateau velocities of 8090 and 8091 in various tempers in comparison with other published data on Al-Li base alloys (3,4,6,7) and results for 7000 and 2000 series alloys. In general for alloys containing > 2 wt% lithium (i.e. Al-Li, Al-Cu-Li, Al-Li-Cu-Mg) plateau velocities are within the range 2-8 x 10^{-9} m/sec. Within this range subtle shifts with temper have been observed (e.g. a reduced velocity for underaged 8090 compared with that of peak aged material in saline environments), but these trends have not been found for all alloy types. The behaviour of these alloys contrasts with that of conventional alloys (e.g. 2024, 7075) and high copper, low lithium alloys (2020) where plateau velocities may decrease by two orders of magnitude as the temper changes from under to peak to overaged. The presence of chloride ions increases crack velocity to a lesser extent than that observed in conventional alloys (18). Furthermore the method of testing, in saline environments, i.e. continuous immersion, alternate immersion or drip feeding does not appear to significantly influence plateau velocity or K_{Iscc}. This contrasts with the result of the initiation studies shown in Table 2.

The variation in K_{Iscc} within the Al-Li-Cu-Mg-Zr alloy system has been determined from V-K curves. No significant trends were observed with either alloy chemistry or ageing practice. A correlation was found between the initial K (the alloys fracture toughness) and K_{Iscc}, Figure 12. This is indicative of a mechanical component in the SC crack growth mechanism.

3.3 Pre-Exposure Embrittlement

The occurrence of reversible pre-exposure embrittlement can be used as a means of indicating if hydrogen embrittlement can cause cracking in a particular alloy system as has been shown for 2000, 5000 and 7000 series alloys. Previous studies (3) on 8090 had indicated the possible role of hydrogen embrittlement in the stress corrosion failure mechanism in Al-Li alloys. In this work, rapid failure in air occurred following pre-exposure to 3.5% NaCl under constant immersion conditions, however, if the pre-exposed samples were held under a vacuum for 28 days no failures were observed on exposure to laboratory air. It is now recognised that the reversible "embrittlement" effect demonstrated by 'recovery' under vacuum may have simply been a purging of the active environment from within the crevice/pit region. In Figure 13 these results are compared with recent data in which the constant immersion environment was changed to synthetic seawater. The extension in time to failure following immersion can be attributed to a buffering action of the synthetic sea water reducing the susceptibility to pitting attack.

Figure 14 shows the reduction in tensile strength (S-T direction) with increasing exposure to 3.5% NaCl under either constant or alternated immersion. The reduction in tensile strength in constant immersion, plateaus after 15 days immersion at a value of 280 MPa. Under alternate immersion condition an equivalent reduction in tensile strength occurs after only 5 complete cycles. The plateau tensile strength, which occurs after 10 cycles, is 200 MPa.

3.4 Surface Attack Morphology

Figure 15 shows the attack morphology developed under alternate and constant immersion for 8090 in under aged, peak aged, isothermally over aged and duplex aged tempers. In the overaged conditions (isothermal or duplex aged) general corrosion attack occurred in both alternate or continuous immersion. The attack occurred on both high and low angle grain boundaries to an approximately equivalent extent. Some preferential attack occurred on intermetallic particles (presumably iron-rich) strung along the high angle grain boundaries. There was no apparent enhancement of attack in the stressed regions of the specimens. In the peak aged temper attack is concentrated at high angle grain boundaries with little attack occurring on the sub-structure. In continuous immersion all grain boundaries were attacked to an equivalent extent whereas in alternate immersion attack appeared to concentrate on certain grain boundaries. In the under aged condition the degree of attack at grain boundaries was increased, alternate immersion tending to promote attack at high angle grain boundaries. In both the under aged and peak aged condition the stressed regions showed some evidence indicative of stress corrosion cracking in the alternate immersion samples.

Figure 16 shows that an Al-Li-Cu-Zn-Mg-Zr aloys in the peak aged condition is virtually immune to surface attack in either continuous or alternate immersion.

4. DISCUSSION

The overriding conclusion from the results of this work is the significant difference between SCC initiation and propagation behaviour in Al-Li base alloys. Initiation characteristics of the alloys are dependent on alloy chemistry, alloy temper and the type of saline test environment whereas crack propagation behaviour appears less sensitive to changes in these parameters.

The important results from the alloy chemistry/crack initiation studies were that susceptibility is highly dependent on copper content for alloys containing 2 to 2.5 wt% lithium in the absence of zinc. In addition this susceptibility can be moderated by over ageing especially for alloy with copper contents below about 2.5 wt%. Within the Al-Li-Cu-Mg system increasing copper relative to magnesium also increases susceptibility to SCC. Crack initiation did not occur for alloys tested without copper under conditions of alternate or continuous immersion. This suggests that the presence of one of the copper containing precipitate phases, e.g. T_1 (Al_2CuLi) or S' (Al_2CuMg) or I-phase ($Al_6Cu(LiMg)_3$), may be responsible for promoting crack initiation rather than δ (AlLi) of the Al-Li system or T-phase (Al_2MgLi) of the Al-Mg-Li system.

The results of the studies on the influence of temper, cold working and ageing practice may be summarised as follows:-

a) Over aged tempers tend to have δ-phase precipitates on sub-grain boundaries and within the matrix (3,17). This leads to a more uniform general corrosion in preference to localised attack of grain boundary precipitates by reducing the potential difference between the copper rich grain boundary particles and the matrix i.e. an increased number of anodic/cathodic sites within the individual grains. This is confirmed by the surface attack morphology studies.

b) In the peak aged condition anodic dissolution of grain boundary phases is enhanced, as evidence from the surface attack studies. This is because the potential difference between the matrix and the grain boundary is increased; δ-phase is absent from the matrix and sub-boundaries and there are preferential anodic sites available at grain boundaries. It is thus proposed that initiation behaviour is controlled by the anodic potential between the grain boundary and the matrix. In the peak aged condition this will be controlled by the degree of S' phase within the matrix. Reducing the ageing temperature refines S' precipitates but will not significantly alter the volume fraction. Hence the modest improvement in SCC initiation behaviur with the lower ageing temperatures. Further evidence for this anodic dissolution mechanism is the effect of the copper to magnesium ratio on SCC, low ratios of Cu/Mg promoting S' precipitation.

The volume fraction of S' is reduced if the alloy is not cold worked prior to ageing and as expected SCC susceptibility increases (Table 3).

c) In under aged tempers S' precipitation within the matrix is reduced, therefore the driving force for anodic dissolution of the, albeit reduced area fraction, of grain boundary precipitates is enhanced (cf Figure 13). The dissolution of the grain boundary precipitates is also exacerbated by localised stress resulting from the alloys propensity to deformed by planar slip due to shearing of coherent ordered δ' precipitates. The fact that cracks readily nucleate in this temper even though the rapidly dissolving grain boundary precipitates are widely spaced may be due to either a solute segregation of active species to the grain boundary during the quench (precursors to the grain boundary precipitates) or hydrogen embrittlement. As yet there is insufficient evidence to support either hypothesis.

The effect of zinc in moderating the detrimental effect of copper cannot as yet be satisfactorily explained. The surface morphology attack studies indicate that the material is much less sensitive to corrosion attack than the zinc free alloys studied in either continuous or alternate immersion. It may well be that the precipitation of a zinc rich phase, both within the matrix and on sub-grain boundaries (17) reduces the potential difference between the matrix and high angle grain boundaries.

Initiation of environmental cracking was significantly easier under conditions of alternate immersion rather than continuous immersion. In fact cracks were only reproducably initiated under continuous immersion when surface stability was sufficiently reduced by a solution pH shift and inhibitors preventing crack blunting. These results are mirrored by the surface attack found between continuous and alternate immersion for under aged and peak aged tempers. These effects have been rationalised as being due to localised solute chemistry changes (3).

In considering the crack propagation behaviour, the plateau velocities of Al-Li-Cu-Mg-Zr alloys in saline environments do not significantly shift with alloy chemistry, ageing practice or test environment. This contrasts with the initiation behaviour discussed above. Similar plateau velocities have been observed in Al-Cu-Li (6,7) and Al-Li (4) alloys; the latter being virtually immune to SCC initiation. It may well be that electrolyte chemistry within a propagating SCC crack is controlling propagation velocity (19). Previous work (3) has shown that restricted geometries and limited electrolyte volume promote SCC initiation which has been attributed to the development of transitory alkaline environments. However, in the case of a propagating crack an acidic environment is present at the crack tip (19). It is thus likely that it is this acidic environment that is controlling crack propagation behaviour c.f. the similarity in behaviour of material in constant immersion with that of AI.

An increased knowledge of the electrochemical environment at the head of a propagating stress corrosion crack and its relationship with alloy chemistry, alloy microstructure and the bulk test environment is required to fully understand the mechanistics of SCC crack propagation.

K_{1scc} of the alloys appears to depend only on the magnitude of the alloys fracture toughness consistent with the view that there is a mechanical factor influencing crack propagation.

5. CONCLUSIONS

1. The SCC initiation behaviour of Al-Li base alloys is dependent on alloy chemistry, alloy temper and test environment.

2. Increased copper contents, under ageing and alternate immersion environments increase susceptibility to SCC initiation.

3. In contrast the propagation behaviour is relatively insensitive to changes in alloy chemistry, alloy temper or test environment. However, there is clearly a mechanical effect to be considered.

4. Both the initiation and propagation results on 8090 compare favourably with those of conventional alloys such as - 2024-T351, 2014-T651, 7075-T651).

5. In considering the overall SCC performance of Al-Li base alloys it must be recognised that although, in certain tempers, crack initiation may readily occur, these cracks rapidly blunt out and do not propagate.

6. ACKNOWLEDGEMENT

This paper is published with the permission of Alcan International Limited.

7. REFERENCES

1. Aluminium-Lithium Alloys II Ed. T.H. Sanders and E.A. Starke TMS-AIME 1984.
2. Aluminium-Lithium Alloys III Ed. C. Baker, P.J. Gregson, S.J. Harris and C.J. Peel, Inst. Metals 1986.
3. N.J.H. Holroyd, A. Gray, G.M. Scamans and R. Hermann in Ref 2 p.310.
4. L. Christodoulou, L. Struble and J.R. Pickens in Ref 1 p.561.
5. E.I. Meletis - 'Stress corrosion cracking properties of 2090 Al-Li alloy'. Proc. Int. Conf. on Fatigue Corrosion Cracking. Salt Lake City, UT 1985.
6. J.G. Rinker, M. Marek and T.H. Sanders, Mat. Sci. Eng. 64, p.203, 1984.
7. A.K. Vasudevan, P.R. Ziman, S.C. Jha and T.H. Sanders in Ref 2 p.303.
8. E.L. Colvin, S.J. Murtha and R.K. Wyss. Aluminium Alloys - Physical and Mechanical Properties. Ed. E.A. Starke and T.H. Sanders pub. EMAS 1986, Vol.3, p.1853.
9. E.I. Meletis, J.M. Sater and T.H. Sanders in Ref 8, Vol.2, p.1157.

10. C.J. Peel, B. Evans, C. Baker, D.A. Bennett, P.J. Gregson and H.M. Flower in Ref 1, p.363.
11. N.J.H. Holroyd and D. Hardie, Corrosion Science 23, p.527, 1983.
12. H.R. Smith and D.E. Piper, Boeing Report D6-24872, 1970.
13. R.F, Ashton, D.S. Thompson, E.A. Starke and F.S. Lin, in Ref 2 p.66.
14. W.S. Miller, J. White, M.A. Reynolds, D.S. McDarmaid and G.M. Starr, Proc. 4th Int. Conf. on Al-Li Alloys - Paris, June 1987. J. de Physique in Press.
15. J. White and W.S. Miller as Ref 14.
16. M. Mamoud as Ref 14.
17. A. Gray as Ref 14.
18. T.J. Summerson and D.O. Sprowls in Ref 8, Vol.3, p.1575.
19. N.J.H. Holroyd, M.R. Jarrett, A. Gray, R. Hermann, AIME Fall Meeting, Orlando 1986 - to be submitted to Corrosion (1987).

TABLE 1 - Chemical Analysis of Experimental Aluminium-Lithium Base Alloys in wt%

Alloy	Li	Mg	Cu	Zr	Fe	Si
Al-Li	2.8	-	-	-	0.07	≤ 0.03
Al-Li-Zr	2.8	-	-	0.16	0.09	0.03
Al-Mg-Li	2.4	3.8	-	0.10	0.06	0.08
Al-Mg-Li-Mn (0.25 Mn)	2.1	3.7	-	0.12	0.06	0.08
Al-Cu-Li	2.2	0.1	2.7	0.14	0.07	0.04
Al-Cu-Li-Mg	2.2	0.4	2.5	0.14	0.06	0.03
Al-Li-Cu-Mg (8090)	2.4	0.61	1.15	0.12	0.09	0.05
Al-Li-Cu-Mg (8091)	2.6	0.59	2.03	0.14	0.06	0.04
Al-Li-Cu-Zn-Mg Zn = 1.01	2.51	1.18	0.81	0.12	0.09	0.04
Al-Zn-Li-Cu-Mg Zn = 3.06	2.41	1.54	0.93	0.11	≤ 0.05	≤ 0.05

TABLE 2 - Preliminary Threshold Stress Levels for 8090 Alloy Compared with Commercial Alloys. Note the Data for 8090 is for a Limited Batch Size and should not be Considered as a Design Value. Upward Shifts in the Threshold Stress are Anticipated.

Alloy	Temper	S-T Threshold Stress 20 Day Life MPa	K_{1c} ST-L MPa$\sqrt{m}$
8090	Under aged	60	27.0
2024	T351	50	24.0
8090	Peak aged	115	20
2014	T651	50	18
7050	T7651	150	20
8090	Over aged	240	12
7075	T7351	240	24

TABLE 3 - Effect of Cold Deformation on the SCC Performance of 8090

	Time to Failure (hr) ST-Orientation 240 MPa				
	Plate			Forgings	
% Strain	0	2.5	3-10	0	2.5
	48	150	160	70	≥ 250

TABLE 4 - Tension-bar Test Data 8090-T651 at 240 MNm^{-2}

Test Environment	SCC Life (Days)
3.5% NaCl A.I.	2.5
3.5% NaCl C.I.	≥ 40
2% NaCl/0.5% Na_2CrO_4 (pH = 3)	15
3.5% NaCl, pH = 3	14, ≥ 30, ≥ 30
3.5% NaCl, pH = 9	2.5, ≥ 30, ≥ 30

TABLE 5 - Stress Corrosion Crack Plateau Velocities and K_{1scc} for Al-Li Base Alloys

Alloy	Temper	Environment	Plateau Velocity m/sec	Stress Intensity range MPa√m
8090	Under aged	) 95% RH	6×10^{-9}	18-30
	Peak aged	)	3×10^{-9}	13-20
	Over aged	) 40°C	1.4×10^{-9}	10-12
8090	Under aged	) 35 g/l	$2\text{-}7 \times 10^{-9}$	15-34
	Peak aged	) NaCl	$4\text{-}8 \times 10^{-9}$	8-16
	Over aged	) AI	4×10^{-9}	6-12
8091	Peak aged	) 35 g/l NaCl	4×10^{-9}	6-10
	Over aged	) AI	4×10^{-9}	7-12
2020	Under aged	) 35 g/l NaCl	2×10^{-8}	11-12
(Al-4.45Cu-1.21Li)	Peak aged	) AI	1×10^{-11}	11-12
(Ref. 6)				
Al-2.9Cu2.2Li	Under aged	) 35 g/l NaCl	6×10^{-9}	8-40
(Ref. 7)	Peak aged	) AI	3×10^{-9}	7-15
	Over aged	)	6×10^{-9}	4-8
Al-2.8%Li	Under aged	) 35 g/l NaCl.CI	2×10^{-9}	8-16
(Ref. 4)	Peak aged	) Chromate Inhibited	2×10^{-9}	8-16
2024	T351	) 35 g/l NaCl	6×10^{-9}	13-27
2024	T851	) AI	$3\text{-}14 \times 10^{-10}$	15-18
7075	T651	)	2×10^{-8}	13-19
7075	T7351	)	5×10^{-10}	18-23

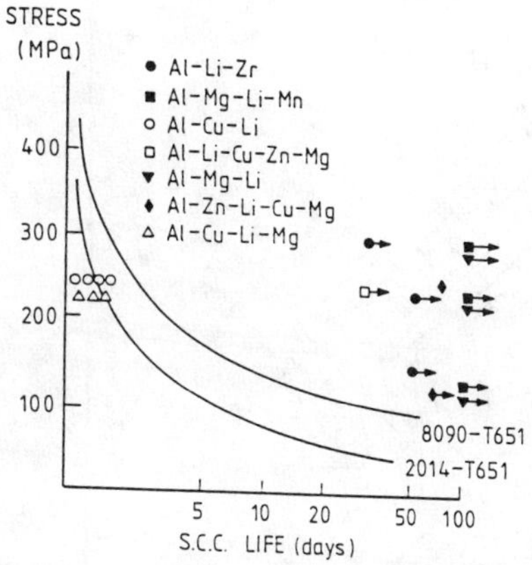

Figure 1. Effect of alloy chemistry on the SCC initiation behaviour of Al-Li base alloys.

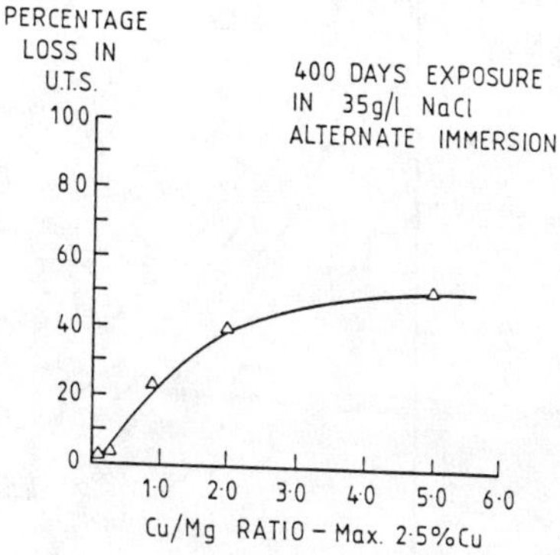

Figure 3. The influence of Cu/Mg ratio on stress corrosion characteristics. Residual strength after 400 days alternate immersion. Longitudinal orientation.

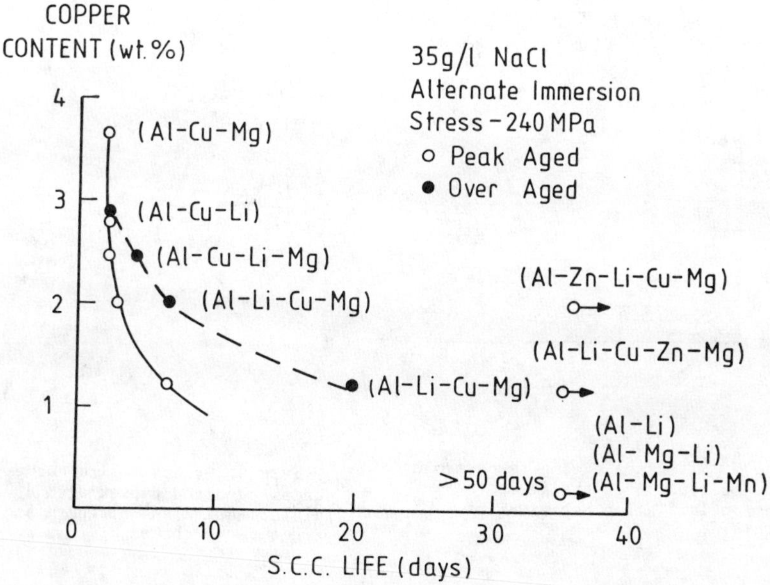

Figure 2. The effect of copper content on the stress corrosion characteristics of various Al-Li base alloys, in the peak and over aged tempers.

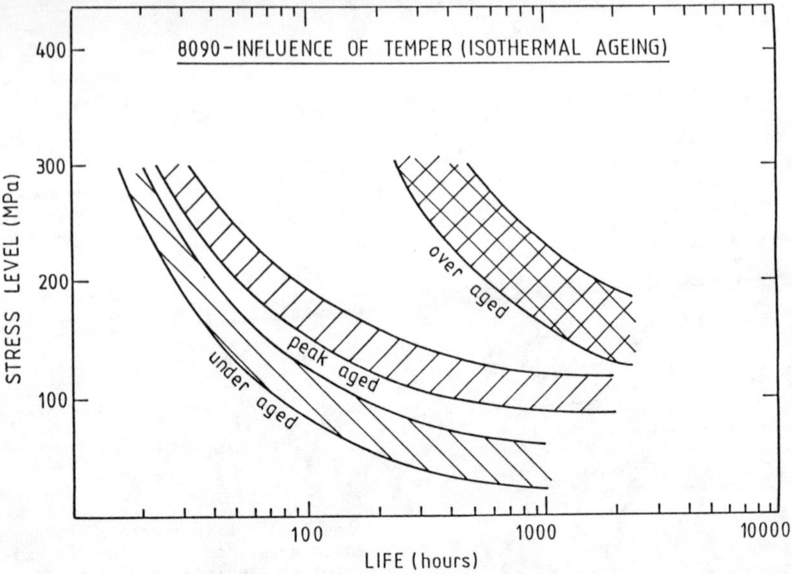

Figure 4. The influence of temper on the stress-life curve of 8090 plate (ST orientation).Under aged = 4h at 190 °C ; Peak aged = 16 h at 190°C ; Overaged = 96 h at 190°C.

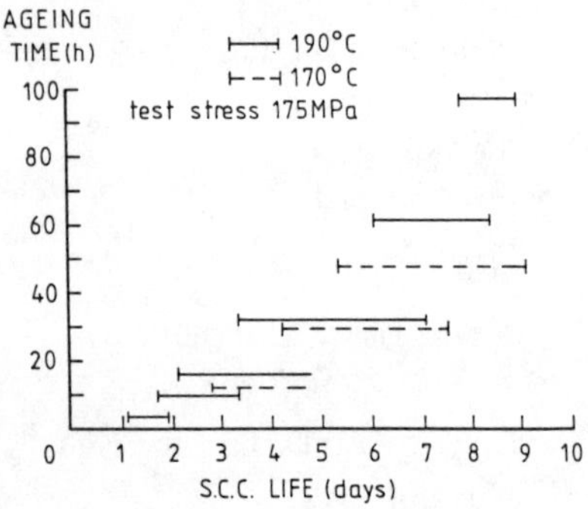

Figure 5. Effect of ageing temperature (170 °C or 190 °C) on the stress corrosion initiation behaviour of 8090 plate.

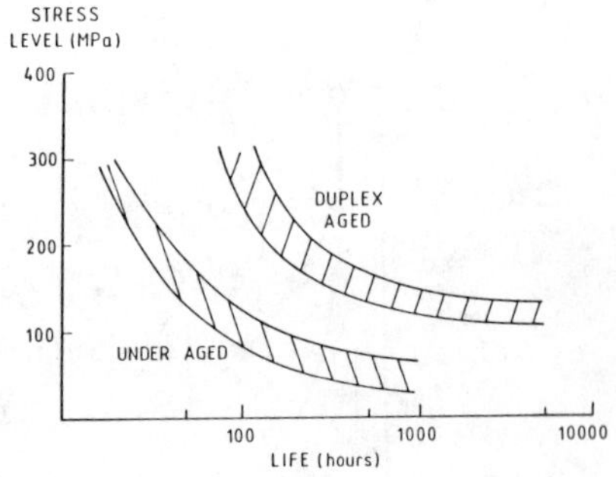

Figure 6. Stress - life curves for under aged and duplex aged 8090 plate. Equivalent strength and fracture toughness.

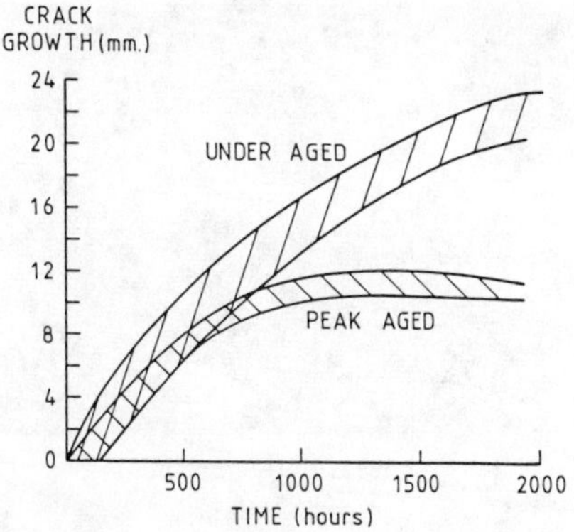

Figure 7. Crack extension v. exposure time curves for under and peak aged 8090 50 mm plate. Drip feed 3.5% NaCl.

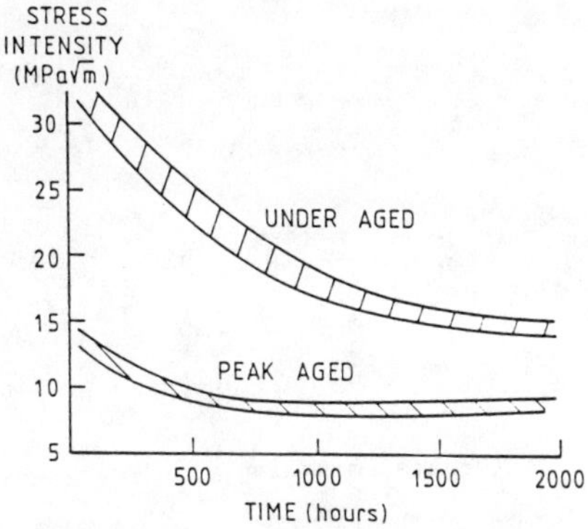

Figure 8. Change in stress intensity with time for under and peak aged 8090 50mm plate. Drip feed 3.5% NaCl.

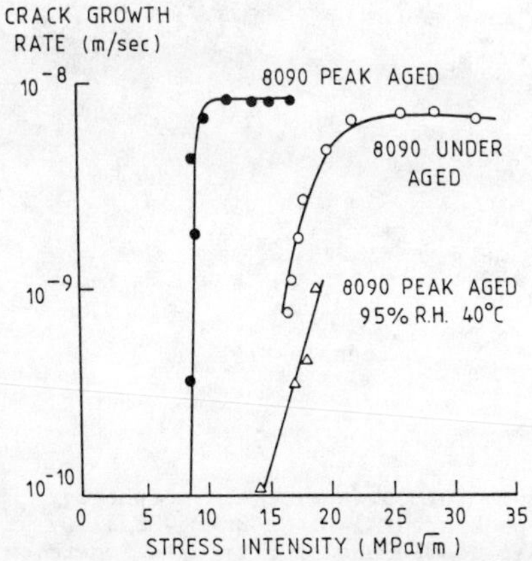

Figure 9. V - K curves for under and peak aged 50mm 8090 plate. Drip feed 3.5% NaCl.

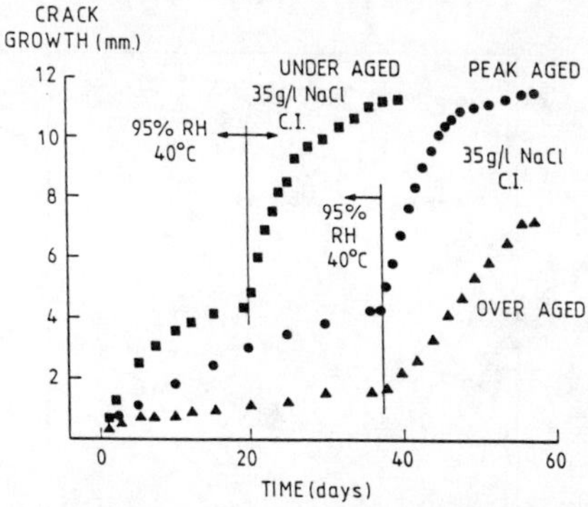

Figure 10. Crack extension v. time curves for 8090 plate in various tempers with change in environment from 95% RH (40 °C) to 3.5%NaCl continuous immersion.

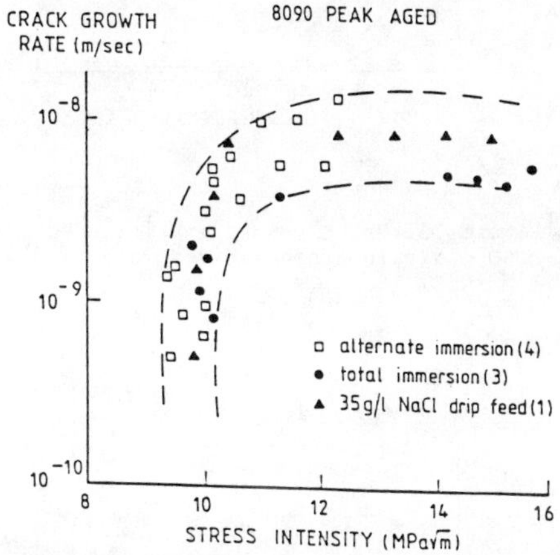

Figure 11. V -K curves for 8090 plate in various saline environments.

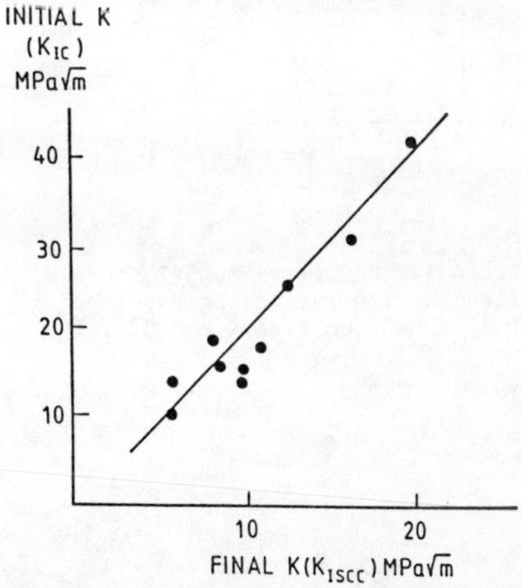

Figure 12. Variation in $K_{I}scc$ with the initial stress intensity $K_{I}c$.

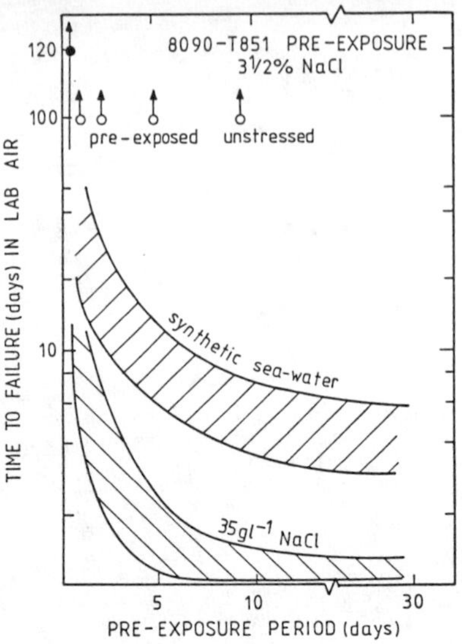

Figure 13. Pre-exposure embrittlement of 8090 plate in synthetic sea water and 3.5% NaCl.

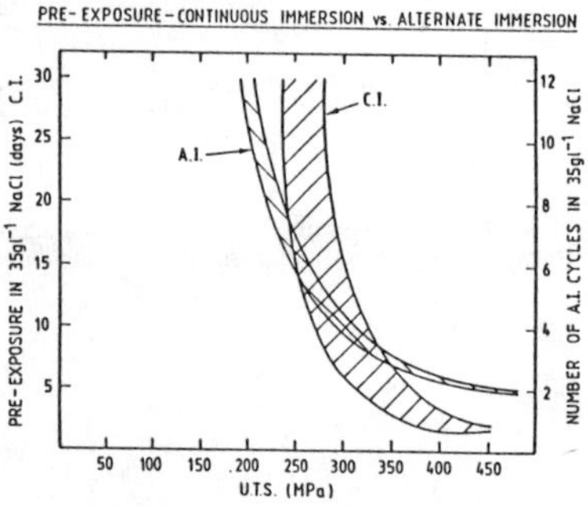

Figure 14. Reduction in S-T tensile strength with increasing exposure to 3.5% NaCl under A.I and C.I..

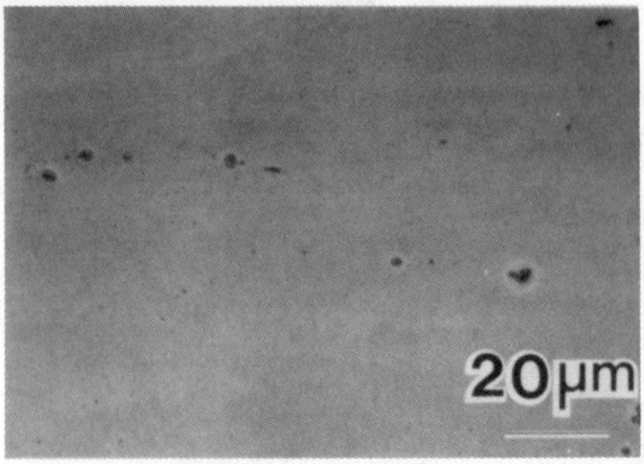

Constant Immersion
3 Hours.

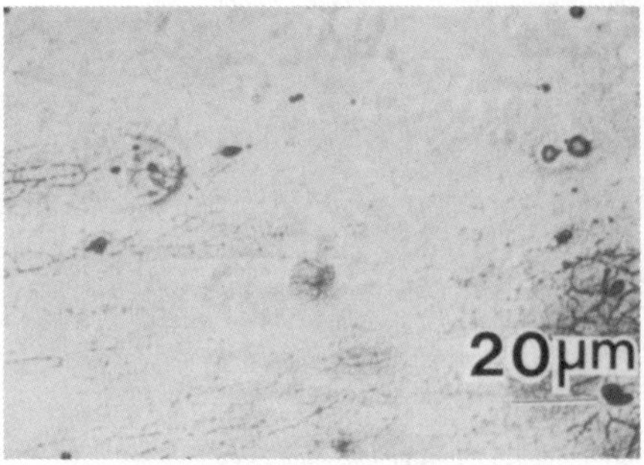

Alternate Immersion (18 cycles

- 3 hours in environment)

Locally Stressed

ıgure 16. Surface attack morphology ɔf peak aged Al-Li-Cu-Zn-Mg-Zr alloy under continuous or alternate immersion in 3.5% NaCl (3 hours exposure).

Constant Immersion
3 Hours.

Alternate Immersion (18 cycles - 3 hours in environment)

Locally Stressed

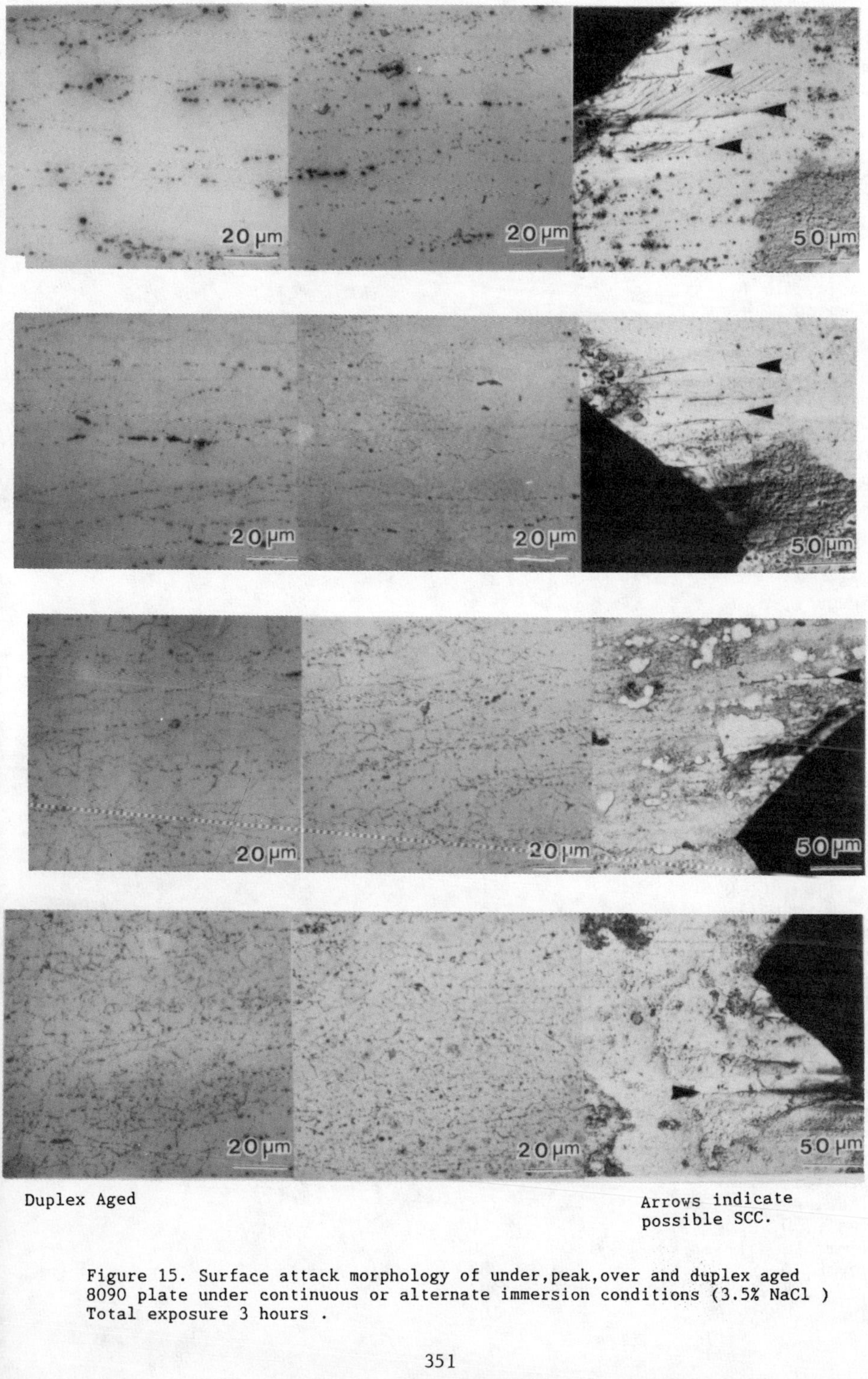

Duplex Aged

Arrows indicate
possible SCC.

Figure 15. Surface attack morphology of under,peak,over and duplex aged 8090 plate under continuous or alternate immersion conditions (3.5% NaCl) Total exposure 3 hours .

AUTHORS INDEX

SUBJECT INDEX

1st International SAMPE Metals and Metals Processing Conference

Volume 1, 1987